Advances in Magnesium Research: New Data

Advances in Magnesium Research: New Data

P.J. Porr
M. Nechifor
J. Durlach

ISBN 2-7420-0606-0

Éditions John Libbey Eurotext
127, avenue de la République
92120 Montrouge, France
Tel: 01 46 73 06 60
e-mail: contact@jle.com
http://www.jle.com

Editor: Maud Thévenin

Foreword

During the last 5 decades the crucial role of electrolytes in life processes has been more and more emphasized by fundamental and clinical researchers. Among these electrolytes magnesium occupies a very special place. Its role as a catalyzer in more than 350 enzymatic reactions, in different membrane processes, its involvement in protein, carbohydrate and lipid metabolism, even in oxidative stress etc. account for its importance in many different fields of human pathology: cardiology, neurology, psychiatry, endocrinology, immunology, obstetrics, pediatrics, gastroenterology a.o.

A significant step in magnesium research was taken in Europe in the sixties with the development of veritable "schools" with J. Durlach in Paris, H. G. Classen in Stuttgart, J. Szántay in Cluj. Within a few years the field of research on magnesium widened considerably, it takes place a real extension of consistent researches about magnesium in nearly all countries of the continent. The international Society for the Development of Research on Magnesium (SDRM) was then created by Jean Durlach, whose aim was to coordinate research and to help magnesium researchers share their findings by fostering every 3rd year both International Symposia (the last one in Australia in 2003, the next one in Japan in 2006) and European Congresses. The last of these (8th European Magnesium Congress) was held in Cluj in 2004 and equaled the high scientific level of the previous ones. More than 60 very interesting plenary lectures, oral communications and posters, not only from Europe, but also from 3 other continents, provided a framework for extensive discussions and interaction between speakers and young researchers in this exciting field of research.

We are assembling in this volume most of the contributions to the Congress which will henceforth be available to an interested audience. We are pleased to express our deep appreciation to all the authors for their excellent work.

Last but not least, we wish to thank the prestigious John Libbey Eurotext for its most useful cooperation

The Editors

Contents

I. Opening lecture

Clinical forms of magnesium depletion with hypofunction of the biological clock

J. Durlach

Président SDRM, Université Pierre et Marie Curie (UPMC) Paris VI, 4 Place Jussieu, Paris, France

Abstract. Mg depletion is a type of Mg deficit due to a dysregulation of the Mg status. It cannot be corrected through nutritional supplementation only, but requires the most specific correction of the dysregulating mechanism. Among those, biological clock dysrhythmias are to be considered. The aim of this study is to analyze the clinical forms of Mg depletion with hypofunction of the Biological Clock (hBC). hBC may be due to either **Primary** disorders of BC [Suprachiasmatic Nuclei (SCN) and pineal gland (PG)] or **Secondary** with homeostatic response [reactive **Photophobia** (Pφ)] to light neurostimulating effects [Nervous Hyper Excitability (NHE)]. The symptomatology is mainly **diurnal** and observed during fair weather (**Spring, Summer**). The elective marker of hBC is represented by a decrease in melatonin and in its metabolites in various fluids. The clinical forms of NHE due to Mg depletion with hBC are central and peripheral. The central forms associate anxiety, headaches and dyssomnia. The peripheral manifestations are neuromuscular: photosensitive epilepsia mainly. Three chronopathological forms of Mg depletion with hBC have been highlighted: 1. **Headaches with Pφ** : mainly migraine 2. **Sudden Infant Death Syndrome** (SIDS) 3. **Multiple Sclerosis** (MS). - **Headaches with Pφ**, *migraine particularly.* These cephalalgias are diurnal with Pφ and are aggravated during the fair seasons (particularly during midnight sun - summer). Migraine is their typical form with its dishabituation to visual stimuli and its occipital cortex hyperexcitability. Comorbidity with anxiety is frequent. In 2/3 of the cases, it appears first. - **SIDS** might be linked to an impaired maturation of both photoendocrine system and brown adipose tissue. - **MS** may be associated with primary disorders of BC. Clinical forms of Mg depletion with hBC in M.S. present diurnal exacerbations and relapses during fair seasons. They have been underestimated because they disagree with the dogma of the « *latitude gradient* », presently questioned. Comorbidities with anxiety and migraine are frequent. hBC may be **treated** by using **darkness therapy with a balanced Mg status**. Absolute light deprivation should only be used only in acute indications and is time-limited. Partial substitutive therapy and chromatotherapy have not been validated yet and are still uncertain.

Keywords: Magnesium, Chronobiology, Anxiety, Migraine, Sudden Infant Death, Multiple Sclerosis

The biological clock (BC) and the magnesium status are strongly correlated. The efficiency of the biological clock represented by suprachiasmatic nuclei and pineal gland is related to the quality of magnesium status [1].

A body of evidence has already stressed the difference between two types of magnesium *deficit*:

– *Deficiency* linked to an insufficient intake which may be corrected, over a long period of time, through a physiological nutritional oral magnesium supplementation, and

– *Depletion* due to a dysregulation of the magnesium status which cannot be corrected through nutritional supplementation only, but requests the most specific correction of the dysregulating mechanism. There exist as many clinical forms of magnesium depletion as many possibilities of the dysregulation of the magnesium status. But in both clinical therapeutics and in animal experiment, the dysregulating mechanisms of magnesium depletion associate a reduced magnesium intake to various

types of stress [2]. Among these, are biological clock dysrhythmias. This allows to distinguish the different manifestations of chronopathological forms of magnesium depletion among various pathologies including migraine, fatigue, fibromyalgia, dyssomnia, epilepsia and even sudden infant death syndrome [1, 3, 4].
The aim of the present study is to analyze the clinical forms of magnesium depletion with hypofunction of the Biological Clock (hBC). hBC may be due to either **Primary** disorders of BC [Suprachiasmatic Nuclei (SCN) and pineal gland (PG)] or **Secondary** to light neurostimulating effects, with their homeostasic response [reactive **Photophobia** (Pφ)].

Clinical Forms of magnesium depletion with hypofunction of the biological clock

The biological clock hypofunction may be due to either primary disorders of BC or secondary homeostasic reactive response to *light hypersensitivity* linked to light neurostimulating effects. The organism responds to the pathogenic effect of this hypersensitivity by protective reactive *photophobia*, whose mechanism is still unclear [5].
The clinical characteristics of these secondary forms of chronobiological Nervous HyperExcitability (NHE) are of circadian as well as of seasonal type: the symptomatology is mainly *diurnal* and observed *in spring and summer*, when light hyperstimulation is obviously maximum during daylight or during the fair seasons. The main biological characteristic is represented by *a decrease in melatonin* (*or in its* metabolite) *levels* in various fluids which has been previously reported as the elective marker of the biological clock [1, 3].
The clinical forms of Nervous HyperExcitability (NHE) are both central and peripheral.

- The central forms associate psychic, algic and dyssomnic manifestations:
  1. Anxiety as manifested from generalized anxiety (TAG) to panic attacks (PA) [6].
  2. Diurnal cephalalgia with photophobia aggravated during the fair seasons (and mainly during the polar summer [7]) whose type is migraine with its occipital cortex hyperexcitability [5, 8-10].
  3. Dyssomnia mainly represented by the delayed sleep phase syndrome (DSPS) observed for instance in jet lag, night work disorders, or insomnia of elderly patients [1, 3, 5, 11] with sometimes inappropriate behaviour [12]. Some chronopathological forms of sudden infant death syndrome (SIDS) may be also associated here [1, 4].
- The central and peripheral manifestations are neuromuscular; mainly represented by photosensitive epilepsia, which may be either generalized or focal, authentified through EEG with intermittent light stimulation (ILS) with its corresponding form observed among TV viewers and video game players [3, 13-15]. Some migraine equivalents may be associated in this context. It is noteworthy that paradoxically epileptic activity may be induced by a light hyperstimulation... Or its suppression as well [3, 16].

The nervous form of chronopathological magnesium depletion may appear clinically as chronic fatigue syndrome (CFS) [17, 18] or as fibromyalgia [3, 19].
All the clinical magnesium depletion forms with biological clock hypofunction may coexist with the same chronobiological characteristics (mainly a decrease in melatonin or in its metabolite levels), the main comorbidity being represented by the association migraine-epilepsia [3].
Three chronopathological forms of magnesium depletion with hBC will be highlighted: headaches with photophobia (mainly migraine), some clinical forms of sudden infant death syndrome and of multiple sclerosis.

Headache with photophobia, migraine particularly

The most typical form of biorhythm-type cephalalgia is **MIGRAINE** headache with photophobia [7, 8, 20-26]. These migraineurs have headaches during artic summers with continuing bright light (particularly during midnight-sun summer) and are in good condition during artic winters [7].

Nature of the magnesium deficit in migraine

Nervous hyperexcitability in migraineurs may be linked to magnesium deficit.

When chronic primary magnesium deficiency coexists with migraine, it only constitutes a decompensatory factor whose control with simple oral nutritional magnesium supplementation should help in migraine therapy as *an adjuvant treatment*: **magnesium deficiency** *does not constitute the cause for migraine « per se »* [27, 28].

Clinical studies on magnesium status in migraineurs have shown heterogeneous and inconstant decreases in extra or intracellular, total or ionized magnesium concentrations in serum, saliva, erythrocyte, mononuclear cells, thrombocyte, even in brain. Positive therapeutic response to oral physiological load is unreliable. These data agree with some dysregulation of the magnesium status in migraine that is to say **magnesium depletion**. The importance of the chronopathologic disorder in the aetiopathogenic mechanisms of the migraine, magnesium depletion should be highlighted [1, 3, 27-44].

Migraine and photic dishabituation

Habituation is a physiological phenomenon characterized by a decrease of the responses to repetitive stimuli: habituation is considered to be a protective mechanism against overstimulation.

Dishabituation, by contrast, is a process that liberates the nervous system from the habituation process. The dishabituation stimuli act as **sensitization** or **potentiation** processes which rely on a dysfunctionning of cortical information processing that might result from the high level of cortical arousal with increased energy demands and hypofunction of subcortico-cortical pathways. The cortical arousal level depends on the actions of various neurotransmitters from the brainstem projecting to the cortex. 5HT (serotonin) acts as a gain control between a noradrenergic, unspecific, facilitating system and a cholinergic, specific, inhibitory system.
Dishabituation may finally induce **generalization**: the alteration being able to involve other stimuli or to invade other substrates [45-49].

Migraine and magnesium depletion by photic sensitization

Migraine may be considered as the type of a chronopathological form of Mg depletion with hypofunction of the biological clock.
Subcortical and cortical genetic factors further the development of a **magnesium depletion** caused by a deficient magnesium dietary intake (**magnesium deficiency**) plus a **photic stress** - mainly circadian (**diurnal**) and seasonal (during « **fair**» **seasons**) - with reactive **photophobia** and **more or less generalized** - sensory, cognitive, painful (trigeminal particularly), alimentary - **sensitization**. It induces cortical and subcortical dysexcitability and spreading oligemia and depression in clinical forms with aura.
Dynamic study of dishabituation shows, **during the interictal period** of migraine, that response to various stressful stimuli overloads a metabolic strain on the brain of migraineurs. This increases the energy demands, triggers the activation of the trigemino-vascular system and leads to **migraine attack**.
During the ictal period, dishabituation and a higher level of cortical arousal can be normalized.

And *later regained* in the next cycle of the disease.

This new physiopathological data on migraine reveal that a shift in the brain metabolic homeostasis could be the main factor for migraine attacks.
The importance of photic dishabituation in **headaches with reactive photophobia** has been shown in studies of visual evoked potentials particularly not only for **migraine** (its typical form) but also in all **headaches with light hypersensitivity**. Between migraine and this type of headache, many authors have suggested **a continuum** of varying severity [1, 3, 39, 42-66] (*Figure 1*).

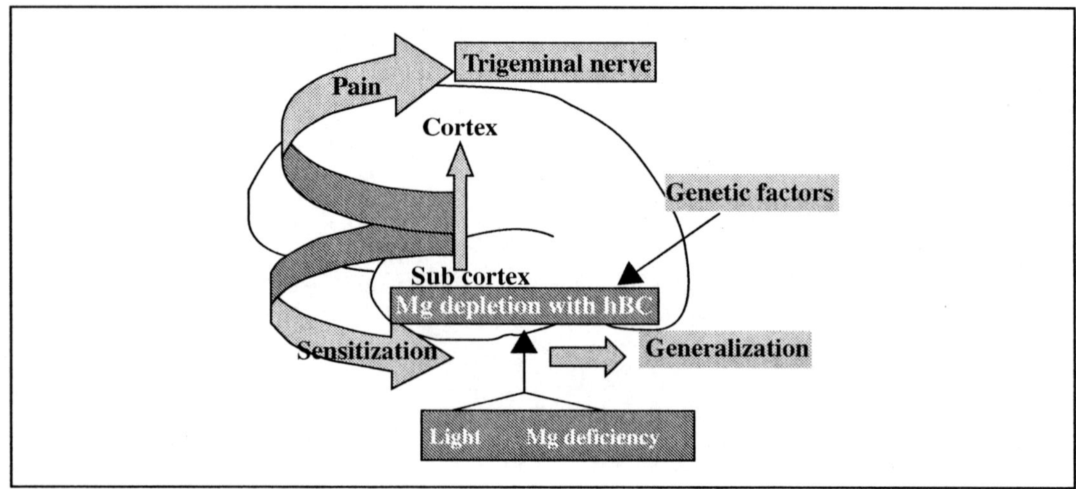

Figure 1. Migraine due to magnesium depletion with hypofunction of biological clock by photic sensitization. Subcortical and cortical genetic factors further the development of **a magnesium depletion** caused by a deficient magnesium dietary intake (**magnesium deficiency**) plus a **photic stress** [mainly circadian: (**diurnal**) and seasonal: (during «**fair**» **seasons**)] with reactive **photophobia** and **more or less generalized** [sensory, cognitive, painful (trigeminal particularly), alimentary] **sensitization**. It induces cortical and subcortical dysexcitability and spreading oligemia and depression in clinical form with aura. Between migraine and headaches with light hypersensitivity there is a continuum of varying severity.

Sudden Infant Death Syndrome (SIDS)

SIDS may be due to gestational Mg deficit: Mg deficiency or various forms of Mg depletion [67].

– SIDS may be caused by the fetal consequences of maternal *Mg deficiency* through an impaired control of Brown Adipose Tissue (BAT) thermoregulation, a mechanism leading to *a modified temperature set point*. SIDS may result from dysthermias: hypo- or hyperthermic forms. A possible prevention could rest on simple nutritional maternal Mg supplementation [4, 67].

– Various stresses in pregnant women or in the infant may convert a simple Mg deficiency into *Mg depletion*: stress in baby care such as bedding in prone position, environmental factors such as parental smoking, but the role of **chronopathological stress** appears to be too often neglected although it constitutes a clinical form of *primary hypofunction of the biological clock* [with its anatomical and clinical stigma such as reduced production of melatonin (↓ MT) and of its urinary metabolite: **6** **S**ulfatoxy-**M**ela**t**onin (↓ 6SMT)]. SIDS might be linked to an impaired maturation of both photoneuroendocrine system and brown adipose tissue (BAT) [4, 67].

Multiple Sclerosis (MS)

Some clinical forms of multiple sclerosis may be associated with primary disorders of magnesium status and of the biological clock (BC).

Magnesium depletion in Multiple Sclerosis

Several markers of the magnesium status have been studied in Multiple Sclerosis patients: decreased ionized magnesium concentration with normal total magnesium in plasma and serum, significantly decreased magnesium concentration in erythrocyte, decreased magnesium in brain especially in white matter without a decrease of myelin magnesium and with magnesium normal cerebrospinal fluid (but with decreased magnesium concentration in liver, spleen, heart and lung), low magnesium levels in rains and soils.

These various markers of Mg deficit may not be due to Mg deficiency, but testify to a clinical form of **Mg depletion**. We have highlighted the possible importance of several types of Mg depletion in the aetiopathogenesis of diverse neurodegenerative diseases, particularly of Mg depletion caused by **the association between** a nutritional factor: **insufficient intake of magnesium** (that is to say Mg deficiency) **and neurostress** (*i.e.* organic or inorganic neurotoxin, viral or parasitic neuroinfection, radiation, chronopathological stress...). This study will mainly focus on the importance of the **chronopathological stresses involved in multiple sclerosis** [1, 3, 28, 29, 68-77].

Clinical forms of Multiple Sclerosis with biological clock dysfunction

Multiple Sclerosis (MS) remains a neurological disease of unknown aetiology but it has been recognized since the early nineteenth century that it is a disease with a unique distribution. Rare in equatorial regions, it becomes increasingly common in higher latitudes. « There is an increased risk of acquiring multiple sclerosis in adult life the greater as the distance that a person has lived away from the equator during childhood and adolescence » [78]. The importance of this **latitude gradient** has been stressed in several following epidemiological studies. Various climatic variables which significantly influence the risk of multiple sclerosis when analyzed alone (hours of sunshine particularly) are found when they are adjusted for latitude to be due to their correlation with this variable [79-82].

Poor photo-stimulation (in winter particularly) may be a factor of increased melatonin production, that is to say of a biological clock hyperfunction [83]. Constantinescu *et al.* have shown that luzindole, a melatonin receptor antagonist, suppresses experimental autoimmune encephalomyelitis (a classical model for multiple sclerosis) [84], but these data have not been confirmed by Maestroni [85]. Phototherapy (which suppresses melatonin production namely) seems rational in this clinical form of multiple sclerosis. Its protective effect may be due not only to melatonin suppression but also to depression of immune response, suppression of inflammatory leukotrienes and cytokines and to increased production of vitamine D (through ultra violet radiation particularly) [1, 70, 86-92].

Further research, with investigations of melatonin (MT) and of its urinary metabolite 6 sulfatoxymelatonin (6SMT) production particularly, will be necessary to determine the frequency of this clinical form of Multiple Sclerosis with Biological Clock Hyperfunction (MS with HBC).

Conversely the importance of the clinical form of **Multiple Sclerosis with Biological Clock hypofunction** (MS with hBC) has been better documented. The main marker of the biological clock, the **nocturnal plasma melatonin levels are decreased** in multiple sclerosis patients. Although melatonin levels were unrelated to the patient age and gender, there was a positive correlation with age of onset of symptoms and an inverse correlation with the duration of illness. Physiological and chronobiological factors for decrease in **melatonin production** are similarly *deleterious factors for multiple sclerosis*: neonatal period, puberty, delivery; diurnal, seasonal and climatic photostimulation.

Multiple sclerosis may directly induce lesions of the Biological Clock. Hypothalamic lesions are frequent in multiple sclerosis. "Systematic pathological investigation of the hypothalamus in multiple sclerosis reveals an unexpected high incidence of active lesions that may impact on hypothalamic functioning" [103]. They may concern anterior hypothalamus where the mammalian circadian oscillator (the SupraChiasmatic Nuclei) is located. The prevalence of Pineal Calcification on Computerized Tomography (CT) scan was seen in 100% of studied multiple sclerosis patients. **Hypofunction of Biological Clock** (hBC) may be caused by **Primary alterations of the Biological Clock** and/or due to **Secondary factors of hypofunction**, seasonal and climatic factors particularly [1, 3, 67, 93-104].

The previously described NORTH-SOUTH gradient is inkeeping with a chronopathological form of multiple sclerosis with **Hyper**function of the Biological Clock (HBC), **where the effects of sun light were beneficial.** Low latitude decreases the risk for multiple sclerosis. But conversely many epidemiologic data have shown *numerous exceptions to the theory of the latitude gradient.*

They highlight **the noxious effects of photostimulation** through migration studies, cluster (or high frequency zone) studies, case controls studies. They agree with chronopathological forms of multiple sclerosis with **hypo**function of the Biological Clock (hBC) with a decreased level of melatonin (↓ **MT**).

This climatic factor of **noxious photostimulation** may be associated with synergic effects of diurnal and seasonal sun exposure: **deleterious effects of sunbathing** and of higher photostimulation during fair seasons: **spring and summer**. Similarly **fair skin** was associated with an increased risk for multiple sclerosis [1, 3, 76, 81, 82, 90, 105-117].

Three types of comorbidities should be stressed: anxiety, sleep disorders and migraine.

– *Anxiety disorders* (panic attack and generalized anxiety disorder) are common in multiple sclerosis and frequently overlooked. This is often due to the difficulty differentiating anxiety from personality correlates, or reactive tendencies in patients with chronic neurologic disease. Anxiety in patients with multiple sclerosis is more often comorbid with depression than alone. Anxiety rating scales median score were respectively 18 in the multiple sclerosis patients (14 in chronic rheumatoid disease) and 6 in the healthy controls [3, 96, 118-124].

– *Sleep disturbances* in multiple sclerosis are common but heterogeneous: from hypersomnia to various types of dyssomnias. A number of circadian rhythm sleep disorders may be observed: a delayed sleep phase syndrome is inkeeping with hypofunction of the Biological Clock (hBC). In subgroups of multiple sclerosis patients, sleep latencies may be reduced: the decrease of mean sleep onset latencies agrees with sleep disorders in magnesium deficit. The studies of sleep disturbances disagree for a generalized circadian disturbance in multiple sclerosis patient, but in subgroups, sleep studies show sleep disorders which are inkeeping with a clinical form of multiple sclerosis induced by magnesium depletion with hypofunction of the biological clock [28, 125-131].

– *Migraine* may be associated with multiple sclerosis but its incidence has not been clearly established perhaps because the diagnostic criteria of migraine are difficult to identify among the diverse neurologic symptoms of multiple sclerosis. Watkins *et al.* found a 27% incidence of migraine in a cohort of 100 consecutive multiple sclerosis patients compared to a 12% incidence in a random selection of age-matched controls. These patients were reported also have twice the incidence of migraine in their family members. Besides cooccurence of multiple sclerosis and migraine, a high incidence of family history of migraine in multiple sclerosis patients is also observed. In another study of 104 consecutive patients the incidence of migraine was 8% [137]. Zorzon *et al.* assessed the risk of multiple sclerosis after information related to demographic data, socioeconomic status, education, ethnicity, changes of domicile, migration, occupation, environmental, nutritional and hormonal factors, exposure to various infections agents, vaccination and family history of diseases. In multiple logistic regression analysis, they found four independent risk factors for multiple sclerosis. Migraine was one of these four risk factors and was frequently comorbid with multiple sclerosis. This comorbidity may rely on increased plasma levels of endothelin-1: a potent vasoconstrictor and a mediator in the inflammatory process [through matrix-protease 2 (MMP2) particularly]. These disorders are inkeeping with the well-known similar disturbances due to magnesium deficit, while the noxious effects of sun exposure agrees with a subgroup of multiple sclerosis with hypofunction of the biological clock [3, 20, 28, 75, 100, 108, 132-140].

These various aspects of Biological Clock hypofunction may be treated by darkness therapy [1, 3, 4, 67] (*Figure 2*).

Darkness therapy

The psycholeptic (or sedative) properties of darkness therapy mirror the psychoanaleptic (or stimulant) properties of phototherapy.

Complexity of the mechanisms [1]

– The mechanisms of the action of darkness appear as the reverse of those obtained with bright light where direct cellular effects (membranous and redox) and neural mediated effects intervene.

– Increased production of melatonin (↑MT) constitutes the best marker of darkness, but it is only an accessory mechanism in the action of darkness.

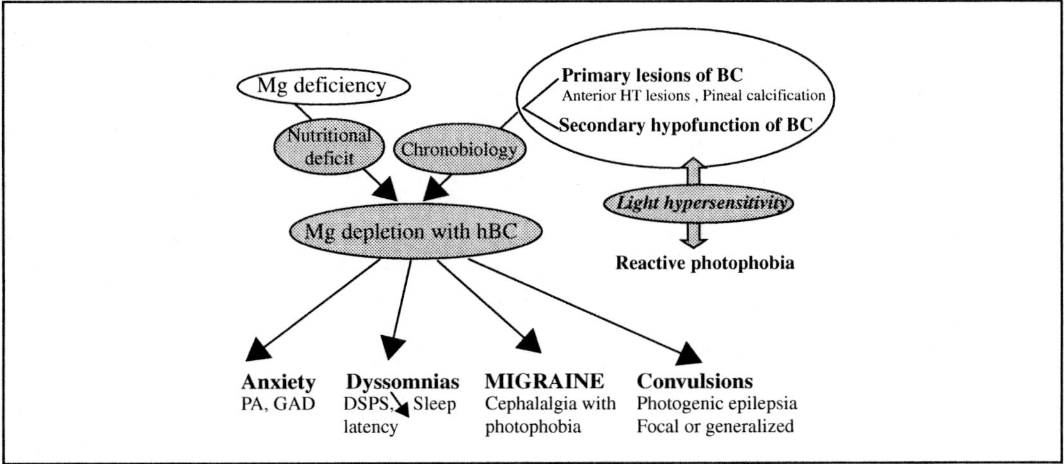

Figure 2. Multiple sclerosis due to magnesium depletion with hypofunction of biological clock. Multiple sclerosis is a heterogeneous disease namely through its etiopathogeny which associate genetic factors and environmental factors (neuroinfections, neurotoxins, radiation, nutritional and chronobiological factors). Among the nutritional and chronobiological etiological forms of multiple sclerosis the importance of multiple sclerosis due to magnesium depletion with hypofunction of biological clock is highlighted. Photic stresses induce **hypofunction of biological clock** which is able to change magnesium deficiency reversible by mere nutritional supplement into **magnesium depletion** not reversible by nutritional supplementation but only by specific chronobiological treatment: *i.e.* darkness therapy. This magnesium depletion with hypofunction of biological clock: - may induce multiple sclerosis, in particular with some specific flickering fields for multiple sclerosis (genetics and environmental: such as neuroinfections, neurotoxins, nutritional factors, radiation...); - but may also induce several photosensitive disorders, anxiety, dyssomnias, MIGRAINE and convulsions, all diseases frequently comorbid with multiple sclerosis.

– The main central neural mechanisms of darkness therapy associate decreased serotonergy (↓5HT) - which could account for the antimigraine effect - and stimulation of inhibitory neuromodulators (↑GABA, ↑TA, ↑kO) and of anti-inflammatory and antioxidative processes - which may induce neural-hypoexcitability (sedative and anticonvulsant effects).

– Humoral transduction may reinforce these last effects by decreasing neuroactive gases (↓CO, ↓NO) through binding of CO with Hb and by increasing melatonin, bilirubin and biliverdin: three antioxidants which have the capacity to quench NO.

Apart from the exception of decreased serotonergy, these effects of darkness are similar to those of magnesium [1].

Methods [1]

Darkness therapy per se

Light deprivation may be obtained by placing the patient in a closed room, in a totally dark environment, with an eye mask on.

This genuine darkness therapy may be used in acute indications, but should be of short duration. It is not compatible with any activity and is frequently associated with induction of bed rest, inactivity and sleep [1, 141].

Relative darkness therapy may be obtained by wearing dark goggles or dark sun glasses but the number of lux passing through is not negligible. This relative darkness therapy may be used as an accessory treatment in the restoration of a light dark schedule: a transition before a totally dark environment [1].

Darkness mimicking agents

Melatonin (MT) is the prototype of darkness mimicking agents. But it appears to be only an accessory factor among the mechanisms of photoperiod actions. Most of the other mechanisms of the effects of darkness have been overlooked, which may account for the controversy around the therapeutic efficiency of MT. Its posology varies from physiological doses (around 0.3 mg) to pharmacological doses (which testify to the weak toxicity of the hormone): usually 3 mg/per dose and per day and even up to 300 mg as a contraceptive. In case of chronopathology, with decreased MT production, MT constitutes a substitutive treatment of its deficiency [1].

Magnetic fields may be used to stimulate the biological clock in a variety of ways to treatment using very weak (picotesla), extremely low frequency (2 to 7 Hz) electromagnetic fields. Transcranial treatment with alternative currents pulsed electromagnetic fields of picotesla flux density may stimulate various brain areas (hypothalamus particularly) and pineal gland (which functions as a magneto receptor). Several studies on its use for treatment of anxiety, migraine and multiple sclerosis particularly [142-147] stressed "the beneficial effects of electromagnetic fields treatment on mood, level of fatigue and cognitive functions with improvements in short and long term memory, alertness, level of energy, concentration, attention, word finding, reading ability, visuospatial and visuoconstructive skills. But the neurological community, the multiple sclerosis organizations and the press remained uninterested in this revolution in multiple sclerosis management" [143]. However a double blind placebo controlled trial has shown that this therapy can alleviate symptoms of multiple sclerosis, although the clinical effects were small [147].

Chromatotherapy uses a short exposure (4 min.) to a precise wavelength spectrum: orange or green for the treatment of hypofunction of biological clock. This method, even successfully used in practice, has not been validated yet [1, 3, 4].

Magnesium, L tryptophan and taurine may also act as darkness-mimicking agents.

– To stimulate the biological clock, it seems well advised to facilitate the neural function of suprachiasmatic nuclei (↑SCN) and the hormonal pineal production (↑MT). The deleterious effects of light and those of magnesium deficiency are often found together and might be partly palliated by **a nutritional magnesium supply** (↑Mg), providing the best possible link between photoperiod and magnesium status [1, 3, 4] (for example the preventive treatment of SIDS must be completed by the prophylactic therapy of its possible chronopathological factor: atoxic nutritional magnesium supplementation for pregnant women and total light deprivation at night for infants over the first year, this latter prescription in agreement with several pioneering studies [4, 67, 148-153]).

Pharmacological use of magnesium is uncertain and apt to induce toxicity. Choice and dose of the Mg salts, oral or parenteral route and indications for the mother or the infant, association with "Mg-fixing agents" remain imprecise [4, 154, 155].

– Supplementation in L tryptophan (↑LTP) may stimulate the tryptophan pathway but may induce toxicity: eosinophilia-myalgia syndrome particularly [1, 3, 4, 156-163].

– Taurine (↑TA) may act as a protective inhibitory neuromodulator which participates in the functional quality of the neural apparatus and in melatonin production and action. Taurine plays a role in the maintenance of homeostasis in the central nervous system, during central nervous hyperexcitability particularly. Taurine a volume-regulating aminoacid is released upon excitotoxicity induced cell swelling. Taurine has an established function as an osmolyte in the central nervous system [1-4, 28, 29, 164-171].

Conclusion

The different clinical forms of magnesium depletion with hypofunction of the biological clock, psychic, algic, dyssomnic, neuromuscular disorders due to either primary disturbances of the biological clock (genetic, ontogenic, infectious or neurodegenerative disorders) or secondary to light hypersensitivity (with reactive photophobia) show the importance of the etiopathogenic chronopathological mechanisms.

Further research will be necessary to determine the place of the various modes of "darkness therapy associated with a balanced magnesium status" in indications as various as prevention of sudden infant death, migraine and multiple sclerosis.

References

1. Durlach J, Pagès N, Bac P, Bara M, Guiet-Bara A. Biorhythms and possible central regulation of magnesium status, phototherapy, darkness therapy and chronopathological forms of Mg depletion. *Magnes Res* 2002; 15: 49-66.
2. Durlach J. Editorial Policy of Magnesium Research: General considerations on the quality criteria for biomedical papers and some complementary guidelines for the contributors of Magnesium Research. *Magnes Res* 1995; 8: 191-206.
3. Durlach J, Pagès N, Bac P, Bara M, Guiet-Bara A, Agrapart C. Chronopathological forms of magnesium depletion with hypofunction or with hyperfunction of the biological clock. *Magnes Res* 2002; 15: 263-8.
4. Durlach J, Pagès N, Bac P, Bara M, Guiet-Bara A. Magnesium deficit and sudden infant death syndrome (SIDS): SIDS due to magnesium deficiency and SIDS due to various forms of magnesium depletion: possible importance of the chronopathological form. *Magnes Res* 2002; 15: 269-78.
5. Main A, Vlakonikolig I, Dowson A. The wavelength of light causing photophobia in migraine and tension-type headache between attacks. *Headache* 2000; 40: 194-9.
6. Keller M, Wiedemann Z, Zihl J. Illumination perception in photophobic patients suffering from panic disorder with agoraphobia. *Acta Psychiatr Scand* 1997; 96: 72-4.
7. Salvesen R, Bakkelund SI. Migraine as compared to other headaches is worse during midnight-sun summer than during polar night. A questionnaire study in an Arctic population. *Headache* 2000; 40: 824-9.
8. Fox AW, Davis RL. Migraine chronobiology. *Headache* 1998; 38: 436-41.
9. Goto Y, Furuta A, Tobimatsu S. Magnesium deficiency differentially affects the retina and visual cortex of intact rats. *J Nutr* 2001; 131: 2378-81.
10. Mulleners WM, Chronicle EP, Vredeveld JW, Koehler PJ. Visual cortex excitability before and after valproate prophylaxis: a pilot study using TMS. *Eur J Neurol* 2002; 9: 35-40.
11. Claustrat B, Brun J, Borson-Chazot F. Mélatonine et rythmes circadiens. *Rev Neurol* 2001; 157(5S): 121-125S.
12. Cohen-Mansfield J, Garfinkel D, Lipson S. Melatonin for treatment of sundowning in elderly persons with dementia: a preliminary study. *Arch Gerontol Geriatr* 2000; 31: 65-76.
13. Parain D. Les épilepsies photosensibles généralisées ou focales. *Rev Neurol* 1998; 154: 757-61.
14. Salas-Puig J, Parra J, Fernandez-Torre JL. Photogenic epilepsy. *Rev Neurol* 2000; 30: S81-S84.
15. Harding GFA. TV can be bad for your health. *Nat Med* 1998; 4: 265-7.
16. Panayotopoulos CP. Fixation-off, scotosensitive and other visual-related epilepsies. In: Zifkin BG, *et al.*, eds. *Reflex epilepsies and reflex seizures: Advances in neurology*. Philadelphia: Lippincott-Raven, 1998: 139-57.
17. Durlach J. Chronic fatigue syndrome and chronic primary magnesium deficiency. *Magnes Res* 1992; 5: 68.
18. Sandrini G, Proietti Cecchini A, Nappi G. Chronic fatigue syndrome: a borderline disorder. *Funct Neurol* 2002; 17: 51-2 (Abstract).
19. Wikner J, Hirsh U, Nettenberg L, Röjdmark S. Fibromyalgia: a syndrome associated with decreased nocturnal MT secretion. *Clin Endocrinol (Oxf)* 1998; 49: 179-83.
20. Vivayan N, Gould S, Watson C. Exposure to sun and precipitation of migraine. *Headache* 1980; 20: 42-3.
21. Blau JN. Migraine pathogenesis: the neural hypothesis reexamined. *J Neurol Neurosurg Psychiatry* 1984; 47: 437-42.
22. Drummond PD. A quantitative assessment of photophobia in migraine and tension headache. *Headache* 1986; 26: 465-9.
23. Woodhouse A, Drummond PD. Mechanisms of increased sensitivity to noise and light in migraine headache. *Cephalalgia* 1993; 13: 417-20.
24. Main A, Dawson A, Gross M. Photophobia and phonophobia in migraineurs between attacks. *Headache* 1997; 37: 492-5.
25. Vingen JV, Sand T, Stovner LJ. Sensitivity to various stimuli in primary headaches: a questionnaire study. *Headache* 1999; 39: 552-8.
26. Rossi LN, Cortinovis I, Menegazzo L, Brunelli G, Bossi A, Macchi M. Classification criteria and distinction between migraine and tension headache in children. *Dev Med Child Neurol* 2001; 43: 45-51.
27. Durlach J, Bac P, Durlach V, Bara M, Guiet-Bara A. Neurotic, neuromuscular and autonomic nervous form of magnesium imbalance. *Magnes Res* 1997; 10: 169-95.
28. Durlach J, Bara M. In: *Le Magnésium en biologie et en médecine*. EM Inter Publ Cachan 2000: 96-9.

29. Durlach J, Bac P, Bara M, Guiet-Bara A. Physiopathology of symptomatic and latent forms of central nervous hyperexcitability due to magnesium deficiency: a current general scheme. *Magnes Res* 2000; 13: 293-302.

30. Ramadan NM, Halvorson H, Vandelinde A, Levine S, Helpern JA, Welsh KMA. Low brain magnesium in migraine. *Headache* 1989; 29: 590-3.

31. Schoenen J, Sianard-Gainko J, Lenaerts M. Blood magnesium levels in migraine. *Cephalalgia* 1991; 11: 97-9.

32. Thomas J, Thomas E, Tomb E. Serum and erythrocyte magnesium concentrations and migraine. *Magnes Res* 1992; 5: 127-30.

33. Gallai V, Sarchielli P, Costa G, Firenze C, Mozucci P, Abbritti G. Serum and salivary magnesium levels in migraine. Results in a group of juvenile patients. *Cephalalgia* 1992; 32: 132-5.

34. Castelli S, Meossi C, Domenici R, Fontana F, Stefani G. Il magnesio nella profilassi della cefaleo primaria e di altri disturbi periodici del bambino. *Ped Med Chir* 1993; 15: 481-8 (Med Surg Ped).

35. Mauskop A, Altura BT, Cracco RQ, Altura BM. Deficiency in serum ionized magnesium but not total magnesium in patients with migraines. *Headache* 1993; 33: 135-8.

36. Gallai V, Sarchielli P, Mozucci P, Abbritti G. Red blood cell magnesium levels in migraine patients. *Cephalalgia* 1993; 13: 74-81.

37. Soriani S, Arnaldi C, de Carlo L, *et al.* Serum and red blood cell magnesium levels in juvenile migraine patients. *Headache* 1995; 35: 14-6.

38. Pfaffenrath V, Wessely P, Meyer C, *et al.* Magnesium in the prophylaxis of migraine: a double blind placebo-controlled study. *Cephalalgia* 1996; 16: 340-6.

39. Aloisi P, Marreli A, Porto C, Tozzi F, Cerone G. Visual evoked potentials and serum magnesium levels in juvenile migraine patients. *Headache* 1997; 37: 383-5.

40. Mishima K, Takeshima T, Shimomura T, *et al.* Platelet ionized magnesium, cyclic AMP and cyclic GMP levels in migraine and tension-type headache. *Headache* 1997; 37: 561-4.

41. Lodi R, Iotti S, Cortelli P, *et al.* Deficient energy metabolism is associated with low free magnesium in the brains of patients with migraine and cluster headache. *Brain Res Bull* 2001; 54: 437-41.

42. Teppert JJ, Rapoport A, Sheftell F. The pathophysiology of migraine. *Neurology* 2001; 7: 279-86.

43. Boska MD, Welch KM, Barker PB, Nelson JA, Schultz L. Contrast in cortical magnesium, phospholipid and energy metabolism between migraine syndromes. *Headache* 2002; 42: 114-9.

44. Bigal ME, Rapoport AM, Sheftell FD, Tepper SJ. New migraine preventive options: an update with pathophysiological considerations. *Rev Hosp Clin Fac Med Sao Paulo* 2002; 57: 293-8.

45. Thompson RF, Spencer WA. Habituation: a model phenomenon for the study of neuronal substrates behaviour. *Psychol Rev* 1996; 73: 16-43.

46. Monnier M, Boehmer A, Scholer A. Early habituation, dishabituation and generalization induced in the visual centres by colour stimuli. *Vision Res* 1976; 16: 1497-504.

47. Schoenen J. Clinical neurophysiology studies in headache: a review of data and pathophysiological hints. *Funct Neurol* 1992; 7: 191-204.

48. Schoenen J. Deficient habituation of evoked cortical potentials in migraine: a link between brain biology, behaviour and trigeminovascular activation? *Biomed Pharmacother* 1996; 50: 71-8.

49. Wang W, Wang GP, Ding XL, Wang YH. Personality and response to repeated visual stimulation in migraine and tension type headache. *Cephalalgia* 1999; 19: 719-24.

50. Ambrosini A, Schoenen J. The electrophysiology of migraine. *Curr Opin Neurol* 2003; 16: 327-31.

51. Marcus DA. Migraine and tension type headaches: the questionable validity of the current classification system. *Clin J Pain* 1992; 8: 28-36.

52. Farkila M. The pathophysiology of migraine. *Ann Med* 1994; 26: 7-8.

53. Spierings ELH. Migraine. Questions and answers. *Merit Publ Internat* 1995: 54-5.

54. de Tomaso M, Sciruicchio V, Guido M, Sasanelli G, Puca F. Steady state visual-evoked potentials in headache: diagnostic value in migraine and tension type headache patients. *Cephalalgia* 1999; 19: 23-6.

55. Evers S, Quibeldey F, Grotemeyer KH, Suhr B, Husstedt IW. Dynamic changes of cognitive habituation and serotonin metabolism during the migraine interval. *Cephalalgia* 1999; 19: 485-91.

56. van Dijk JG. Neurophysiological evidence of increased cortical reactivity in migraine. *Funct Neurol* 2000; 15: 73-7 (Suppt. to n°3).

57. Sand T, Vanagaite Vingen J. Visual long-latency auditory and brainstem auditory evoked potentials in migraine: relation to pattern size, stimulus intensity, sound and light discomfort thresholds and pre-attack state. *Cephalalgia* 2000; 20: 804-20.

58. Sheperd AJ. Visual contrast processing in migraine. *Cephalalgia* 2000; 20: 865-80.

59. Bowyer SM, Aurora SK, Moran JE, Tepley N, Welch KMA. Magnetoencephalographic fields from patients with spontaneous and induced migraine aura. *Ann Neurol* 2001; 50: 582-7.

60. Welch KMA, Bowyer SM, Aurora SK, Moran JE, Tepley N. Visual-stress induced migraine compared to spontaneous aura studied by magnetoencephalography. *J Headache Pain* 2001; 2: S131-S136.

61. Sheperd AJ. Increased visual after-effects following pattern adaptation in migraine: a lack of intracortical excitation. *Brain* 2001; 124: 2310-8.

62. Bäcker M, Sander D, Hammes MG, *et al.* Altered cerebrovascular pattern in interictal migraine during visual stimulation. *Cephalalgia* 2001; 21: 611-6.

63. Legrain V, Janne P, Laloux P, Ossemann M, Dupuis M, Regnaert C. Intérêts cliniques et physiopathologiques des potentiels évoqués cognitifs dans la migraine (Clinical and pathophysiological contribution of event-related potentials used to study migraine headache). *Rev Neurol* 2001; 157: 365-75.

64. Shepperd AJ, Palmer JE, Davis G. Increased visual after-effects in migraine following pattern adaptation extend to stimultaneous tilt illusion. *Spat Vis* 2002; 16: 33-4.

65. De Marinis M, Pujia A, Natale L, D'arcangelo E, Accornero N. Decreased habituation of the R2 component of the blink reflex in migraine patients. *Clin Neurophysiol* 2003; 114: 889-93.

66. Friberg L, Sandrini G, Jänig W, *et al.* Instrumental investigations in primary headache. An updated review and new prospectives. *Funct Neurol* 2003; 18: 127-44.

67. Durlach J, Pagès N, Bac P, Bara M, Guiet-Bara A. New data on the importance of gestational magnesium deficiency. *Magnes Res* 2004; 17: 116-25.

68. Heipertz R, Eickhoff K, Karstens KH. Magnesium and inorganic phosphate content in CSF related to blood brain barrier function in neurological disease. *J Neurol Sci* 1979; 40: 87-95.

69. Moscarello MA, Chia LS, Leighton D, Absolom D. Size and surface charge properties of myelin vesicles from normal and diseased (Multiple Sclerosis) brain. *Neurochem* 1985; 45: 415-21.

70. Hasanen E, Kinnunen E, Alhonen P. Relationship between the prevalence of Multiple Sclerosis and some physical and chemical properties of soil. *Sci Total Environ* 1986; 58: 263-72.

71. Yasui M, Yase Y, Ando K, Adachi K, Mukoyama M, Ohsugi K. Magnesium concentration in brains from Multiple Sclerosis patients. *Acta Neurol Scand* 1990; 81: 187-200.

72. Yasui M, Ota K. Experimental and clinical studies on dysregulation of magnesium metabolism and the aetiopathogenesis of Multiple Sclerosis. *Magnes Res* 1992; 5: 295-302.

73. Altura BT, Bertschat F, Jeremias A, Ising H, Altura BM. Comparative findings on serum IMg^{2+} of normal and diseased human subjects with the NOVA and KONE ISE's for Mg^{2+}. *Scand J Clin Lab Invest* 1994(Suppl 217): 77-81.

74. Slelmasiak Z, Solski J, Jakubowska B. Magnesium concentration in plasma and erythrocytes in Multiple Sclerosis. *Acta Neurol Scand* 1995; 92: 109-11.

75. Durlach J, Bac P, Durlach V, Bara M, Guiet-Bara A. Are age-related neurodegenerative diseases linked with various types of magnesium depletion? *Magnes Res* 1997; 10: 339-53.

76. Johnson S. The possible role of gradual accumulation of Cu, Cd, Pb and Fe and gradual depletion of Zn, Mg, Se, vitamins B2, B6, D and E and essential fatty acids in Multiple Sclerosis. *Med Hypotheses* 2000; 55: 239-41.

77. Bolviken B, Celius EG, Nilsen R, Strand T. Radon: a possible risk factor in Multiple Sclerosis. *Neuroepidemiology* 2003; 22: 87-94.

78. Davenport CB. Multiple Sclerosis from the standpoint of geographic distribution and race. *Arch Neurol Psychiatry* 1922; 8: 51-60.

79. Norman JE, Kurtzke JF, Beebe GW. Epidemiology of Multiple Sclerosis in U.S. veterans: 2. Latitude, climate and the risk of Multiple Sclerosis. *J Chronic Dis* 1983; 36: 551-9.

80. Rosen LN, Livingstone IR, Rosenthal NE. Multiple Sclerosis and latitude: a new perspective and an old association. *Med Hypotheses* 1991; 36: 376-8.

81. Hutter CDD, Laing P. Multiple Sclerosis: Sunlight, diet, immunology and aetiology. *Med Hypotheses* 1996; 46: 67-74.

82. Carlyle IP. Multiple Sclerosis: a geographical hypothesis. *Med Hypotheses* 1997; 49: 477-86.

83. Compston DA, Batchelor JR, Earl CJ, McDonald WI. Factors influencing the risk of Multiple Sclerosis developing in patients with optic neuritis. *Brain* 1978; 101: 495-511.

84. Contantinescu CS, Hilliard B, Ventura E. Luzindole, a melatonin receptor antagonist, suppresses experimental antoimmune encephalomyelitis. *Pathobiology* 1997; 65: 190-4.

85. Maestroni GJM. The immunotherapeutic potential of melatonin. *Exp Opin Invest Drugs* 2001; 10: 467-76.

86. Constantinescu CS. Melanin, melatonin, MSH and the susceptibility to antoimmune demyelinisation: a rationale to light therapy in Multiple Sclerosis. *Med Hypotheses* 1995; 45: 455-8.

87. McMichael AJ, Hall AJ. Does immunosuppressive UV radiation explain the latitude gradient for Multiple Sclerosis? *Epidemiology* 1997; 8: 642-5.

88. Ponsonby AL, McMichael A, van der Mei I. UV radiation and antoimmune disease: insights from epidemiological research. *Toxicology* 2002; 181-182: 71-8.

89. Staples JA, Ponsonby AL, Lim LL, McMichael AJ. Ecologic analysis of some immune-related disorders including type I diabetes in Australia: latitude regional UV radiation and disease prevalence. *Environ Health Perspect* 2003; 111: 518-23.

90. van der Mei IA, Ponsonby AL, Dwyer T, *et al.* Past exposure to sun, skin phenotype and risk of Multiple Sclerosis. *BMJ* 2003; 327: 316.

91. Hayes CE. Vitamin D: a natural inhibitor of Multiple Sclerosis. *Proc Nutr Soc* 2000; 59: 531-5.

92. Zittermann A. Vitamin D in preventive medicine: are we ignoring the evidence? *Br J Nutr* 2003; 89: 552-72.

93. Sandyk R, Awerbuch GI. The pineal gland in multiple sclerosis. *Intern J Neurosc* 1991; 61: 61-7.

94. Sandyk R, Awerbuch GI. Nocturnal plasma MT and MSH levels during exacerbation of multiple sclerosis. *Intern J Neurosc* 1992; 67: 173-86.

95. Sandyk R. Multiple sclerosis: The role of puberty and the pineal gland in its pathogenesis. *Intern J Neurosc* 1993; 68: 209-25.

96. Sandyk R. Nocturnal MT secretion in multiple sclerosis patients with effective disorders. *Intern J Neurosc* 1993; 68: 227-40.

97. Sandyk R, Awerbuch GI. Multiple sclerosis: relationship between seasonal variations of relapse and age of onset. *Intern J Neurosc* 1993; 71: 147-57.

98. Sandyk R, Awerbuch GI. Relationship of nocturnal MT levels to duration and course of multiple sclerosis. *Intern J Neurosc* 1994; 75: 229-37.

99. Sandyk R, Awerbuch GI. The relationship of pineal calcification to cerebral atrophy on CT scan in multiple sclerosis. *Intern J Neurosc* 1994; 76: 71-9.

100. Sandyk R, Awerbuch GI. The cooccurence of multiple sclerosis and migraine headache: the serotoninergic link. *Intern J Neurosc* 1994; 76: 249-57.

101. Sandyk R. Role of the pineal gland in multiple sclerosis: a hypothesis. *J Altern Complement Med* 1997; 3: 267-90.

102. Sakai N, Miyajima H, Shimizo T, Arai K. Syndrome of inappropriate secretion of ADH associated with MS. *Intern Med* 1992; 31: 463-6.

103. Huitinga I, de Groot CJ, van der Valk P, Kamphorst W, Tilders PJ, Swaab DF. Hypothalamic lesions in multiple sclerosis. *J Neuropathol Exp Neurol* 2001; 60: 1208-18.

104. Huitinga I, Erkurt ZA, van Beurden D, Swaab DF. Impaired hypothalamus-pituitary-adrenal axis activity and more severe multiple sclerosis with hypothalamic lesions. *Ann Neurol* 2004; 55: 37-45.

105. Neutel CI. Multiple sclerosis and the Canadian climate. *J Chronic Dis* 1980; 33: 47-56.

106. Bamford CR, Sibley WA, Thies C. Seasonal variation of multiple sclerosis exacerbations in Arizona. *Neurology* 1983; 33: 697-701.

107. Laborde JM, Dando WA, Teetzen ML. Climate, diffused solar radiation and multiple sclerosis. *Soc Sci Med* 1988; 27: 231-8.

108. Harbison JW, Calabrese VP, Edlich RF. A fatal case of sun exposure in a multiple sclerosis patient. *J Emerg Med* 1989; 7: 465-7.

109. O'Reilly MA, O'Reilly PM. Temporal influences on relapses of multiple sclerosis. *Eur Neurol* 1991; 31: 391-5.

110. Kurtzke JF, Delasnerie-Laupretre N. Reflection on the geographic distribution of multiple sclerosis in France. *Acta Neurol Scand* 1996; 93: 110-7.

111. Hayes CE, Cantorna MT, de Luca H. Vitamin D and multiple sclerosis. *PSEBM* 1997; 216: 21-7.

112. Hogancamp WE, Rodriguez M, Weinshenker BG. The epidemiology of multiple sclerosis. *Mayo Clin Proc* 1997; 72: 871-8.

113. Jin Y, de Pedro-Cuesta J, Soderstrom M, Stawiarz L, Link H. Seasonal patterns in optic neuritis and multiple sclerosis: a metanalysis. *J Neurol Sci* 2000; 181: 56-64.

114. Azoulay-Cayla A. La sclérose en plaques est-elle une maladie d'origine virale? (Is multiple sclerosis a disease of viral origin?). *Path Biol* 2000; 48: 4-14.

115. Rosati G. The prevalence of multiple sclerosis in the world: an update. *Neurol Sci* 2001; 22: 117-39.

116. Pugliatti M, Sotgiu S, Solinas G, *et al.* Multiple sclerosis epidemiology in Sardinia: evidence for a true increasing link. *Acta Neurol Scand* 2001; 103: 20-6.

117. Pugliatti M, Sotgiu S, Solinas G, Castiglia P, Rosati G. Multiple sclerosis prevalence among Sardinians: further evidence against the latitude gradient theory. *Neurol Sci* 2001; 22: 163-5.

118. Zivadinov R, Iona L, Monti-Bragadin L, *et al.* The use of the standardized incidence and prevalence rates in the epidemiological studies on multiple sclerosis. *Neuroepidemiology* 2003; 22: 65-74.

119. Feinstein A. O'Connor P, Gray T, Feinstein K. The effects of anxiety on psychiatric morbidity in patients with multiple sclerosis. *Mult Scler* 1999; 5: 323-6.

120. Riether AM. Anxiety in patients with multiple sclerosis. *Semin Clin Neuropsychiatry* 1999; 4: 103-13.

121. Minden SL. Mood disorders in multiple sclerosis: diagnosis and treatment. *J Neurovirol* 2000; 6: S160-S167.

122. Zorzon M, de Masi R, Nasuelli D, *et al.* Depression and anxiety in multiple sclerosis. A clinical and MRI study in 95 subjects. *J Neurol* 2001; 248: 416-21.

123. Defer G. Évaluation neuropsychologique et psychopathologique dans la sclérose en plaques (Neurophysiological and psychopathological assessment in multiple sclerosis). *Rev Neurol* 2001; 157: 1128-34.

124. Léger E, Ladouceur R, Freeston MH. Anxiété et limitation physique: une relation complexe (Anxiety and physical limitation: a complex relation). *Encephale* 2002; 28: 205-9.

125. Rumbach L, Tongio MM, Warter JM, Collard M, Kurtz D. Multiple sclerosis, sleep latencies and HLA antigens. *J Neurol* 1989; 236: 309-10.

126. Taphoorn MJ, van Someren E, Snoek FJ, *et al.* Fatigue, sleep disturbances and circadian rhythm in multiple sclerosis. *J Neurol* 1993; 240: 446-8.

127. Tachibana N, Howard RS, Hirsh NP, Miller DH, Moseley IF, Fish D. Sleep problems in multiple sclerosis. *Eur Neurol* 1994; 34: 320-3.

128. Ferini-Strambi L, Filippi M, Martinelli V, *et al.* Nocturnal sleep study in multiple sclerosis: correlations with clinical and brain Magnetic Resonance Imaging findings. *J Neurol Sci* 1994; 125: 194-7.

129. Aner RN, Rowlands CJ, Perry SF, Reunners JE. Multiple sclerosis with medullary plaques and fatal sleep apnea (Ondine's curse). *Clin Neuropathol* 1996; 15: 101-5.

130. Poirrier P. Photopériode, photothérapie et troubles du rythme veille-sommeil (Photoperiod, phototherapy and wakefulness-sleep rhythm disorders). *Rev Neurol* 2001; 157: S140-S144.

131. Iseki K, Mesaki T, Oka Y, *et al.* Hypersomnia in multiple sclerosis. *Neurology* 2002; 59: 2006-7.

132. Watkins SM, Espir M. Migraine and multiple sclerosis. *J Neurol Neurosurg Psychiatry* 1969; 32: 35-7.

133. Freedman MS, Gray TA. Vascular headache: a presenting symptom of multiple sclerosis. *Can J Neurol Sci* 1989; 16: 63-6.

134. Buchholtz DW, Reich SG. The menagerie of migraine. *Semin Neurol* 1996; 16: 83-93.

135. Haufschild T, Shaw SG, Kesselring J, Flammer J. Increased endothelin-I plasma levels in patients with multiple sclerosis. *J Neuro-Ophtalmol* 2001; 21: 37-8.

136. Evans RW, Rolak IA. Migraine versus multiple sclerosis. *Headache* 2001; 41: 97-8.

137. Fryze W, Zaborski J, Clonkowska A. Pain in the course of multiple sclerosis. *Neurol Neurochir Pol* 2002; 36: 275-84.

138. Pache M, Kaiser HJ, Akhalbedashvili N, *et al.* Extraocular blood flow and endothelin-I plasma levels in patients with multiple sclerosis. *Eur Neurol* 2003; 49: 164-8.

139. Pagès N, Gogly B, Godeau G, *et al.* Structural alterations of the vascular wall in Mg-deficient mice. A possible role of gelatinase A (MMP2) and B (MMP9). *Magnes Res* 2003; 16: 43-8.

140. Zorzon M, Zivadinov R, Nasuelli D, *et al.* Risk factors of multiple sclerosis: a case control study. *Neurol Sci* 2003; 24: 242-7.

141. Polacek L, Stein J. Experiences with sleep therapy of multiple sclerosis. *Cesk Neurol* 1959; 22: 20-9.

142. Mix E, Jensen HL, Lehmitz R, *et al.* Effect of pulsating electromagnetic field therapy on cell volume and phagocytosis activity in multiple sclerosis. *Psychiatr Neurol Med Psychol (Leipz)* 1990; 42: 457-66.

143. Sandyk R. Electromagnetic fields for treatment of multiple sclerosis. *Intern J Neuroscience* 1996; 87: 1-4.

144. Sandyk R. Therapeutic effects of alternating current pulsed electromagnetic fields in multiple sclerosis. *J Altern Complem Med* 1997; 3: 365-86.

145. Richards TL, Lappin MS, Lawrie FW, Stegrauer KC. Bioelectromagnetic applications for multiple sclerosis. *Phys Med Rehabil Clin N Am* 1998; 9: 659-74.

146. Brola W, Wegrzyn W, Czernicki J. Effect of variable magnetic field on motor impairment and quality of life in patients with multiple sclerosis. *Wiad Lek* 2002; 55: 136-43.

147. Lappin MS, Lawrie FW, Richards TL, Kramer ED. Effects of a pulsed electromagnetic therapy on multiple sclerosis. Fatigue and quality of life: a double blind, placebo controlled trial. *Altern Ther* 2003; 9: 38-48.

148. Hedlmaier G, Hoffmann K. Melatonin stimulates growth of brown adipose tissue. *Nature* 1974; 247: 224-5.

149. Lawson K, Daum C. Turkewitz. Environmental characteristics of a neonatal intensive-care unit. *Child Dev* 1977; 48: 1633-9.

150. Mann NP, Haddow R, Strokes L, Rutter N. Effect of night and day on preterm infants in a newborn nurserey: randomized trial. *BMJ* 1986; 293: 1265-7.

151. Auliciems A, Barnes A. Sudden Infant deaths and clear weather in a subtropical environment. *Soc Sci Med* 1987; 24: 51-6.

152. Nelson EAS, Taylor BJ. Climatic and social association with post-neonatal mortality rates in New Zeland. *New Zeland Med J* 1988; 101: 443-6.

153. Blacburn S, Patteson D. Effects of cycled light on activity state and cardiorespiratory function in preterm infants. *J Perinat Neonatal Nurs* 1991; 4: 47-54.

154. Goldberg P, Fleming MC, Picard EH. Decreased relapse rate through dietary supplementation with Ca, Mg and vitamin D. *Med Hypotheses* 1986; 21: 193-200.

155. Durlach J, Durlach V, Bac P, Bara M, Guiet-Bara A. Magnesium and therapeutics. *Magnes Res* 1994; 7: 313-28.

156. Hyypa MT, Jolma T, Riekkinen P, Rinne UK. Effects of L-Tryptophan treatment on central indoleamine metabolism and short lasting-neurologic disturbances in multiple sclerosis. *J Neural Transm* 1975; 37: 297-304.

157. Scott Jr. CF, Cashman N, Spitler LE. Experimental allergic encephalitis: treatment with drugs which alter CNS serotonin levels. *J Immunopharmacol* 1982–1983; 4: 153-62.

158. Heiman-Patterson TD, Bird SJ, Parry GJ, *et al.* Peripheral neuropathy associated with eosinophilia-myalgia syndrome. *Ann Neurol* 1990; 28: 522-8.

159. Arnouts PJ, Colemont LJ, van Outryve MJ, van Moer EM. L-Tryptophan-induced eosinophilia-myalgia syndrome. *J Intern Med* 1991; 230: 83-6.

160. Mayeno AN, Gleich GJ. Eosinophilia-myalgia syndrome and tryptophan production: a cautionary tale. *Trends Biotechnol* 1994; 12: 346-52.

161. Sternberg EM. Pathogenesis of L-tryptophan eosinophilia-myalgia syndrome. *Adv Exp Med Biol* 1996; 398: 325-30.

162. Sandyk R. Tryptophan availability and the susceptibility to stress in multiple sclerosis: a hypothesis. *Int J Neurosci* 1996; 86: 47-53.

163. Gross B, Ronen N, Honigman S, Livne E. Tryptophan toxicity: time and dose response in rats. *Adv Exp Med Biol* 1999; 467: 507-16.

164. Lopez-Colome A, Erlig D, Pasantes-Morales H. Different effects of Ca flux blocking agents on light and K stimulated release of taurine from retina. *Brain Res* 1976; 113: 527-34.

165. Pasantes-Morales H, Ademe RM, Quesada O. Protective effects of taurine on the light induced disruption of isolated frog rot outer segments. *J Neurosci Res* 1981; 6: 337-48.

166. Sturman JA, Lu P, Xu YX, Imaki H. Feline maternal taurine deficiency: effects on visual cortex of the offspring. A morphometric and immunohistochemical study. In: Huxtable R, Michalk DV, eds. *Taurine in Health and Disease*, New York: Plenum Publ, *Adv In Exp Med Biol*, 1994: 359; 369-92.

167. Stover JF, Pleines VE, Morganti-Kossman MC, Kossman T, Lowitzch K, Kempski OS. Neurotransmitters in cerebrospinal fluid reflect pathological activity. *Eur J Clin Invest* 1997; 27: 1038-43.

168. Pasantes-Morales H, Quesada O, Moran J. Taurine: an osmolyte in mammalian tissue. In: Schaffer S, Lombardini JB, Huxtable RJ, eds. *Taurine 3*. New York: Plenum publ. *Adv In Exp Med Biol*, 1998: 442; 209-17.

169. Labiner DM, Yan CC, Weinand ME, Huxtable RJ. Disturbances of aminoacids from temporal lobe synaptosomes in human complex partial epilepsy. *Neurochem Res* 1999; 24: 1379-83.

170. Husy N, Deleuze C, Bres V, Moos FC. New role of taurine as an osmomediator between glial cells and neurons in the rat supraoptic nucleus. In: Della Corte L, Huxtable RJ, Sgaragli G, Tipton KF, eds. *Taurine 4*. New York: Kluwer Ac and Plenum Publ. *Adv In Exp Med Biol*, 2000: 483; 227-37.

171. Garseth M, White LR, Aasly J. Little change in CSF aminoacids in subtypes of multiple sclerosis compared with acute polyradiculoneuropathy. *Neurochem Int* 2001; 39: 111-5.

II. Magnesium and neurosciences

Magnesium: out of sight, out of mind?

R. Vink

Department of Pathology, University of Adelaide, Adelaide, SA 5005, Australia

Abstract. Considerable evidence has accumulated firmly establishing that intracellular calcium plays a critical role in the various pathologies that result in cell death in the central nervous system. In contrast, intracellular magnesium does not have such universal acknowledgement as an important factor in neuronal cell death despite considerable evidence to the contrary. Much of this reluctance toward magnesium is due to the fact that few technologies are able to accurately determine free magnesium concentration. Those technologies that have been successfully used often produce widely varying results between different laboratories. On the other hand, indirect evidence using magnesium as a pharmacotherapy has conclusively demonstrated that it is a neuroprotective agent in such diverse pathologies such as traumatic brain injury, stroke, brain edema, drug induced injury (*eg.*, alcohol, cocaine), migraine, and cerebral palsy, just to name a few. This review will critically analyze the evidence suggesting a ubiquitous role for free magnesium in neuronal cell death, focusing on measurements of free magnesium concentration, its change, and the relationship to functional outcome.

With the publication of the 1916 article describing magnesium sulphate as a clinical anesthetic [1], it has been understood that magnesium may play a major role in brain function. Nonetheless, while the importance of magnesium in brain metabolism and function has been acknowledged, the characterization of this role has essentially not been pursued to the same degree as the body's other major divalent cation, calcium. Indeed, in view of this preferential treatment, magnesium has been termed the forgotten cation. Recent reports of magnesium being the pharmacologic drug of choice in eclampsia [2], a useful agent in preventing myocardial reperfusion injury [3], and implicated in reducing the incidence of cerebral palsy [4] has rekindled interest in using magnesium salts as pharmacologic agents in a diversity of pathophysiological conditions. Moreover, an increasing number of basic science reports are providing direct evidence that magnesium is a critical factor in the brain's response to metabolic and physiologic stress. The possible role that magnesium may play in brain function, particularly with respect to brain injury, and its pharmacologic implications, is the subject of this article.

Free magnesium change in vivo?

Before any factor can be considered regulatory, there must be evidence to demonstrate that its concentration can change, *in vivo*, under physiological conditions. With magnesium, much of the problem has resulted from difficulties in measuring the intracellular free concentration. In this respect, a number of techniques have been developed, but most of these are invasive and, for one reason or another, somewhat inadequate [5]. More recently, the application of magnetic resonance spectroscopy to the study of free magnesium concentration has permitted non-invasive measurements of magnesium and, after extensive evaluation in a number of different physiological situations and comparison to previous methods, there is general agreement that the technique does give a useful approximation of the intracellular free concentration (for review, see [6]). The first demonstration of *in vivo* free magnesium change in brain using magnetic resonance was in traumatic brain injury where declines in free concentration of up to 60% were reported to occur within 30 minutes of the traumatic event [7]. Subsequently, decreases have also been reported in migraine, where the low intracellular free magnesium has been described as the link between the physiologic threshold for migraine and the attack itself

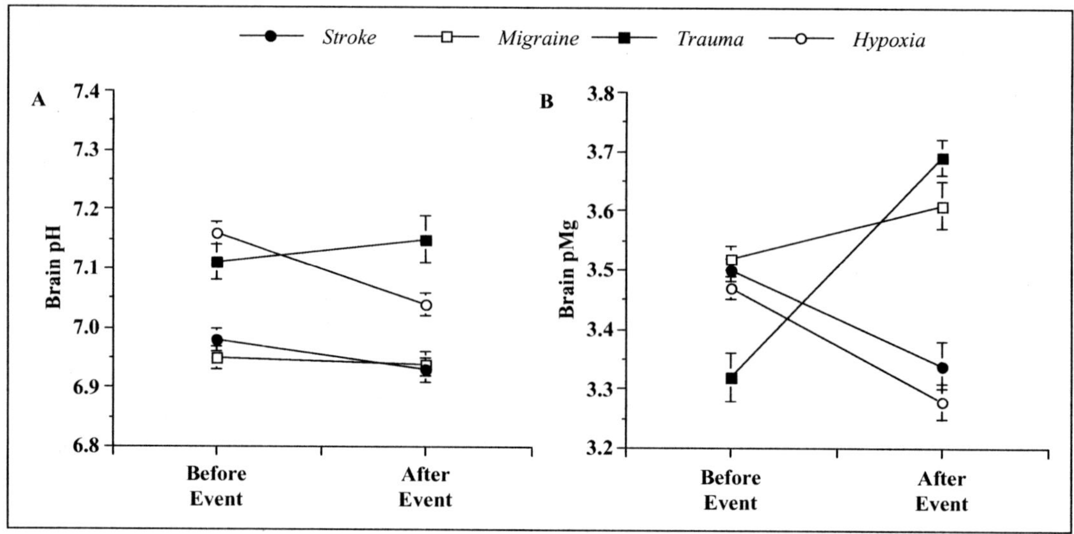

Figure 1. Changes in brain pH (A) and free magnesium concentration (B) following traumatic injury, migraine, ischemia and hypoxia using data derived from previously published material [8, 11-13]. Free magnesium is expressed as pMg (-log (Mg_i)) to simplify comparison to the more familiar pH. Scales are identical. Changes in pMg are significantly greater than changes in pH for each of the conditions. Following trauma and migraine, there are significant declines in free magnesium concentration but no significant change in pH. Conversely, during ischemia and hypoxia, small decreases in intracellular pH are associated with large increases in free magnesium concentration.

[8], and following alcohol exposure [9, 10]. In contrast, increases in the cation have been reported following stroke [11] and hypoxia/ischemia [12]. *Figure 1* summarizes the degree and direction of free magnesium change in some of these conditions.

What determines whether free magnesium increases or decreases in different brain pathologies seems to be dependent on the ATP levels. In conditions where ATP does not decline significantly (trauma, alcohol exposure and migraine), the free magnesium level declines. In contrast, where ATP depletion occurs (stroke and hypoxia/ischemia), there is an immediate increase in free magnesium concentration associated with the liberation of the magnesium cation from the nucleotide. This raises the question of whether magnesium change is an injury factor or a consequence of injury. In the case of ATP depleted conditions, the magnesium change seems more likely a consequence of injury. We are left to conclude that acute magnesium decline may be an injury factor whereas any increase in magnesium should be interpreted as a marker of injury and not as an injury factor in itself.

Magnesium as a neuroprotectant

The finding that brain free magnesium can change *in vivo* lent support to previous findings suggesting a potential neuroprotective effect of the cation, leading to many more studies in this area. The cation has now been described as a possible neuroprotectant in brain and spinal cord injury [13], ischemia [14], neonatal hypoxia/asphyxia [15], maternal hypoxia [16] and peripheral nerve damage [17]. With respect to seizures, magnesium has been shown to have a more pronounced central anticonvulsant effect than phenytoin [18]. Of further significance, the studies examining neuroprotective effects of magnesium in traumatic brain injury have shown that the increase in brain free magnesium in response to parenteral magnesium administration is related to the degree of improvement in neurologic outcome [13]. Indeed, magnesium salts have been shown to have positive effects on both motor outcome and cognitive outcome after trauma [19, 20]. This improvement in functional outcome may be correlated to the protective effects that magnesium has at the cellular level after traumatic brain injury [21, 22]. It would therefore seem that the intracellular free magnesium concentration is critically important to maximal recovery of function. Clearly, when the intracellular free magnesium levels are low, a return to normal concentration is required to promote cell repair.

Mechanisms of protection

It should be stated from the outset that magnesium can potentially be protective at both the intracellular and the extracellular level. Therefore, conditions where intracellular magnesium is not decreased (*eg*, stroke) may benefit from the extracellular effects of the cation, whereas conditions that have an intracellular depletion (*eg*, trauma) will benefit from the combined intracellular and extracellular effects of the cation.

Extracellular

The impetus for a renewed interest in mechanistic aspects of magnesium in the brain injury process can be traced back to the finding that magnesium is a voltage dependent blocker of the n-methyl-D-aspartate (NMDA) channel [23]. This channel is associated with entry of calcium into neurones, an excess of which leads to the initiation of a cascade of autodestruction [24]. By blocking the NMDA channel, magnesium reduces calcium influx into the cell and protects against this injury cascade. This neuroprotective property of magnesium has now been extensively demonstrated both *in vitro* and *in vivo* [15, 23-27]. Interestingly, the block of the NMDA channel increases substantially with age [26] suggesting that the importance of the magnesium block increases as the complexity of the CNS increases. Magnesium also reduces NMDA receptor binding capacity [27] and has been shown to reduce glutamate release [24]. The effects of magnesium on calcium entry into cells are not limited to the NMDA channel. Indeed, the magnesium ion has been described as nature's physiologic calcium antagonist having a direct effect on calcium uptake [28], and this property has been extensively pursued in pharmacologic cardiovascular studies. However, the modulation of ion fluxes is not limited to calcium, with the K^+ channel also being directly regulated by magnesium concentration [28]. Finally, magnesium has been recently shown to attenuate blood brain barrier permeability and reduce vasogenic edema [29], and is well known as a vasodilator that has the ability to inhibit cerebrovasospasms [9].

Intracellular

Through its effects on the Na^+/K^+ ATPase, magnesium can be considered as having an indirect effect on membrane potential. Similarly, magnesium's effects on receptor sensitivity are not limited to extracellular effects on the excitatory amino acids or ligand gated ionotropic receptors. Modulatory effects on receptor binding and activity having been described for the seven-transmembrane receptors (metabotropic receptors) including serotonergic and opioid receptors. Specifically, intracellular magnesium is required for the activation of G-proteins and therefore the coupling of G-protein-linked receptors to their effector mechanisms [30].

Magnesium is an essential cofactor in all energy producing and consuming reactions, and therefore changes in the cation concentration will have profound consequences for all bioenergetic reactions, including those of oxidative phosphorylation. It is an integral component of cell membranes (the membrane is in fact thought to be a free magnesium buffer) and as such has been demonstrated to be involved in the process of lipid peroxidation and generation of free radicals [31]. Its critical role in RNA aggregation and protein synthesis has been well described, as has its role on the replication of DNA itself [32]. Recent evidence also supports a role in apoptosis. Specifically, magnesium has been shown to inhibit the opening of the mitochondrial permeability transition pore [33] as well as the induction of DNA fragmentation factor. The work of Delivoria-Papadopolous and colleagues [34] has also demonstrated that magnesium alters the balance between pro-apoptotic bax and the anti-apoptotic bcl-2, thus inhibiting the activation of caspases. Finally, the cation is thought to regulate the processing of amyloid precursor protein away from the neurotoxic β amyloid form and toward the neuroprotective soluble APPα form [35]. This potentially has implications for Alzheimer's Disease.

What regulates magnesium?

This is perhaps the central dilemma. Unlike calcium concentration, factors controlling intracellular magnesium concentration are still largely unknown. There is evidence now appearing suggesting specific magnesium transporters. There is also evidence appearing that suggests that receptor-mediated stimuli are directly connected to the regulation of intracellular magnesium [36]. Unfortunately, the progress that is being made is limited, again because of the technical difficulties associated with "seeing" magnesium changes. Which brings us back to the title of this short paper. Because we currently can't "see" magnesium, we should not leave it out of mind. With a better understanding of its role and regulation, magnesium may represent as important a therapeutic target as calcium.

Acknowledgements

RV is supported, in part, by the Australian National Health and Medical Research Council.

References

1. Peck CH, Meltzer SJ. Anesthesia in human beings by intravenous injection of magnesium sulfate. *JAMA* 1916 ; 67 : 1131-3.

2. Neilson JP. Magnesium sulphate : the drug of choice in eclampsia. *BMJ* 1995 ; 311 : 702-3.

3. Steurer G, Yang P, Rao V, Mohl W, Glogar D, Smetana R. Acute myocardial infarction, reperfusion injury, and intravenous magnesium therapy - basic concepts and clinical implications. *Am Heart J* 1996 ; 132(Suppl) : 478-81.

4. Nelson KB, Grether JK. Can magnesium sulfate reduce the risk of cerebral palsy in very low birthweight infants? *Pediatrics* 1995 ; 95 : 263-9.

5. Alvarez-Leefmans FJ, Giraldez F, Gamino SM. Intracellular free magnesium in excitable cells : its measurements and its biological significance. *Can J Physiol Pharmacol* 1987 ; 65 : 915-25.

6. London RE. Methods for measurement of intracellular magnesium : NMR and fluorescence. *Annu Rev Physiol* 1991 ; 53 : 241-58.

7. Vink R, McIntosh TK, Demediuk P, Weiner MW, Faden AI. Decline in intracellular free magnesium concentration is associated with irreversible tissue injury following brain trauma. *J Biol Chem* 1988 ; 263 : 757-61.

8. Ramadan NM, Halvorson H, Vande-Linde A, Levine SR, Helpern JA, Welch KMA. Low brain magnesium in migraine. *Headache* 1989 ; 29 : 416-9.

9. Altura BM, Altura BT, Gupta RK. Alcohol intoxication results in rapid loss in free magnesium in brain and disturbances in brain bioenergetics : relation to cerebrospasm, alcohol-induced strokes, and barbiturate anesthesia-induced deaths. *Magnes Trace Elem* 1992 ; 10 : 122-35.

10. Mullins PGM, Vink R. Chronic alcohol exposure decreases brain intracellular free magnesium concentration in rats. *Neuroreport* 1995 ; 6 : 1633-6.

11. Helpern JA, Van de Linde AMQ, Welch KMA, *et al.* Acute elevation and recovery of intracellular Mg^{2+} following human focal cerebral ischemia. *Neurol* 1993 ; 43 : 1577-81.

12. Williams GD, Smith GD. Application of the accurate assessment of intracellular magnesium and pH from the 31P shifts of ATP to cerebral hypoxia-ischemia in neonatal rat. *Magn Reson Med* 1995 ; 33 : 853-7.

13. Vink R, Cernak I. Regulation of brain intracellular free magnesium following traumatic injury to the central nervous system. *Front Biosci* 2000 ; 5 : 656-65.

14. Izumi Y, Roussel S, Pinard E, Seylaz J. Reduction of infarct volume by magnesium after middle cerebral artery occlusion in rats. *J Cereb Blood Flow Metab* 1991 ; 11 : 1025-30.

15. Hoffman DJ, Marro PJ, McGowan JE, Mishra OP, Delivoria-Papadopoulos M. Protective effect of MgSO4 infusion on NMDA receptor binding characteristics during cerebral cortical hypoxia the newborn piglet. *Brain Res* 1994 ; 644 : 144-9.

16. Hallak M, Hotra JW, Kupsky WJ. Magnesium sulfate protection of fetal rat brain from severe maternal hypoxia. *Obstet Gynecol* 2000 ; 96 : 124-8.

17. Greensmith L, Mooney EC, Waters HJ, Houlihanburne DG, Lowrie MB. Magnesium ions reduce motor neuron death following nerve injury exposure to n-methyl-D-aspartate. *Neuroscience* 1995 ; 68 : 807-12.

18. Mason BA. Magnesium is more efficacious than phenytoin in reducing n-methyl-D-aspartate seizures in rats (1994). *Am J Obstet Gynecol* 1994 ; 171 : 999-1002.

19. McIntosh TK, Faden AI, Yamakami I, Vink R. Magnesium deficiency exacerbates and pretreatment improves outcome following traumatic brain injury in rats : ^{31}P magnetic resonance spectroscopy and behavioural studies. *J Neurotrauma* 1988 ; 5 : 17-31.

20. Vink R, O'Connor CA, Nimmo AJ, Heath DL. Magnesium attenuates persistent functional deficits following diffuse traumatic brain injury in rats. *Neurosci Lett* 2003 ; 336 : 41-4.

21. Bareyre FM, Saatman KE, Raghupathi R, McIntosh TK. Magnesium attenuates persistent functional deficits following diffuse traumatic brain injury in rats. *J Neurotrauma* 2000 ; 17 : 1029-39.

22. Saatman KE, Bareyre FM, Grady MS, McIntosh TK. Acute cytoskeletal alterations and cell death induced by experimental brain injury are attenuated by magnesium treatment and exacerbated by magnesium deficiency. *J Neuropathol Exp Neurol* 2001 ; 60 : 183-94.

23. Mayer ML, Westbrook GL, Guthrie PB. Voltage-dependent block by Mg^{2+} of NMDA responses in spinal cord neurons. *Nature* 1984 ; 309 : 261-3.

24. Rothman SM, Olney JW. Glutamate and the pathophysiology of hypoxic-ischemic brain damage. *Ann Neurol* 1986 ; 19 : 105-11.

25. Frandsen A, Schousboe A. Effect of magnesium on NMDA mediated toxicity and increases in (Ca2+)i and cGMP in cultured neocortical neurones : evidence for distinct regulation of different responses. *Neurochem Int* 1994 ; 25 : 303-8.

26. Strecker GJ. Blockade of NMDA activated channels by magnesium in the immature rat hippocampus. *J Neurophysiol* 1994 ; 72 : 1538-48.

27. Hallak M, Berman RF, Irtenkauf SM, Janusz CA, Cotton DB. Magnesium sulphate treatment decreases N-Methyl-D-Aspartate receptor binding in the rat brain - an autoradiographic study. *J Soc Gynecol Invest* 1994 ; 1 : 25-30.

28. Agus ZS, Morad M. Modulation of cardiac ion channels by magnesium. *Annu Rev Physiol* 1991 ; 53 : 299-307.

29. Esen F, Erdem T, Aktan D, *et al.* Effects of magnesium administration on brain edema and blood brain barrier breakdown after experimental traumatic brain injury in rats. *J Neurosurg Anesthesiol* 2003 ; 15 : 119-25.

30. Birnbaumer L, Abramowitz J, Brown AM. Receptor-effector coupling by G proteins. *Biochim Biophys Acta* 1990 ; 1031 : 163-224.

31. Gunther T, Hollriegl V, Vormann J, Bubeck J, Classen HG. Increased lipid peroxidation in rat tissues by magnesium deficiency and vitamin E depletion. *Magnes Bull* 1994 ; 16 : 38-43.

32. Terasaki M, Rubin H. Evidence that intracellular magnesium is present in cells at a regulatory concentration for protein synthesis. *Proc Natl Acad Sci USA* 1985 ; 82 : 7324-6.

33. Halestrap AP, Kerr PM, Javadov S, Woodfield KY. Elucidating the molecular mechanism of the permeability transition pore and its role in reperfusion injury of the heart. *Biochim Biophys Acta* 1998 ; 1366 : 79-94.

34. Ravishanker S, Ashraf QM, Fritz K, Mishra OP, Delivoria-Papadopoulos M. Expression of bax and bcl2 proteins during hypoxia in cerebral cortical nuclei of newborn piglets : effect of administration of magnesium sulfate. *Brain Res* 2001 ; 18 : 23-9.

35. Van den Heuvel C, Finnie JW, Blumbergs PC, *et al.* Upregulation of neuronal amyloid precursor protein (APP) and APP mRNA following magnesium sulphate therapy in traumatic brain injury. *J Neurotrauma* 2000 ; 17 : 1041-53.

36. Wolf FI, Francesco AD, Covacci V, Cittadini A. cAMP activates magnesium efflux via the Na/Mg antiporter in ascites cells. *Biochem Biophys Res Commun* 1994 ; 202 : 1209-14.

Variations of magnesium concentrations in psychosis

M. Nechifor[1], C. Văideanu[2], I. Mîndreci[3], C. Borza[2]

1. Dept. of Pharmacology, University of Medicine and Pharmacy "Gr. T. Popa" Iasi, Romania ; 2. Psychiatry Hospital "Socola", Iasi, Romania ; 3. Biophysics Dept, University of Medicine and Pharmacy "Gr. T. Popa" Iasi, 700115, Romania

Psychoses are characterized by disturbances in thought processing and behavior leading to a loss of contact with reality. These diseases are different clinical and pathogenic entities. Major depression, schizophrenia (with different forms) and bipolar psychosis affect an important number of persons. Only schizophrenia prevalence is 1.4-4.6/1000 inhabitants. Psychosis has a chronic evolution characterized by relapses. Pathogenic mechanisms of psychosis are incompletely known. Magnesium and other cations play complex roles at CNS level. There are data about a possible involvement of intra- and extracellular changes of cations in the pathogeny of psychosis.

Schizophrenia

Schizophrenia is a severe psychiatric disease characterized by progressive disintegration of personality and relations between patient and the social environment. The subtypes of schizophrenia are: paranoid type; disorganized type; catatonic type; undifferentiated type; residual type. The main clinical symptoms are: delusions, hallucinations disorganized or catatonic behavior, negative symptoms (*e.g* affective flattening, alogia, avolition) (DSM-IV).

Data regarding changes in plasma and cellular magnesium level in schizophrenia are different and sometime discordant. Some authors found a raised Mg level in patients with schizophrenia ([1] and others).

Banki *et al.*, 1985 [2] and Gattaz *et al.*, 1983 [3] found a raised Mg level in CSF. Widner *et al.*, 1993 [4] consider that hypermagnesemia might be involved in decreasing catecholaminergic neurotransmission in some parts of the brain in these patients. On contrary other authors found a decreased magnesium plasma level in schizophrenic patients [5-7]. In some studies there are significant changes of magnesium plasma level in schizophrenic patients *vs* normal persons. If data regarding changes in plasma magnesium concentrations are quite heterogenic and sometimes contradictory, studies showing a raised copper level and a zinc deficiency are much more consonant [8, 9]. Also data regarding the influence of some antipsychotic drugs on Mg plasma levels in schizophrenia are different.

Jabotinsky-Rubin *et al.*, 1993 [10] have shown that haloperidol decreases the (high) level of magnesium in patients with schizophrenia without changing plasma calcium and phosphorus concentrations. Ang *et al.*, 1993 [11] have considered that hypomagnesemia is one of the causes of resistance to antipsychotics in some patients.

Diselectrolytic disturbances from hemodyalisis are considered by some authors [12] as being involved in psychic symptoms that appear in such conditions (loss of memories, personality disturbances, irritability, etc.).

We searched plasma and erythrocite concentrations of magnesium in a group of 56 adult patients (32 women, 24 men) with paranoid schizophrenia (after DSM-IV) admitted with psychotic attack in Clinical Psychiatric Hospital "Socola" Iasi and that did not get any antipsychotic treatment before admittance. The group of patients had an average age 38 years (18-65 years). In the group were included patients there were hospitalized at least 30 days. There were not include patients with chronic renal failure, cardiac failure, hepatic cirrhosis, chronic ethanol intake, malabsorbtion syndrome, treatment with diuretics or with magnesium or other cations containing drugs.

There were determined plasma and intraerythrocite magnesium concentration using spectrophotometry with atomic absorbtion at admittance before treatment and after 3 weeks of treatment.

Patients were treated with haloperidol, 8 mg/kg/day (patients) and risperidone (Rispolept[R]) 6 mg/kg/d *per os*, 21 days (in 2-3 equal doses day). There were determined plasma and erythrocytic magnesium concentrations also to a healthy control patients with the same age and gender as schizophrenic patients. Our data have shown that plasma magnesium is not significantly changed in patients with schizophrenia at admittance versus control group, but erythrocytic magnesium is significantly decreased (4.82 ± 0.22 mg/dl in schizophrenic patients before treatment *versus* 5.82 ± 0.11 mg/dl in control group, $p < 0.05$) (*figure 1*). Both treatments with haloperidol, respectively risperidone, increase moderately but significantly, erythrocytic magnesium concentration (*figure 2*). This was positively correlated with improving clinical status of the patients.

A typical neuroleptic (haloperidol) and an atypical one (risperidone) increases both intraerythrocite magnesium level [13]. We consider that increasing intra erythrocytic magnesium acts as a Ca^{2+} antagonist at the level of calcium channels coupled NMDA receptors. By glutamate stimulation of these receptors is explained some of the symptoms during acute schizophrenia (agitation, panic, delirium). Yamakura *et al.*, 1998 [14] have shown that haloperidol and other butirophenones besides actions on dopamine receptors have a direct action on calcium channels coupled with NMDA receptors. We consider that this action might be enhanced by magnesium. Risperidone decreases glutamate-NMDA binding [15]. This means that in this case decreasing of some symptoms from acute schizophrenia involved decreasing activation of calcium channels bound with NMDA receptors. This effect is enhanced by increased concentrations of magnesium. We believe that increase of Mg^{2+} cell level is important for antipsychotic action of haloperidol and risperidone. The effect of magnesium could be produced by:

- Decreasing neuronal response to glutamate stimulation of NMDA receptors,
- Decreasing the presynaptic release of some excitatory aminoacids,
- Decreasing formation of peroxidic radicals,

The raise of magnesium concentration can reduce anxiety, hallucinations and agitation.

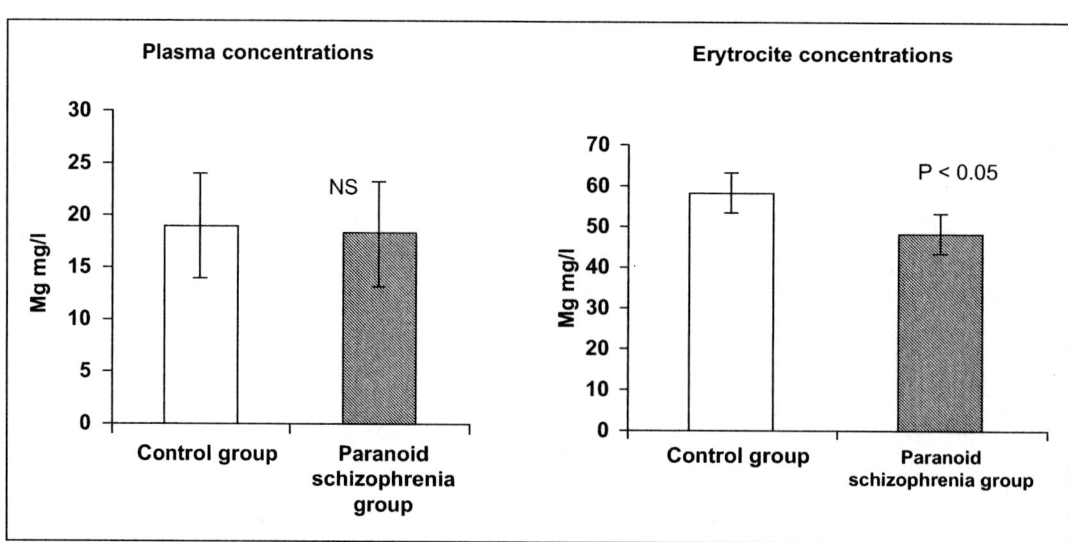

Figure 1. Plasma and erythrocytic magnesium concentrations in patients with acute episode of paranoid schizophrenia before therapy.

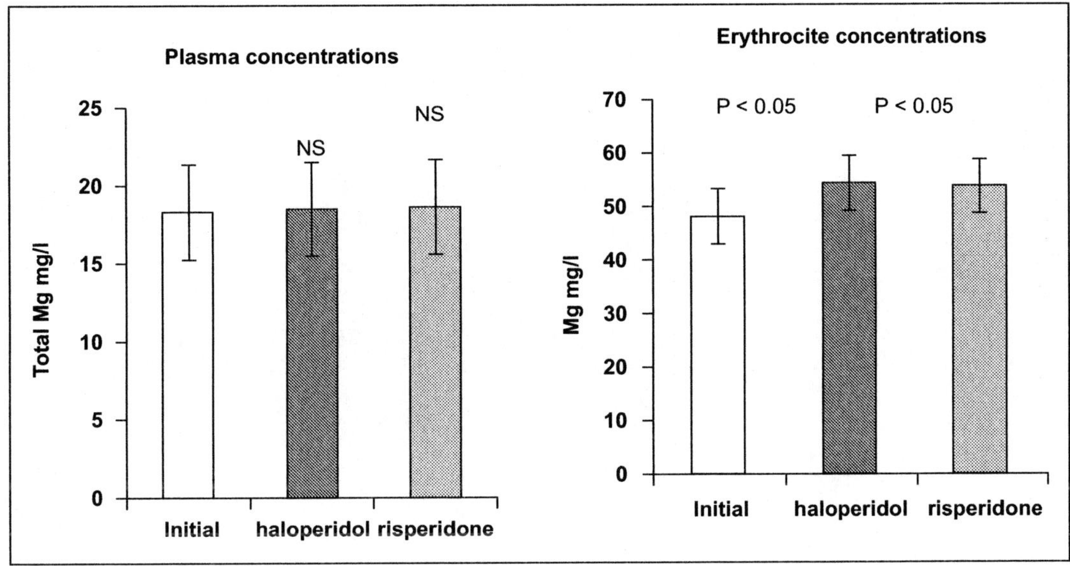

Figure 2. The influence of haloperidol therapy (H), 8 mg/d or risperidone (R), 6 mg/d on plasma and erythrocytic magnesium concentrations in patients with paranoid schizophrenia.

Bipolar disorders (BD)

Many studies have shown there are changes of plasma concentrations of some cations in BD. Lower values of plasma Mg level in BD versus healthy control were observed by Herzberg and Herzberg, 1997 [16] and Durlach, 1980 [17]. Some studies correlate psychotic agitation in mania with statistical increases in serum total calcium and phosphorus [18] but other authors deny this correlation. Kumar and Kurup, 2002 [19] consider that decreasing activity of Na^-K^- ATP-ase plays a main role in the pathogeny of schizophrenia and manic depressive psychosis. This decreases neuronal transmembranary ionic changes and led to increasing intracellular concentrations of calcium and decreasing intracellular concentrations of magnesium. This fact is in agreement with data about decreasing intracellular magnesium concentrations in these patients. We consider that decreasing Mg concentrations plays a role in the pathogenic mechanism of BD also by increasing glutamate action and entrance in high quantity of Ca^{2+} into the neuron following a low Mg^{2+} concentration.

Other authors sustain that in BD there are no significant changes of Mg concentrations in CSF and in blood plasma and neither an influence of lithium therapy on the magnesium levels [20].
Mg administration was proved useful in therapy-resistant maniac agitation [21]. Magnesium effect is different from calcium that in I.V. administration slightly intensifies agitation in some patients with maniacal psychosis [22].

Association of magnesium medicines (Magnesiocard^R) for long time (32 weeks) with the lithium therapy in patients with BP increased therapeutic efficiency of its administration [23]. It was proved that Li^+ administration determines increasing intracellular magnesium concentrations. There is a competition between Li^+ and Mg^{2+} for some binding sites from some cells [24, 25]. We consider this increase in free intraneuronal Mg^{2+} as one of the antimaniacal mechanisms of action of lithium (in agreement with other authors - Amari *et al.*, 1999 [26]). By depressing Na+- K+ ATP-ases activity, Li+ might play a role in modulation of magnesium intracellular concentrations.

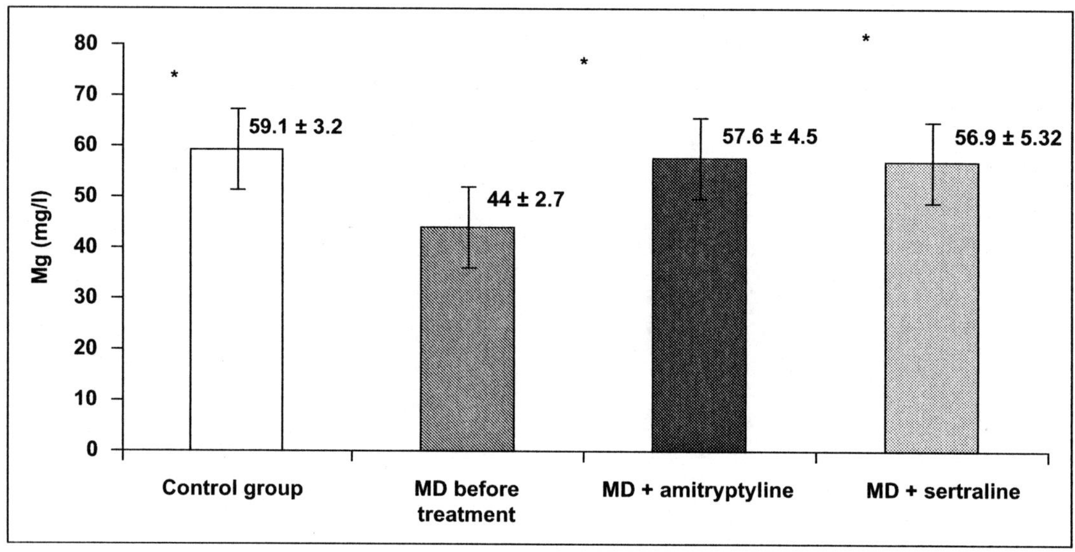

Figure 3. Erythrocytic magnesium concentration in patients with major depression (MD) before and after the treatment. * $p < 0.01$ *vs* MD group before treatment.

Major depression (MD)

MD is a psychosis characterized by intra- and extracellular variations of magnesium and other cations. Existent data are contradictory. Levine *et al.*, 1999 [27] found raised Ca/Mg ratio in the moment of admittance of acutely depressed patients both in serum and CSF. Young *et al.*, 1996 [28] consider there are not significantly changes in plasma magnesium concentrations and Ca/Mg ratio is not significantly changed in patients with major depression *vs* healthy controls. Frazer *et al.*, 1983 [29] have found total plasma Mg higher in patients with MD than in healthy control. Widmer *et al.*, 1995 [4] found that middle and highly depressed patients had higher erythrocyte and plasma magnesium than lowly depressed patients or healthy controls. The same authors consider it can not be established any correlation between plasma levels of Na^+, K^+ or Ca^{2+} and intensity of depression. Kamei *et al.*, 1998 [30] consider there are significantly changes only in erythrocyte Ca^{2+} in MD without significantly changes in Mg^{2+} concentrations. Other authors didn't find changes in the Mg concentrations and serum ratio Ca^{2+}/Mg^{2+} in patients with MD versus normal subjects [28].

Serum copper was found increased in patients with MD [31] and in the same time zinc plasma level was decreased [32]. Data regarding the influence of some antidepressant medications on plasma and cellular concentrations on some cations are scarcely. It was shown on an experimental model of depression in rat – not identical with human depression - that imipramine and cytalopram decreased moderately copper level in the brain [31].

We searched in a group of 52 patients with MD (Hamilton score > 17) the influence of treatment with amitriptyline (3 x 25 mg/day *per os* at least 28 days) or sertraline (Zoloft^R- 150 mg/day *per os* at least 28 days).

Our data have shown in patients with MD in the moment of the admittance (before antidepressive therapy) the plasma magnesium level decreased not significantly statistic (20.2 ± 2.1 mg/l in MD group *versus* 22.9 ± 1.9 mg/l in healthy controls). If consider only patients with severe MD (Hamilton score > 17) than decreasing plasma magnesium is significantly statistic though moderately (22.9 ± 1.9 mg/l in healthy control *versus* 17.8 ± 2.2 mg/l in patients with MD, $p < 0.05$) [33]. Decreasing plasma zinc level and increasing plasma copper are significantly versus healthy control (Cu^{2+} 0.86 ± 0.05 mg/l in MD patients *vs* 0.63 ± 0.01 mg/l in healthy controls, $p < 0.05$). Our data show that erythrocyte magnesium is significantly decreased in patients with MD *vs* control (44.0 ± 2.7 mg/l in MD

group *versus* 59.1 ± 3.2 mg/l in control group). Our results about a lower level of erythrocyte magnesium in patients with severe MD are correlated with Levine *et al.*, 1999 [27] that report a higher Ca/Mg ratio in cerebrospinal fluid in the moment of hospitalization during MD crisis.
Our results show that amitriptyline and sertraline increases significantly erythrocyte magnesium level (*figure 3*) and serum zinc and decreases serum copper [34]. This correlates with an improving in clinical symptoms. Increasing cell concentrations of magnesium might contribute to decreasing anxiety in these patients, in accordance with DeSouza *et al.*, 2000 [35] and Hantouche *et al.*, 1998 [36] data. We consider variations in the magnesium concentrations and other bivalent cations too (*e.g.* zinc and copper) play an important role, even in not fully understood, in the pathogeny of some psychosis.

References

1. Chugh TD, Dhingra RK, Gulati RC, Bathla JC. Magnesium in schizophrenia. *Indian J Med Res* 1973 ; 61(7) : 998-1001.

2. Banki CM, Vojnik M, Papp Z, Balla KZ, Arato M. Cerebrospinal fluid magnesium and calcium related to amine metabolites, diagnosis, and suicide attempts. *Biol Psychiatry* 1985 ; 20(2) : 163-71.

3. Gattaz WF, Kattermann R, Gattaz D, Beckmann H. Magnesium and calcium in the CSF of schizophrenic patients and healthy controls: correlations with cyclic GMP. *Biol Psychiatry* 1983 ; 18(8) : 935-9.

4. Widmer J, Henrotte JG, Raffin Y, Bovier P, Hilleret H, Gaillard JM. Relationship between erythrocyte magnesium, plasma electrolytes and cortisol, and intensity of symptoms in major depressed patients. *J Affect Disord* 1995 ; 34(3) : 201-9.

5. Kirov GK, Tsachev KN. Magnesium, schizophrenia and manic-depressive disease. *Neuropsychobiology* 1990 ; 23(2) : 79-81.

6. Kanofsky JD, Sandyk R. Magnesium deficiency in chronic schizophrenia. *Int J Neurosci* 1991 ; 61(1-2) : 87-90.

7. Levine J, Rapoport A, Mashiah M, Dolev E. Serum and cerebrospinal levels of calcium and magnesium in acute versus remitted schizophrenic patients. *Neuropsychobiology* 1996 ; 33(4) : 169-72.

8. Pfeiffer CC, Iliev V, De Gara M, Smyrl EG, Cawley JC. Blood histamine levels, basophil counts, and trace metals in the schizophrenias. *Psychopharmacol Bull* 1971 ; 7(3) : 37.

9. Olatunbosun DA, Akindele MO, Adadevoh BK, Asuni T. Serum copper in schizophrenia in Nigerians. *Br J Psychiatry* 1975 ; 127 : 119-21.

10. Jabotinsky-Rubin K, Durst R, Levitin LA, *et al.* Effects of haloperidol on human plasma magnesium. *J Psychiatr Res* 1993 ; 27(2) : 155-9.

11. Ang AW, Ko SM, Tan CH. Calcium, magnesium, and psychotic symptoms in a girl with idiopathic hypoparathyroidism. *Psychosom Med* 1995 ; 57(3) : 299-302.

12. Bustamante J. Nephropathy due to analgesics : interstitial nephritis caused by phenacetin in nine patients. *Med Clin (Barc)* 1979 ; 73(7) : 277-80.

13. Richelson E. Neuroleptic binding to human brain receptors : relation to clinical effects. *Ann N Y Acad Sci* 1988 ; 537 : 435-42.

14. Yamakura T, Sakimura K, Mishina M, Shimoji K. Sensitivity of the N methyl D aspartate receptor channel to butyrophenones is dependent on the epsilon2 subunit. *Neuropharmacology* 1998 ; 37(6) : 709-17.

15. Tarazi FI, Baldessarini RJ, Kula NS, Zhang K. Long-term effects of olanzapine, risperidone, and quetiapine on ionotropic glutamate receptor types : implications for antipsychotic drug treatment. *J Pharmacol Exp Ther* 2003 ; 306(3) : 1145-51.

16. Herzberg L, Herzeberg B. Mood change and magnesium. A possible interaction between magnesium and lithium? *J Nerv Ment Dis* 1977 ; 165(6) : 423-6.

17. Durlach J, Durlach V, Bac P. Magnesium and therapeutics. *Magnes Res* 1994 ; 7 : 313-28.

18. Carman JS, Post RM, Runkle DC, Bunney Jr. WE, Wyatt RJ. Increased serum calcium and phosphorus with the 'switch' into manic or excited psychotic state. *Br J Psychiatry* 1979 ; 135 : 55-61.

19. Kumar AR, Kurup PA. Inhibition of membrane Na+-K+ ATPase activity : a common pathway in central nervous system disorders. *J Assoc Physicians India* 2002 ; 50 : 400-6.

20. George MS, Rosenstein D, Rubinow DR, Kling MA, Post RM. CSF magnesium in affective disorder : lack of correlation with clinical course of treatment. *Psychiatry Res* 1994 ; 51(2) : 139-46.

21. Heiden A, Frey R, Presslich O, Blasbichler T, Smetana R, Kasper S. Treatment of severe mania with intravenous magnesium sulphate as a supplementary therapy. *Psychiatry Res* 1999 ; 89(3) : 239-46.

22. Carman JS, Wyatt RJ. Calcium : bivalent cation in the bivalent psychoses. *Biol Psychiatry* 1979 ; 14(2) : 295-336.

23. Chouinard G, Beauclair L, Geiser R, Etienne P. A pilot study of magnesium aspartate hydrochloride (Magnesiocard) as a mood stabilizer for rapid

cycling bipolar affective disorder patients. *Prog Neuropsychopharmacol Biol Psychiatry* 1990 ; 14(2) : 171-80.

24. Layden BT, Abukhdeir AM, Williams N, *et al.* Effects of Li (+) transport and Li (+) immobilization on Li (+)/Mg (2+) competition in cells : implications for bipolar disorder. *Biochem Pharmacol* 2003 ; 66(10) : 1915-24.

25. Abukhdeir AM, Layden BT, Minadeo N, Bryant FB, Stubbs Jr. EB, Mota de Freitas D. Effect of chronic Li+ treatment on free intracellular Mg2+ in human neuroblastoma SH-SY5Y cells. *Bipolar Disord* 2003 ; 5(1) : 6-13.

26. Amari L, Layden B, Nikolakopoulos J, *et al.* Competition between Li+ and Mg2+ in neuroblastoma SH-SY5Y cells : a fluorescence and 31P NMR study. *Biophys J* 1999 ; 76(6) : 2934-42.

27. Levine J, Stein D, Rapoport A, Kurtzman L. High serum and cerebrospinal fluid Ca/Mg ratio in recently hospitalized acutely depressed patients. *Neuropsychobiology* 1999 ; 39(2) : 63-70.

28. Young LT, Robb JC, Levitt AJ, Cooke RG, Joffe RT. Serum Mg2+ and Ca2+/Mg2+ ratio in major depressive disorder. *Neuropsychobiology* 1996 ; 34(1) : 26-8.

29. Frazer A, Ramsey TA, Swann A, *et al.* Plasma and erythrocyte electrolytes in affective disorders. *J Affect Disord* 1983 ; 5(2) : 103-13.

30. Kamei K, Tabata O, Muneoka K, Muraoka SI, Tomiyoshi R, Takigawa M. Electrolytes in erythrocytes of patients with depressive disorders. *Psychiatry Clin Neurosci* 1998 ; 52(5) : 529-33.

31. Schlegel-Zawadzka M, Krosniak M, Nowak G. Brain copper levels after antidepressant treatment. In : Collery P, Bratter P, Negrelli de Bratter V, Khassanova L, Etienne JC, eds. *Metal Ions in Biology and Medicine.* Paris : John Libbey, Eurotext, 1998 : 703-6.

32. Maes M, De Vos N, Demedts P, Wauters A, Neels H. Lower serum zinc in major depression in relation to changes in serum acute phase proteins. *J Affect Disord* 1999 ; 56(2-3) : 189-94.

33. Nechifor M, Vaideanu C, Boisteanu P, Mindreci I, Cuciureanu R, Nechifor C. Changes in Mg2+ and other cations plasma concentration in patients with major depression. In : Escanero JE, Alda JO, Guerra M, Durlach J, eds. *Advances in Magnesium Research-Physiology, Pathology and Pharmacology.* Zaragosa : Prensas Universitaria de Zaragoza, 2003 : 177-81.

34. Nechifor M, Văideanu C, Palamaru I, Borza C, Mândreci I. *The Influence Of Some Antipsychotic Drugs On Erythrocyte And Plasma Magnesium Level And Of Other Bivalent Cations. 10th International Magnesium Symposium, Cairns, Australia.* 2003 ; (Abstracts Book, 3).

35. De Souza MC, Walker AF, Robinson PA, Bolland K. A synergistic effect of a daily supplement for 1 month of 200 mg magnesium plus 50 mg vitamin B6 for the relief of anxiety-related premenstrual symptoms : a randomized, double-blind, crossover study. *J Womens Health Gend Based Med* 2000 ; 9(2) : 131-9.

36. Hantouche EG, Guelfi JD, Comet D. Alpha-beta L-aspartate magnesium in treatment of chronic benzodiazepine abuse : controlled and double-blind study versus placebo. *Encephale* 1998 ; 24(5) : 469-79.

Serum magnesium concentrations in migraine with aura

M. Cojocaru[1], I.M. Cojocaru[2], C. Musuroi[2], M. Botezat[2], L. Lazăr[2], A. Drută[2]

1. Central Clinical Laboratory Colentina ; 2. "Carol Davila" University of Medicine and Pharmacy, Clinic of Neurology, Colentina Clinical Hospital, Bucharest, Romania

Summary. The aim of this study was to clarify the relationship between serum magnesium concentrations and migraine with aura. The study included forty patients with migraine with aura (27 women and 13 men), mean age 52 ± 5 years, and 18 healthy, sex- and age-matched subjects as controls. The serum magnesium concentrations were evaluated by Vitros 750 XRC, Johnson & Johnson kit (Ortho Clinical Diagnostics). Data are expressed as arithmetic mean (x) ± standard deviation (SD). The unpaired Student's t-test was used for statistical comparison between mean values of magnesium and considered statistically significant at $p < 0.05$. Serum magnesium concentrations of the migraine patients and controls were 0.65 ± 0.07 mmol/l and 0.89 ± 0.08 mmol/l, respectively ($p < 0.001$). Serum magnesium concentrations were below the normal reference range in 47.4% of the migraine patients and 12.7% of the control subjects. The reasons of magnesium deficiency in migraine attacks are not yet clear. The role of various effects of low magnesium levels in the development of migraines with aura remains to be elucidated. Low magnesium levels may potentiate the sensitivity of the N-methyl-D-aspartate (NMDA) receptor to glutamic acid (Glu), a neuroexcitatory amino acid in the brain. Our results suggest that magnesium may be involved in migraine pathophysiology. We feel that a trial of oral magnesium supplementation can be recommended to a majority of migraine sufferers. Further clinical and experimental studies will be necessary to elucidate the role of magnesium in the pathogenesis of migraine.

Magnesium is mainly an intracellular cation, with less than 1% of total body content present in the extracellular fluids. The Mg concentration in serum represents not more than 0.3% of total body Mg. Nevertheless, serum or plasma Mg measurement is the most readily available and widely used test of Mg status. Of the total Mg in serum, around 55% is present as free ionised $Mg2^{+}$ as superscript, 15% is complexed to anions (*e.g.* bicarbonate, citrate, sulfate) and 30% is bound to proteins, mainly albumin. Magnesium is believed to be important neuroprotective agent.

Migraine is taken to mean: repeated attacks of moderate to intense, throbbing or pounding headaches, usually on one side of the head and often accompanied by nausea, vomiting and hypersensitivity to sound and light. Even moderate physical exertion can make the headache noticeably worse.

The management of migraine ideally requires that the factors that cause the attacks are identified and, if possible, eliminated. However, this is not easy because attacks can be initiated by a variety of triggers, making complete avoidance impossible [1, 2].

In the last few years a fundamental role for magnesium in establishing the threshold for migraine attacks and involvement in the pathophysiologic mechanisms related to its onset has become evident [3].

The aim of this study was to assess the magnesium levels of serum in patients with migraine with aura (classic migraine), during attack, compared to the serum magnesium levels of healthy control group.

Subjects

According to the International Headache Society (IHS) classification criteria 40 patients with migraine with aura (27 women and 13 men) and 18 healthy controls matched for age and sex participated in the study. The mean age (range) of the patients and controls was 52 ± 5 years (36-64) and 53 ± 5 years (38-63), respectively. The patients with migraine with aura were recruited from the Clinic of Neurology-Colentina Clinical Hospital.
Patients with classic migraine present visual warnings, known as an aura or prodrome, before the actual headache. In addition to visual disturbances, the aura may include smells, unusual odours or sensations of tingling or numbness. Migraine patients are defined as those who have at least two attacks with aura or at least five attacks without aura. An attack can occur at any time of the day or night but the onset is most frequent on rising.
Patients who had previously used migraine prophylactics, drugs including Mg-salts and had a systemic disease were excluded from the study.

Materials and Methods

Serum is the preferred specimen. Plasma may be used, but avoid using EDTA as anticoagulant, as it will give lower test results.
The serum Mg levels of all patients were evaluated during migraine attacks. Venous blood samples from the patients were drawn in tubes (vacutainer, Becton Dickinson).
Quantitative analysis of Mg in serum was done by VITROS 750 XRC, Johnson & Johnson kit (Ortho Clinical Diagnostics). All samples were analysed in duplicate and repeated if the difference between individual values relative to the mean was > 10%.
Data processing and statistical analysis were done using Excel 2002. Normally distributed data were expressed as arithmetic mean (x) ± standard deviation (SD). Differences between groups were evaluated using unpaired Student's t-test and considered statistically significant at $p < 0.05$.

Results

In comparison with normal subjects, migraine patients had lower levels of serum magnesium. Mean ± SD serum Mg concentration of the patients with migraine and the controls were 0.65 ± 0.07 and 0.89 ± 0.08 mmol/l, respectively ($p < 0.001$) (*Figure 1*).
In 47.4% of the patients with migraine and 12.7% of the control subjects serum Mg concentrations were below the normal reference range of 0.75 to 0.95 mmol/l.
Although serum Mg-levels during attacks were lower in patients with more frequent attacks, the longer duration of attacks and more severe migraine.

Discussion

A migraine attack usually lasts between 4 to 72 hours and produces the following symptoms: a moderate to severe pounding headache on one side of the head, aggravated by movement and accompanied by symptoms such as nausea, vomiting and hypersensitivity to light or sound.
Although our understanding of the pathophysiology of migraine has increased in recent years, its precise mechanism is still far from resolved [4].
Mg is important in vasoconstriction, spreading cortical depression, central neurotransmitter dysfunction and platelet hyperaggregation, which are all considered possible mechanisms immediately preceding or occurring in migraine attacks [5].
Similar to findings from other authors, the mean serum Mg concentration of patients with migraine was significantly lower than in controls [6].
This study showed significant decrease of Mg levels of serum during attacks.

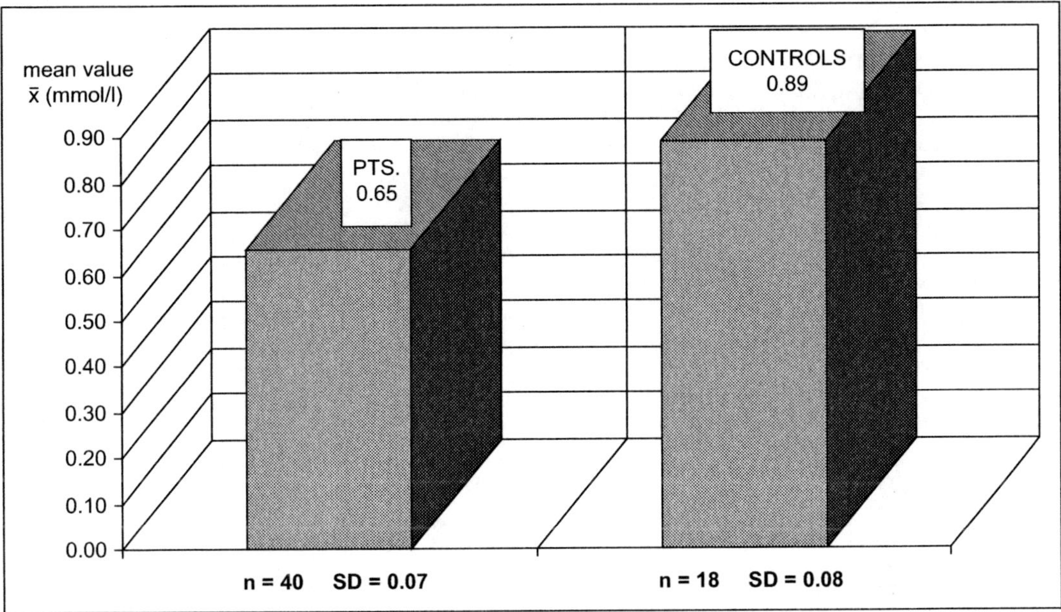

Figure 1. Mean value of serum magnesium in patients with migraine with aura and in controls ($P < 0.001$).

The reasons for the high prevalence of Mg deficiency in patients with migraine are not clear, but may include increased urinary loss, lower dietary intake, or impaired absorption of Mg compared to healthy individuals.

The importance of magnesium in the pathogenesis of migraine headaches is clearly established by a large number of clinical and experimental studies. However, the precise role of various effects of low magnesium levels in the development of migraines remains to be discovered. Magnesium concentration has an effect on serotonin receptors, nitric oxide synthesis and release, NMDA receptors, and a variety of other migraine related receptors and neurotransmitters. The available evidence suggests that up to 50% of patients during an acute migraine attack have lowered levels of ionized magnesium [7].

Decreased serum and intracellular levels of magnesium have been reported in patients with migraine. It has been suggested that magnesium may play an important role in the attacks and pathogenesis of headaches. It is suggested that reduced platelet ionized magnesium in patients with tension-type headache is related to abnormal platelet function, and that increased platelet cyclic AMP in patients with migraine is related to alteration of neurotransmitters in the platelet [5].

The migraine patients studied had a reduced mononuclear magnesium content compared to age-matched healthy control subjects. The authors say that the lower magnesium content in mononuclear cells could indirectly indicate the reduction of brain magnesium concentration, which has recently been demonstrated in the course of migraine [8].

The authors believe that low red blood cell magnesium levels could be a peripheral expression of the reduced brain magnesium concentration observed in migraine patients [9].

Serum magnesium levels tended to be further reduced during attacks which could be an expression, at the peripheral level, of reduced cerebral magnesium levels which would contribute, at least in part, to defining the threshold for migraine attacks [10].

It was suggested that magnesium may be useful in the prevention of migraine attacks [11].

The findings support the hypothesis of a magnesium deficit in people suffering from migraine and raise the problem of the relationship between migraine and other pathologies, including chronic magnesium deficit, latent tetany due to magnesium deficit, mitral valve prolapse, and allergy [12].

The Thomas's study shows that migraine patients have a magnesium deficit, which, while not constant, is a frequent occurrence [13].
Another study indicates that low intracellular brain magnesium concentration may be the link between the physiologic threshold for migraine and the attack itself [14].
Welch *et al.* propose that patients who suffer from migraine with aura have a susceptibility to spontaneous neuronal discharges and subsequent spreading depression, hypersusceptibility is supported by increased turnover of high-energy phosphates, low intracellular Mg2+ and large amplitude depolarizing waves on magnetoencephalography [14].
It is hypothesized that reduction might be due to an abnormal regulation of intracellular magnesium possibly reflecting at the periphery changes observed in the brain of migraineurs [6].
Changes in neuroexcitatory amino acids and magnesium, which may reflect a predisposition of the migraine patient, notably those having attacks with aura, will develop spreading depression [1].
Low magnesium levels may potentiate the sensitivity of the N-methyl-D-aspartate (NMDA) receptor to glutamic acid (Glu), a neuroexcitatory amino acid in the brain. Our results suggest that magnesium may be involved in migraine pathophysiology.
These results may support the use of Mg in prophylactics of migraine [15-17].

Conclusion

Low plasma magnesium concentration is a highly specific indicator of poor magnesium status.
We have demonstrated that low Mg status is common in patients with migraine with aura.
Magnesium seems to play significant role in the pathogenesis of migraine.
If serum Mg is low, an intervention to increase dietary intakes of Mg may be beneficial.
Placebo controlled, double blind and randomised studies should be performed to evaluate the effects of Mg replacement therapy on migraine and to determine the place of Mg in migraine therapy.

References

1. Ferrari MD. Biochemistry of migraine. *Pathol Biol* 1992 ; 40(4).

2. Rude RK. Magnesium deficiency : a cause of heterogeneous disease in humans. *J Bone Miner Res* 1998 ; 13 : 749-58.

3. Mauskop A, Altura BM. Role of magnesium in the pathogenesis and treatment of migraines. *Clin Neurosci* 1998 ; 5(1) : 24-7.

4. Welch KM, Barkley GL, Tepley N, Ramdan NM. Central neurogenic mechanisms of migraine. *Neurology* 1993 ; 43(6 Suppl 3) : S21-S25.

5. Mishima K, Takeshima T, Shimomura T, *et al.* Platelet ionized magnesium, cyclic AMP, and cyclic GMP levels in migraine and tension-type headache. *Headache* 1997 ; 37(9) : 561-4.

6. Schoenen J, Sianard Gainko J, Lenaerts M. Blood magnesium levels in migraine. *Cephalalgia* 1991 ; 11 : 97.

7. Mauskop A. Intravenous magnesium sulfate relieves migraine attacks in patients with low serum ionized magnesium levels : A pilot study. *Clin Sci* 1995 ; 89 : 633-6.

8. Gallai V, Sarchielli P, Morucci P, Abbritti G. Magnesium content of mononuclear blood cells in migraine patients. *Headache* 1994 ; 34(3) : 160-5.

9. Gallai V, Sarchielli P, Morucci P, Abbritti G. Red blood cell magnesium levels in migraine patients. *Cephalalgia* 1993 ; 13(2) : 81-94.

10. Gallai V, Sarchielli P, Coata G, Firenze C, Morucci P, Abbritti G. Serum and salivary magnesium levels in migraine. Results in a group of juvenile patients. *Headache* 1992 ; 32(3) : 132-5.

11. Taubert K. Magnesium in migraine. Results of a multicenter pilot study. *Fortschr Med* 1994 ; 112(24) : 328-30.

12. Thomas J, Thomas E, Tomb E. Serum and erythrocyte magnesium concentrations and migraine. *Magnes Res* 1992 ; 5(2) : 127-30.

13. Thomas J, Tomb E, Thomas E, Faure G. Migraine treatment by oral magnesium intake and correction of the irritation of buccofacial and cervical muscles as a side effect of mandibular imbalance. *Magnes Res* 1994 ; 792 : 123-7.

14. Welch KM, Barkley GL, Ramadan NM, D'Andrea G. NMR spectroscopic and magnetoencephalographic studies in migraine with aura : support for the spreading depression hypothesis. *Pathol Biol* 1992 ; 40(4) : 349-54.

15. Trauninger A, Pfund Z, Koszegi T, Czopf J. Oral magnesium load test in patients with migraine. *Headache* 2002 ; 42 : 114-9.

16. Durlach J, Durlach V, Bac P, *et al.* Magnesium and therapeutics. *Magnesium Res* 1994 ; 7 : 313-28.

17. Seref D, Okay V, Babur D, Mehmet AT. Efficacy of intravenous magnesium sulfate in the treatment of acute migraine attacks. *Headache* 2001 ; 41 : 171-7.

III. Magnesium and cardiovascular diseases

Protective effects of magnesium and orotic acid on the cardiovascular system and brain functions

H.G. Classen

Head of Pharmacology and Toxicology of Nutrition, University of Hohenheim, Fruwirthstrasse 16, D-70593 Stuttgart, Germany

Summary. Hypomagnesaemia is present at serum concentrations < 0.76 mmol Mg/L; suboptimal levels are < 0.80 mmol/L. Optimal concentrations are > 0.80 mmol/L, an evidence-based lower limit of the reference interval is around 0.85 mmol/L, according to Elin. Low plasma Mg concentrations indicate Mg deficiency and are associated with increased cardio- and cerebrovascular risk and various functional disturbances impairing the quality of life. They usually respond to oral Mg supplements. Mg depletion (Durlach) results when cells cannot maintain physiological Mg concentrations and require Mg-fixing agents in addition to Mg. Orotic acid, a key intermediate in the biosynthesis pathway of pyrimidines, stimulates the synthesis of proteins, phospholipids and ATP and hence provides binding-sites for Mg. Mg-orotate hence is a suitable salt also for the treatment of Mg depletion.

Keywords: Mg deficiency, Mg depletion, orotic acid, epidemiology

Beneficial effects of parenteral Mg injections on cardiovascular disturbances were first presented as observational single-case studies. In 1935, Zwillinger reported on the attenuation of digitalis-induced ventricular arrhythmias by Mg-sulfate [1], and in 1959, Parsons *et al.* [2] and Schreiber [3] successfully treated coronary spasms and stenocardic attacks with Mg injections. At the same time, Durlach [4] used oral Mg supplements for the treatment of functional, central-nervous disturbances, summarized as "*spasmophilie constitutionelle idiopathique*", later confirmed by Fehlinger [5]. Unfortunately, results of these and numerous additional studies are mixed together even today, and positive as well as negative data are used as proof that Mg has great, or only minor or even any importance for cardio- and cerebrovascular diseases. To dispel this confusion it is indispensable to differentiate between physiological and pharmacological effects of Mg [6]: Pharmacological Mg effects can be described as "nature's physiologic calcium blocker" [7] resulting, independently of the starting situation, when serum Mg concentrations are increased above 1.10 mmol/L [8]. Hence the results of i.v.-intervention studies like LIMIT-2, ISIS-4 or MAGIC do not allow any conclusion concerning the prognostic importance of the preceding Mg status (deficient or sufficient) for the development of disease. From a (patho-)physiological point of view it is necessary to consider Mg input (usually via food and water) and output (losses via faeces, urine, sweat a.s.o.).

Deficiency results when input < output. Serum Mg, although representing only 0.3% of total body Mg, is a valuable indicator: concentrations < 0.76 mmol/L indicate Mg deficiency, concentrations < 0.80 mmol/L are suboptimal and optimal levels are around 0.85 mmol/L [9], [10]. Intracellular Mg is > 90% bound to ligands, mostly phosphates. Mg depletion results when insufficient amounts of ligands are available, *e.g.* due to excessive consumption of energy-rich phosphates. Under these conditions Mg-fixing agents must be administered in addition to Mg [6].

Mg deficiency is obligatorily associated with secondary electrolyte disturbances, *e.g.* intracellular loss of potassium and accumulation of sodium and calcium. These complex alterations readily explain the associations between Mg deficiency and disease.

Prognostic importance of serum Mg

In studies on 199 patients with chronic congestive heart failure, all receiving diuretics and digoxin, Gottlieb *et al.* reported low serum Mg (<0.80 mmol/L) in 19%; these patients developed more frequent ventricular complexes, more episodes of ventricular tachycardia and had significantly worse survival rates after 1 and 2 years than subjects with serum Mg levels of 0.80 to 1.05 mmol/L [11]. In the FRAMINGHAM heart study, 3,327 healthy residents, aged 44 years, revealed higher prevalence rates of ventricular arryhthmias at lower serum Mg and K; a decrease of serum Mg by 0.08 mmol/L was associated with a 20% greater odds of VPCs [12]. In the NHANES I study, 10-year follow-up data on 492 subjects with and 7,759 subjects without cardiovascular events, adjusted relative risk for coronary heart and vascular disease events (deaths and hospitalisation) fell to 68% as serum Mg rose from < 0.81 to 0.87 mmol/L [13]. 19-year follow-up data on 12,500 subjects revealed serum Mg < 0.80 mmol/L in 23% of the sample; the highest age-adjusted ischemic heart disease and all-cause mortality rates occurred in this subgroup.

Mortality decreased to 79%, 66%, and 69% when serum Mg increased to 0.84, 0.89 and > 0.89 mmol/L [14]. In the ARIC study, 4-year follow-up data of 15,248 subjects aged between 45 and 64 years were collected. Serum Mg was lower in subjects with cardiovascular disease, hypertension and diabetes, and carotid wall thickness significantly increased by 0.0119 mm for each 0.1 mmol/L decrease in serum Mg [15]. The PROMISE study on 1,068 patients with congestive heart failure was biased by the fact that subjects with serum K < 3.5 mmol were excluded (although Mg deficiency is associated with secondary electrolyte disturbances!). Nevertheless, serum Mg < 0.75 mmol/L occurred in 19% of the remainder who exhibited more frequently ventricular couplets; however, there was no difference in survival time, probably due to the exclusion criteria [16]. In 957 patients undergoing coronary bypass graft surgery low serum Mg increased risk of death or myocardial infarction at one year [17], accordingly, Mg infusions reduced postoperative arrhythmic events [18]. A meta-analysis of 20 studies on totally 1,220 patients revealed a dose-dependent reduction of systolic and diastolic blood pressure following oral Mg supplementation [19]. These results are supported by data of Singh who kept 400 high-risk individuals during 10 years either on usual, or on a Mg-rich diet.

In the “usual” diet group serum Mg was lower and sudden deaths were 1.5 fold more common than in the Mg-rich diet group [20]. It is noteworthy that serum Mg has also prognostic importance for neurological events: a 20-months follow-up study on 323, 68-years old patients with symptomatic peripheral artery disease revealed that patients with serum Mg < 0.76 mmol/L exhibited a 3.3 fold increased risk for neurological events in comparison to subjects with levels > 0.84 mmol/L [21].

Cardiovascular and brain functions

Evidence is given that neural activity and ventricular fibrillation are closely associated [22], frequency of sudden cardiac death, for example, peaks in the early morning [23]. Symptoms of the tetanic syndrome and functional neurovegetative disorders are associated with low serum Mg, oral Mg supplements offer relief as shown by Durlach, Fehlinger, Schimatschek *et al.* and Seelig *et al.* [5,6,24,25]. Sleep disorders are associated with Mg deficiency [26] and have been improved with Mg supplements [27,28]. Increasing evidence is given that postoperative pain perception is related to Mg metabolism since Mg infusions reduce pain, the consumption of analgesics and opioids, and improve well-being [29–31]. Alltogether these data point to the fact that extracellular Mg status not only modifies the reactivity of Ca-entry channels of excitable tissues, but simultaneously central-nervous NMDA-receptor activity.

Magnesium depletion and orotic acid

In 1969, Meerson reported that orotic acid supports adaptive reactions via activated synthesis of nucleic acids (RNA), proteins, phospholipids, glycogen and ATP under hypoxic conditions [32]. In recently infarcted dog hearts subjected to cardioplegic arrest, pre-treatment with orally administered orotic acid significantly increased myocardial concentrations of ATP and glycogen thus providing binding-sites for Mg [33]. In fact, Golf *et al.* [34] as well as Branea *et al.* [35] reported beneficial effects of Mg-orotate undergoing coronary artery bypass surgery. In conclusion, orotic acid should be listed among and used as "*magnésium fixateur*" [6].

Conclusion

Convincing experimental [36], epidemiological and clinical evidence is provided that Mg deficiency and depletion, as severest form, must be listed among important cardiovascular and cerebrovascular risk factors. Therefore optimal serum Mg concentrations of > 0.80 mmol/L should be provided, preferentially by oral supplements, for prevention of disease and fatalities and for improving quality of life.

References

1. Zwillinger L. Über die Magnesiumwirkung auf das Herz. *Klin Wschr* 1935 ; 14 : 1429-33.
2. Parsons RS, Butler TC, Sellars EP. The treatment of coronary artery disease. *Med Procedure* 1959 ; 5 : 487-8.
3. Schreiber G. Erfahrungen mit Magnesiumsulfat-Injektionen bei der Behandlung von stenocardischen Anfällen. *Münchner Med Wschr* 1959 ; 101(41) : 1796-8.
4. Durlach J. Traitement pratique de la spasmophilie constitutionnelle idiopathique. *Vie Med* 1961 ; 42 : 1067-71.
5. Fehlinger R. Therapy with magnesium salts in neurological diseases. *Magnesium-Bull* 1990 ; 12 : 35-42.
6. Durlach J. *Magnesium in Clinical Practice.* London : John Libbey, 1988.
7. Iseri LT, French JH. Magnesium. Nature´s physiologic calcium blocker. *Am Heart J* 1984 ; 108 : 188-93.
8. Classen HG, Schimatschek HF, Spiessmann B. Pharmacology of orally administered magnesium salts with special reference to acid-base status. In : Rayssiguier Y, Mazur A, Durlach J, eds. *Advances in Magnesium Research.* Eastleigh : John Libbey, 2001 : 459-63.
9. Spätling L, Classen HG, Külpmann WR, *et al.* Kardiovaskuläres Risiko korreliert mit Serummagnesium. *MMW Fortschr Med* 2000 ; 142 : 441-2.
10. Elin RJ. Laboratory evaluation of chronic latent magnesium deficiency. In : Rayssiguier Y, Mazur A, Durlach J, eds. *Advances in Magnesium Research.* Eastleigh : John Libbey, 2001 : 233-9.
11. Gottlieb SS, Baruch L, Kukin ML, Bernstein JL, Fisher ML, Packer M. Prognostic importance of the serum magnesium concentration in patients with congestive heart failure. *J Am Coll Cardiol* 1990 ; 16 : 827-31.
12. Tsuji H, Venditti FJ, Evans JC, Larson MG, Levy D. The associations of levels of serum potassium and magnesium with ventricular premature complexes (the Framingham Heart Study). *Am J Cardiol* 1994 ; 74 : 232-5.
13. Gartside PS, Glueck CJ. The important role of modifiable dietary and behavioral characteristics in the causation and prevention of coronary heart disease hospitalization and mortality : The prospective NHANES I follow-up study. *J Am Coll Nutr* 1995 ; 14 : 71-9.
14. Ford ES. Serum magnesium and ischaemic heart disease : findings from a national sample of US adults. *Internat J Epidemiol* 1999 ; 28 : 645-51.
15. Ma J, Folsom AR, Melnick SL, *et al.* Associations of serum and dietary magnesium with cardiovascular disease, hypertension, diabetes, insulin, and carotid arterial wall thickness : The ARIC study. *J Clin Epidemiol* 1995 ; 48 : 927-40.
16. Eichhorn EJ, Tandon PK, DiBianco R, *et al.* Clinical and prognostic significance of serum magnesium concentration in patients with severe chronic congestive heart failure : The PROMISE study. *J Am Coll Cardiol* 1993 ; 21 : 634-40.
17. Booth JV, Phillips-Bute B, McCants CB, *et al.* Low serum magnesium level predicts major adverse cardiac events after coronary artery bypass graft surgery. *Am Heart J* 2003 ; 145 : 1108-13.
18. Speziale G, Ruvolo G, Fattouch K, *et al.* Arrhythmia prophylaxis after coronary artery bypass grafting : regimens of magnesium sulfate administration. *Thorac Cardiovasc Surg* 2000 ; 48 : 22-6.
19. Jee SH, Miller ER, Guallar E, Singh VK, Appel LJ, Klag MJ. The effect of magnesium supplementation on blood pressure : a meta-analysis of randomized clinical trials. *Am J Hypertens* 2002 ; 15 : 691-6.
20. Singh RB. Effect of dietary magnesium supplementation in the prevention of coronary heart disease and sudden cardiac death. *Magnesium Trace Elem* 1990 ; 9 : 143-51.
21. Amighi J, Sabeti S, Schlager O, *et al.* Low serum magnesium predicts events in patients with advanced atherosclerosis. *Stroke* 2004 ; 35 : 22-7.
22. Lown B, Verrier RI. Neural activity and ventricular fibrillation. *N Engl J Med* 1976 ; 294 : 1165-71.
23. Guagnano MT, Davi G, Sensi S. Morning sudden cardiac death. *Int J Immunopathol Pharmacol* 2000 ; 13 : 55-60.

24. Schimatschek HF, Classen HG, Baerlocher K, Thöni H. Hypomagnesiämie und funktionell-neurovegetative Beschwerden bei Kindern : Eine Doppelblindstudie mit Magnesium L-Aspartat-Hydrochlorid. *Der Kinderarzt* 1997 ; 28 : 196-203.

25. Seelig MS, Rosanoff A. *The Magnesium Factor*. New York : Penguin Group Inc, 2003.

26. Popoviciu L, Bagathai J, Buksa C, *et al.* Clinical and polysomnographic researches in patients with sleep disorders associated with magnesium deficiencies. In : Lasserre B, Durlach J, eds. *Magnesium – A Relevant Ion*. London : John Libbey, 1991 : 353-65.

27. Held K, Antonijevic IA, Kunzel H, *et al.* Oral Mg supplementation reverses age-related neuroendocrine and sleep EEG changes in humans. *Pharmacopsychiatry* 2002 ; 35 : 135-43.

28. Murck H, Steiger A. Mg reduces ACTH secretion and enhances spindle power without changing delta power during sleep in man – possible therapeutic implications. *Psychopharmacology (Berl)* 1998 ; 137 : 247-52.

29. Kara H, Sahin N, Ulusan V, Aydogdu T. Magnesium infusion reduces perioperative pain. *Eur J Anaesthesiol* 2002 ; 19 : 52-6.

30. Levaux C, Bonhomme V, Dewandre PY, Brichant JF, Hans P. Effect of intra-operative magnesium sulphate on pain relief and patient comfort after major lumbar orthopaedic surgery. *Anaesthesia* 2003 ; 58 : 131-5.

31. Unlugene H, Gunduz M, Ozalevli M, Akman H. A comparative study on the analgesic effect of tramadol, tramadol plus magnesium, and tramadol plus ketamine for postoperative pain management after abdominal surgery. *Acta Anaesthesiol Scand* 2002 ; 46 : 1025-30.

32. Meerson FZ. The myocardium in hyperfunction, hypertrophy, and heart failure. *Circ Res* 1969 ; 24(25) : 1-163.

33. Newman AJ, Chen X-Z, Rabinov M, Williams JF, Rosenfeldt FL. Sensitivity of the recently infarcted heart to cardioplegic arrest. *J Thorac Cardiovasc Surg* 1989 ; 97 : 593-604.

34. Golf S, Haase G. Der Einfluss von Magnesiumorotat auf den antiarrhythmischen Therapiebedarf bei supraventrikulären Arrhythmien nach aortokoronarer Bypassoperation. *Herz/Kreislauf J* 1996 ; 13 : 44-54.

35. Branea I, Gaita D, Dragulescu SI, *et al.* Assessment of treatment with orotate magnesium in early postoperative period of patients with cardiac insufficiency and coronary artery by-pass grafts (ATOMIC). *Rom J Int Med* 1999 ; 37 : 287-96.

36. Vormann J, Fischer G, Classen HG, Thöni H. Influence of decreased and increased magnesium supply on the cardiotoxic effects of epinephrine in rats. *Arzneimittel-Forsch/Drug Res* 1983 ; 33 : 205-10.

Drug treatment in acute myocardial infarction the position of magnesium - a critical review

R.H. Smetana

University Clinic of Vienna, Department of Internal Medicine IV, Vienna, Austria

Summary. According to the ACC/AHA guidelines for the management of patients with acute myocardial infarction the recommendation is as follows: antithrombotic and anticoagulant therapy; intravenous nitroglycerin for the first 24-48 hours in patients with acute myocardial infarction, prolonged use over 48 hours in patients with acute myocardial infarction, prolonged use over 48 hours in case of recurrent angina; reperfusion therapy such as acute PTCA or thrombolysis, in addition to these agents/procedures other *routineously* applied drugs such as beta-blockers are an established regimen; calcium channel blockers such as verapamil or diltiazem may be given to patients in whom β-adrenergic receptor blockers are ineffective or contraindicated. In borderline patients the new generation of short-life beta-blockers such as esmolol could also serve as an appropriate substance. The use of intravenous magnesium could be the ideal agent in case that conventional antianginal therapy is not applicable due to aggravation of hypotension. Furthermore, several large trials have shown essential benefit for intravenous magnesium therapy with regard to reduction of life-threatening arrhythmias and left ventricular failure. ACE-inhibitors in the early post-infarction phase are indicated in extended anterior wall infarction and are generally useful. Therefore, the optimal therapy regimen in acute myocardial infarction requires careful titration of antianginal and hypotensive drugs in order to establish equilibrium of pain relief and sufficient myocardial perfusion for the benefit of the patient. Future prospects are focussed on intravenous short-acting beta-blockers and intravenous magnesium.

Blood pressure and diastolic coronary blood flow predominantly determine coronary perfusion. During acute coronary syndromes and myocardial infarction *malperfusion* and consequently even necrosis of ischemic myocardium takes place. The idea, that sympathetic activity and myocardial norepinephrine release play a significant role in the vicious cycle leading to myocardial damage has been established during the last years.

Physiological and pathophysiological cardiac mechanisms

The activation of the sympathetic nervous system has to be considered as a compensatory mechanism to maintain perfusion pressure to vital organs and enhance myocardial contractility. Nevertheless, one of the most important actions of catecholamines is to alter the cardiac electrolyte metabolism. Adrenaline stimulates the beta-adrenergic receptors, and this leads to increased intracellular cyclic AMP, which activates the membrane-bound, magnesium and ATP dependent sodium-potassium-pump. This action is a physiological mechanism of the vital myocardium that turns into pathophysiological pattern in case of ischemic myocardium.

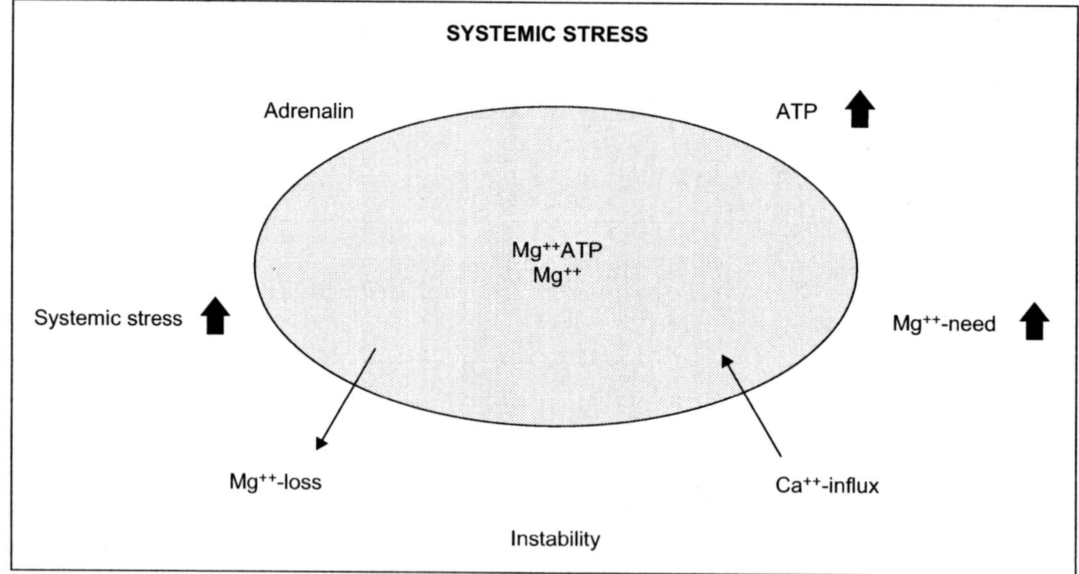

Figure 1. Pathomechanism of systemic stress.

Catecholamines and Systemic Stress

In relation to the duration and severity of coronary artery disease (CAD) stress and magnesium deficiency are mutually enhancing. Thus, a vicious cycle may develop resulting in adrenergic overstimulation. The enhanced catecholamine output in CAD requires an increased magnesium supply to maintain intracellular energy-bound metabolic processes. Therefore, physical and psychological stress increase the amount of magnesium required. The catecholamines cause movement of magnesium from intra- to extracellular compartments. Consequently, the electric membrane becomes unstable and membrane permeability increases, followed by cellular loss of magnesium and potassium and an increased risk of developing ventricular ectopic beats. The loss of myocardial magnesium is associated with the uptake of calcium that proceeds cardiac damage by calcium accumulation (*Figure 1*).

Acute myocardial infarction

In acute myocardial infarction (AMI) the restoration of coronary blood flow by pharmacological or mechanic intervention is essential for the survival of the ischemic myocardium. Early cardiac reperfusion, however, may be followed by complications as myocardial stunning, arrhythmia and even lethal cell injury. The possible pathomechanisms underlying reperfusion injury according to the present knowledge are: Oxygen radicals generated by the restoration of blood flow, ischemia-induced activation of the complement system, and disturbance of calcium homeostasis in addition to oxidative stress and complement-mediated damage in the myocardium. The development of the mechanical stunning of the heart is related to cytosolic calcium overload at the onset of reperfusion.

State of the art therapeutic regimen

Today's standard therapy of acute coronary syndromes and acute myocardial infarction consists of a variety of drugs, some of which have pressure-decreasing abilities, which might be more harmful in the acute setting than of benefit for the patient. To enlighten the background of these antianginal and vasodilatory drugs, a review of the current literature was undertaken with special respect to the ACC/AHA guidelines for the management of patients with AMI.

Standard therapy today includes antithrombotic management with predominantly aspirin, but also other new drugs such as ticlopidine or clopidogrel, anticoagulation with heparins and Glycoprotein IIb/IIa inhibitors and vasodilatory drugs such as nitroglycerin and magnesium. Furthermore beta-blockers, ACE inhibitors, diuretics in case of cardiogenic failure or positive inotrope substances such as the various catecholamines like dopamine are part of the standard medical therapy. Further treatment may include thrombolytic therapy or acute catheterization.
In the following paragraphs the mechanisms and base of evidence for the single substances which might cause hypotension will be reviewed.

Nitroglycerin

The primary mechanism of action is vasodilation, which is attributable primarily to the relaxation of smooth muscle cells in veins, arteries and arterioles taking advantage of the different sensitivities of these cells in the different tissues (veins > arteries > arterioles). The metabolic conversion of organic nitrates to the endothelium-derived relaxing factor nitric oxide is the main mechanism of action and an important factor in the endogenous modulation of vascular tone. The reduction in pre-load combined with the after load reduction resulting from arterial vasodilation decrease cardiac workload and cardiac oxygen demand. By vasodilation of coronary vessels and the coronary vascular bed perfusion can be improved and ischemia reduced. Still there is no evidence of a reduction in mortality by the application of nitrates in acute coronary syndromes. Nitroprussid even seems to endanger coronary perfusion by hypotension. If nitrates are not titrated intravenously in a very careful manner, *e.g.* decrease in arterial pressure by no more than 10% or mean blood pressure < 80 mmHg, hypotension and consequently hypoperfusion of the jeopardized myocardium and aggravation of myocardial oxygen demand by reflex tachycardia are risked.

Beta-adrenoreceptor Blockers

Beta-blockers are recommended as an agent of first choice in the absence of contraindications within 12 hours after onset of symptoms. Beta-blockers diminish catecholamine induced cardiovascular hyperreactivity, their ability to improve outcome in patients with acute coronary syndromes might be due to the reduction of heart rate, a lowering of systemic arterial pressure and negative inotropy, all of which lead to an decrease in oxygen demand. Immediate application reduces the extent of the infarcted area, concomitant complications as well as the rate of reinfarction in patients receiving thrombolysis. Contraindications such as heart rate less than 60 bpm, systolic arterial pressure less than 100 mmHg, heart failure, second- or third grade AV block or lung disorders have to be strictly obeyed. Patients with moderate LV failure (the presence of bibasilar rates without evidence of low cardiac output) can be treated if titration is performed carefully. Meta-analysis showed a 12% reduction in short term mortality if beta-blockers are applied in the early hours of the event; the reduction of mortality is independent of the classical risk factors. The intravenous application of the new generation of beta-blockers with an extremely short half life such as esmolol widens the range of beta-blocker interventions and therefore is a promising perspective in treatment of AMI.

Amiodarone

This substance commonly also counted as a beta-blocker is a complex antiarrhytmic drug combining the abilities of noncompetitive β-adrenoceptorblockade, calcium channel blockade, blockade of sympathetic efferents and class I antiarrhythmics. It is approved for the treatment and prophylaxis of recurrent ventricular fibrillation and hemodynamically destabilizing ventricular tachycardia. Side effects may include such as hypotension, bradycardia and AV-block.

ACE-Inhibitors

ACE-inhibitors are about equally effective as beta-blockers, calcium-channel blockers or diuretics in hypertension, but without the negative effects on lipid metabolism and insulin sensitivity. In patients with extended anterior myocardial infarction or signs of threatening heart failure in the absence of significant hypotension or known contraindications this medication is of benefit if applied in the first 24 hours after onset of symptoms. Solely oral ACE-inhibitors seem to be beneficial; in the only study showing no benefit but even a slight tendency towards damage by an early hypotensive reaction intravenous enalaprilat was used.

Calcium Channel Blockers

There is currently no class I evidence for the use of calcium-antagonists in the setting of acute myocardial infarction. New studies though suggest that Verapamil given to patients without heart failure in the late phase of myocardial infarction reduces mortality and morbidity significantly. A multicenter trial will tempt to prove the benefit in patients with manifest heart failure, especially when given in combination with ACE-inhibitors.

Magnesium

The application of magnesium in acute myocardial infarction has currently no class I evidence although several large trials have shown essential benefit for intravenous magnesium therapy with regard to reduction of life-threatening arrhythmia and left ventricular failure. Magnesium acts as cofactor for over 350 enzymes and metabolic processes in the human body. Its prevalence as a cofactor of the cellular energy metabolism as magnesium-ATP is of obvious importance. During ischemia caused by impaired coronary blood flow, the need for magnesium augments, leading to increased utilization of intracellular reserves. As the equilibrium of the electrolyte-fluxes is very sensitive to changes of the ionic stability, an increased influx of calcium is taking place. Growing instability of the cell membrane, finally leading to membrane leakage and loss of magnesium into the extracellular compartment follows this disturbance of cellular homeostasis. These incidents are followed by increasing loss of function of the myocyte, leading to development of systemic stress and to a further rise in catecholamine output. A vicious cycle consisting of increased catecholamine stimulation, extended activation of the energy metabolism of the cell and therefore augmented need for magnesium develops, finally leading to the total loss of cell function and cell lysis. Magnesium may therefore play a pivotal role in the prevention of reperfusion injury by membrane stabilization and reduction of calcium overload in the myocardial mitochondria and by conservation of ATP as magnesium ATP consequently, combined with the mechanism of improved coronary blood flow by vasodilatation. The use of intravenous magnesium in AMI is still a stimulating option in the scope of future strategies in therapeutic regimen, in spite the MAGIC trial could not confirm its efficacy.

Conclusion

The optimal therapy regimen in acute myocardial infarction requires careful titration of antianginal and hypotensive drugs in order to establish equilibrium of pain relief and sufficient myocardial perfusion for the benefit of the patient. A concertant action in finding complementary therapeutic concepts is the goal for the next millennium. Magnesium as a therapeutic agent has similar characteristics like beta-blockers and calcium-channel-blockers, furthermore it has membrane-stabilizing effects. Magnesium applicated intravenously has pharmacological effects. Anyway, in treatment of acute AMI, Magnesium has to be given before pharmacological or mechanical reperfusion in order to prevent calcium accumulation and reperfusion injury. The common procedure of invasive cardiac intervention (angioplasty, stenting) and intravenous magnesium application could become a gold standard in treatment of acute AMI [1–10].

References

1. Ryan JR, Anderson JL, Antmann EM, *et al.* ACC/AHA Guidelines for the Management of Patients with Acute Myocardial Infarction. *JACC* 1996 ; 28 : 1328-428.
2. Ryan TJ, Antmann EM, Brooks NH, *et al.* 1999 Update : ACC/AHA Guidelines for the Management of Patients with Acute Myocardial Infarction. *JACC* 1999 ; 34 : 890-911.

3. Jugdutt BI. Effects of nitrates on myocardial remodeling after acute myocardial infarction. *Am J Cardiol* 1996 ; 77(13) : 17C-23C.

4. Klein W, Eber B, Gasser R, *et al.* Current Aspects of ACE inhibitor therapy from the cardiologic viewpoint. *Wi Med Wochenschr* 1996 ; 146(11) : 221-4.

5. Jespersen CM. Verapamil in acute myocardial infarction. The rationales of the VAMI and DAVIT III trials. *Cardiovasc Drugs Ther* 1999 ; 13(4) : 301-7.

6. Maggioni AP, Latini R. How to use ACE-inhibitors, beta-blockers, and newer therapies in AMI. *Am Heart J* 1999 ; 138 : 183-7.

7. De Vreede JJM, Gorgels APM, Verstraaten GMP, Vermeer F, Dassen WRM, Wellens HJJ. Did prognosis after acute myocardial infarction change during the past 20 years? A meta-analysis. *J Am Coll Cardiol* 1991 ; 18 : 698-706.

8. Rogers WJ. Contemporary management of acute myocardial infarction. *Am J Med* 1995 ; 157 : 195-206.

9. Glogar DH, Yang P, Steurer G. Management of acute myocardial infarction : Evaluating the past, practicing in the present, elaborating the future. *Am Heart J* 1996 ; 132 : 465-70.

10. Antman EM, Smetana RH, *et al.* The MAGIC Trial Investigators. Early administration of intravenous magnesium to high-risk patients with acute myocardial infarction in the Magnesium in Coronaries (MAGIC) Trial : a randomized controlled trial. *Lancet* 2002 ; 360 : 1189-96.

The relationships between magnesium deficiency and cardio-cerebrovascular risk factors

I.A. Gutiu[1], L.I. Gutiu[2]

1. Dept. of Medical Emergencies, "Saint Mary" Hospital; University of Medicine and Pharmacy "Carol Davila", Bucharest, Romania ; 2. Dept of Neurology, Central Military Hospital; University of Medicine and Pharmacy "Carol Davila", Bucharest, Romania

Some studies shown that, between the magnesium deficiency and increased risk of cardio-cerebrovascular disease exists a relationship [1].

The mechanism is still unclear. However, it was demonstrated that magnesium deficiency (even only extra cellular) may induce a nonspecific inflammation characterized by: activation of leucocytes and macrophages, freeing of cytokines and free radicals, increased oxidation of lipids and LDL; all are basic mechanisms in atherogenesis [2].

These aspects generated by magnesium deficiency (demonstrated in animal experiments) are more difficultly to demonstrate in clinical research. However, some clinical studies confirmed the existence of an interrelationship between the cardio-cerebrovascular risk factors (CVRF) and magnesium deficiency.

In actual conception, CVRF can be classified in traditional risk factors such as dyslipidaemia, smoking, arterial hypertension, diabetes, and new risk factor such as low level, nonspecific, chronic inflammation, immunity alterations, infections, etc. [3-6] (*Tables I, II*).

On this background we established the objectives of work, as in *Table III*.

Table I. Traditional Cardiovascular Risk Factors.

Modifiable	Non-modifiable
- Dyslipidemia	- Age
- Smoking	- Sex
- Hypertension	- Family history of premature CHD
- Diabetes mellitus	
- Obesity	
- Dietary factors	
- Thrombogenic factors	
- Sedentary lifestyle	

Table II. Some New Cardiovascular Risk Factors.

Infections
- *Chlamydia pneumoniae*
- *Helicobacter pylori*
- *Cytomegalovirus*
- Oral germs from dental plaque
- Other: *Herpesviruses, Hepatitis virus A, B, C, etc.*
- Infectious burden

Chronic nonspecific systemic inflammation
- ↑ Fibrinogenaemia, ↑ CRP, ↑ IL-6

Autoimmunity
- Detection of antibodies against oxidized LDL, HSP-60 etc.

Table III. Objective of this lecture.

- To discuss certain relationships between serum magnesium and cardio-cerebrovascular risk factors;
- To appreciate the role of magnesium in atherogenesis in coronary artery disease and stroke patients;
- To insist on the relationships between magnesium level and the new cardiovascular risk factors as inflammation and infections;
- For this purpose, we selected a number of cardiovascular patients and analyzed the cardiovascular risk factors in relationship with blood magnesium level.

Material and methods

We studied the patients admitted in hospitals and investigated by the physicians. In the protocol we introduced a number of new cardiovascular risk factors as C reactive protein (CRP), serum fibrinogen. We included in study 139 patients mainly women (61%), admitted for the consequences of atherosclerosis (coronary disease or stroke) (*Table IV*).
The main characteristics of study group are synthesized in *Table V*.

The data of the patients was analyzed statistically using dispersion analysis ANOVA on quartiles of serum magnesium levels.

Table IV. Clinical Characteristics of the Study Group.

No of patients	139
Sex (F/M)	85/54 (61%/39%)
History of Myoc. Infarction	60 (47%)
Angina pectoris	31 (25%)
Stroke	36 (28%)

Table V. Baseline Characteristics of Study Population [1].

Number:	139
Sex:	M = 54 (39%); F = 85 (61%)
Age (years)	67 +/- 9
Height (m)	1.62 +/- 0.09
Weight (kg)	69 +/- 14
Cholesterol (magnesium %)	201 +/- 57
Triglyceride (magnesium %)	129 +/- 79
Glucose (magnesium %)	114 +/- 42
Smoking index	8.1 +/- 20.2
Arterial pressure (mmHg)	153 +/- 27
WBC	6470 +/- 2213
Fibrinogen (magnesium %)	440 +/- 101
C Reactive Proteine+ (%)	37
Calcium (magnesium %)	9.3 +/- 0.9
Magnesium (magnesium %)	2.2 +/- 0.7
HDL	40.1 +/- 16.6
LDL	143.6 +/- 58.8
Creatinine (magnesium %)	1.01 +/- 0.2

Results

All significant data are presented in *Table VI*. We retain only cardiovascular risk factors with significant difference between first and last quartiles of serum magnesium level.

On data of the *Table VI* we conclude that hypomagnesaemia is more frequently met in the patients with cerebro-cardiovascular disease who have dyslipidaemia, serum glucose increased, higher level of serum CRP and increased body mass index (BMI).

These all risk factors are gathered in *Table VII*.

Table VI. The Main Cardiovascular Risk Factors Analysed by ANOVA dispersion test on serum Magnesium Quartiles [1] and found having statistical significance in relationship with magnesium level.

Variables (risk factors)	Magnesium level Quartile 1 (< = 1.7 magnesium)	Magnesium level Quartile 2-3 (> 1.7 < 2.2 magnesium)	Magnesium level Quartile 4 (> = 2.2 magnesium)	P[a]
Cholesterol	189.9 +/- 53.4	193.2 +/- 44.4	230.2 +/- 62.2	**< 0.005**
HDL-C	30.2 +/- 10.4	43.8 +/- 17.9	44.8 +/- 14.9	**< 0.001**
B.M.I.	26.8 +/- 4.7	25.8 +/- 4.6	23.7 +/- 4.5	**< 0.047**
Glucose level	126.2 +/- 30.7	108.9 +/- 37.8	108.6 +/- 34.0	**< 0.04**
CRP (%)	43.3	38.7	36.0	**< 0.01**
Calcium	9.08 +/- 1.29	9.15 +/- 0.96	9.94 +/- 0/56	**< 0.002**

[a] Statistical significance of difference between quartile 1 and quartile 4 serum magnesium level.

Table VII. Synthesis of the significant cardiovascular risk factors.

- Cholesterol
- HDL-C
- CRP ↑ (%)
- BMI
- Glucose level
- Calcium

Discussions

We observed the following anomalies of lipids in hypomagnesaemia study group:
- Decrease of HDL-C,
- Decrease of total cholesterol,
- Increase of blood glucose,
- Increase of BMI.

More important is the HDL-C that have a demonstrated protective role in atherogenesis and, now, is subject of large prophylactic study. The decrease of total cholesterol in patients with low magnesium show that, in these cases with CAD, stroke, and a long evolution of atherosclerosis, may be more important intervention of other risk factors implied in plaque instability, especially manifested in this group, as new cardiovascular risk factors: inflammation, infections, etc.
The link between hypomagnesaemia and dyslipidaemia is presented in *Table VIII*, as conclusions of a number of published works [7-9].

Concerning the relationships between hypomagnesaemia and nonspecific chronic inflammation in atherogenesis we argue with the conclusions of more studies which analyzed these aspects, synthesized in *Table IX*.

In this work we noted an increase of some nonspecific inflammation in the hypomagnesaemia group of patients as high CRP level, versus normal or high blood magnesium patients.

Table VIII. Dyslipidaemia and Hypomagnesaemia.

Supposed intervention of Magnesium in cardiovascular disease
- In ventricular arrhythmias
- In cardiac failure
- In atherogenesis on:
  - Platelet activity
  - Blood pressure
  - Oxidative stress
  - Lipid peroxidase
  - Catecholamine reduction
  - Coronary dilatation
  - Reperfusion syndrome

Table IX. Nonspecific chronic inflammation and hypomagnesaemia-experimental evidence of Magnesium intervention in atherogenesis on inflammation link [1, 2, 11].

- Experimental evidence of magnesium intervention in atherogenesis on inflammation link
- Freeing of substance P, ↑ interleukines, PMN activation
- Increase of phagocytosis
- Freeing of inflammation mediators as: cytokines, prostaglandines, leucotrienes, etc.
- Endothelial dysfunction: ↓ Nitric oxide, ↑ Insulin resistance, possible initiating process of atherosclerosis, etc.
- Possible intervention on immunologic system: for example, ↑ of C3 complement, ↓ innate immune cellular protection, etc.

Conclusions

– Our study says that hypomagnesaemia have more influence in CVD or stroke in the presence of a chronic inflammation syndrome possibly generated by some infections with large extension in population (see Helycobacter pylori case or gum chronic infections, etc.) [4, 7].

– The mechanisms are listed but we note as more possible the concurrent intervention of magnesium on: a) ↑ phagocytosis; b) freeing of cytokines: interleukines, prostaglandines, leucotrienes, etc.; c) endothelial dysfunction: ↓ nitric oxide, ↑ insulin resistance (in our patients, in hypomagnesium group, serum glucose is significant increased), etc.

– We note that this is a preliminary study and our results are obtained on a relative small number of patients.

– But, these relationships seem very logically and the study must continue.

– The role of CRP in this discussion is linked to very large and important trials demonstrating the importance of CRP determination for the plaque instability appreciation. Our work seems to demonstrate the role of hypomagnesaemia in low level inflammation process in patients with atherosclerosis consequences [3, 4].

References

1. Altura B, Brust M, Bloom S, *et al.* Magnesium dietary intake modulates blood lipid levels and atherogenesis. *Oroc Natl Acad Sci* 1990 ; 87 : 1840-4.

2. Bussiere FI, Gueux E, Rock E, Mazur A, Rayssiguier Y. Protective effect of calcium deficiency on the inflammatory response in magnesium deficient rats. *Eur J Nutr* 2002 ; 41 : 197-202.

3. Pfeilschifter J, Loditz R, Pfohl M, Schatz H. Changes in proinflammatory cytokine activity after menopause. *Endocr Rev* 2002 ; 23(1) : 90-119.

4. Kiechl S, Egger G, Mayr M, *et al.* Chronic infections and the risk of carotid atherosclerosis. Prospective results from a large population study. *Circulation* 2001 ; 193 : 1064-70.

5. Haraszty VI, Zambon JJ, Trevisan M, Zeid M, Genco RJ. Identification of periodontal pathogens in atheromatous plaques. *Periodontal* 2000 ; 7(10) : 1554-60.

6. Li X, Kolltveit KM, Tronstadt L, Olsen I. Systemic diseases caused by oral infection. *Clin Microb Rev* 2000 ; 13(4) : 547-58.

7. Jing MA, Folsom AR, Melnick SL, *et al.* Association of serum and dietary magnesium with cardiovascular disease, hypertension, diabetes, insulin, and carotid arterial wall thickness : the ARIC study. *J Clin Epidemiol* 1995 ; 19 : 258-72.

8. Sherer T, Bitzur R, Cohen H, *et al.* Mechanism of action of the anti-atherogenic effect of magnesium. lessons from a mouse model. *Magnes Res* 2001 ; 14(3) : 173-9.

9. Stanton MF, Lowenstein FW. Serum magnesium in women during pregnancy, while taking contraceptives, and after menopause. *J Am Coll Nutr* 1987 ; 6(4) : 313-9.

10. Bureau I, Anderson RA, Armand J, *et al.* Trace mineral status in post menopausal women : impact of hormonal replacement.

11. Teragawa H, Kato M, Matsura H, Kajiama G. Magnesium causes nitric oxide independent coronary artery dilatation in humans. *Heart* 2001 ; 86 : 212-6.

Advances in Magnesium Research: New Data 2006: 55-60

Role of magnesium in essential hypertension in teenagers

G. Sur, O. Maftei

2nd Pediatric Clinic, Cluj-Napoca, Romania

Abstract. Magnesium is involved in the pathogenesis of essential hypertension by decreasing the sympathetic reactivity, cardiac excitability and vascular tone, contractility and reactivity. The authors of this study proposed to assess the reduction of blood pressure values after magnesium supplementation in hypomagnesemic teenagers with essential hypertension. In this retrospective study across 3 years, 17 patients with primary hypertension were included. Those with magnesium deficit received Magnerot for 2 months in association with a salt-restrictive diet and physical exercises. After magnesium supplementation, both systolic and diastolic pressure readings decreased and the clinical status improved. The role of magnesium supplementation in magnesium-deficient adolescents with high blood pressure seems to be beneficial, but health-promoting lifestyle changes are also important.

Magnesium plays an important role in the physiology of the cardiovascular apparatus and the pathogenesis of cardiovascular disease [1]. Magnesium is involved in more than 300 essential metabolic reactions, including ion transport systems that are implicated in the conduction of nerve impulses, the muscle contraction and the normal rhythm of the heart [1, 2].

Magnesium is a potent vasodilator. Magnesium acts in a way related to calcium channel blocker drugs (physiologic calcium channel blocker) [3]. When administered in combination, synthetic calcium channel blockers and magnesium are synergistic in the treatment of hypertension [3]. Blood pressure correlates positively with intracellular calcium concentration and negatively with magnesium concentration [3, 4].

Magnesium also decreases sympathetic nervous system activity, a fact that contributes to relaxing smooth muscle in blood vessels, subsequently reducing the blood pressure [1]. Plus, magnesium supplementation lowers LDL-cholesterol levels and raises HDL-cholesterol levels, reducing the risk for cardiovascular disease [4].

In the table below, we present the clinical effects of suppressed intracellular free magnesium and/or elevated cytosolic free calcium not only in hypertension, but also in other conditions associated with it [5] (*Table I*).

Magnesium-deficient individuals present: increased stress susceptibility, cramps, muscle tremors, sleep disturbances, poor concentration, irritability, nausea, constipation, cardiac arrhythmia, rise in blood pressure [6].

The National Heart, Lung and Blood Institute included in the classification of hypertension one more category: pre-hypertension, defined as consistent systolic blood pressure (SBP) readings of 120-139 mmHg or diastolic blood pressure (DBP) of 80-89 mmHg (*Table II*).

Objective

The aim of the present study was to assess the potential benefits of magnesium repletion in hypomagnesemic patients with high blood pressure.

Material and method

We made a retrospective study over a period of 3 years (January 2001- December 2003) on 17 patients aged 14-18 years with essential hypertension, admitted in 2nd Pediatric Clinic of Cluj-Napoca, Romania. To all these subjects we had excluded an underlying disease as a cause of secondary hypertension.
According to the 7th Report of Joint National Committee, 12 of the study patients were pre-hypertensives, 4 had stage 1 hypertension being on beta-blocker medication (Metoprolol 50 mg/day) and 1 had stage 2 hypertension being treated concurrently with a beta-blocker (Metoprolol 50 mg/day) and a diuretic (Indapamide 2.5 mg/day) (*Table III*).
Most of our subjects had low serum magnesium, only 3 pre-hypertensive patients had a normal plasma level of magnesium. All those teenagers who had hypomagnesemia were given Magnerot orally 3 times a day, daily for a period of 8 weeks. One tablet of Magnerot comprises 500 mg of magnesium orotate, containing 32.8 mg of magnesium, equivalent of 2.7 mEq (1.35 mmol), though our patients received 8.1 mEq (4.05 mmol) magnesium/day.
Health-promoting lifestyle changes were recommended to all the patients, including: reduction in sodium intake, weight reduction, physical activity, discouragement of smoking, restriction in alcohol intake and avoiding stress.
We checked blood pressure values and serum magnesium level at baseline, after 4 weeks and 8 weeks, respectively.

Table I. Main effects of magnesium.

Cell type	Clinical expression
Vascular smooth muscle cell	Vasoconstriction, HBP
Fat or skeletal muscle cell	Insulin resistance
Pancreatic beta cell	Hyperinsulinemia
Cardiac myocyte	Left ventricular hypertrophy
Platelet	Increased platelet aggregation
Renal tubular cells	Increased sodium retention, calcium excretion
Central and peripheral sympathetic neurons	Increased catecholamine release, increased sympathetic tone

Table II. 7th Report of Joint National Committee on Prevention, Detection, Evaluation and Treatment of High Blood Pressure (2003).

	SBP (mmHg)	DBP (mmHg)
Normal	< 120 and	< 80
Pre-hypertension	120-139 or	80-89
Stage 1 hypertension	140-159 or	90-99
Stage 2 hypertension	≥ 160 or	≥ 100

Table III. The repartition of our study patients.

	Total	Serum level of magnesium		
		Low	Normal	
Pre-hypertension	12	9	3	
Stage 1 hypertension	4	4	-	← β-blocker
Stage 2 hypertension	1	1	-	← β-blocker+ diuretic
Total	17	14	3	
		↑ Magnesium		

Results

After 2 months of magnesium supplementation, salt-restrictive diet and physical exercise, systolic blood pressure decreased respectively by 1.6 mmHg in pre-hypertensive group with normal serum magnesium, 5 mmHg in pre-hypertensive group with low serum magnesium, 7.5 mmHg in stage 1 hypertensive group and 10 mmHg in stage 2 hypertensive group (*Figure 1*).

Diastolic blood pressure fell by 1.67 mmHg in patients who did not receive any additional magnesium and more in "magnesium group", respectively 4.4 mmHg in pre-hypertensive group with low serum magnesium, 7.5 mmHg in stage 1 hypertensive group and 5 mmHg in stage 2 hypertensive group (*Figure 2*).

After magnesium supplementation, serum magnesium was higher. Changes in magnesium level averaged 0.07 mmol/l in pre-hypertensive group, 0.05 mmol/l in stage 1 hypertensive group and 0.06 mmol/l in stage 2 hypertensive group, after 8 weeks of treatment (*Figure 3*).

Magnesium supplement was well tolerated and more over, magnesium-deficient children improved their clinical status: no palpitations, no sleep disturbances and lower irritability compared with "magnesium group", the group of subjects who did not receive any additional magnesium, did not experience these positive effects.

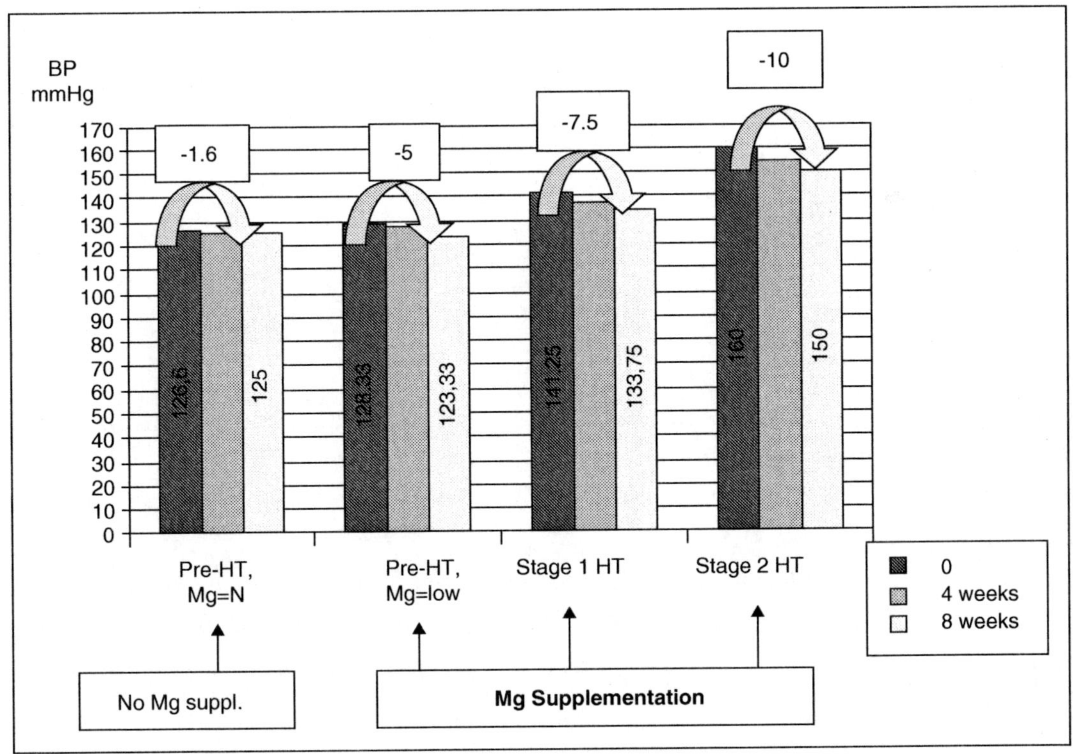

Figure 1. Mean values for SBP readings.

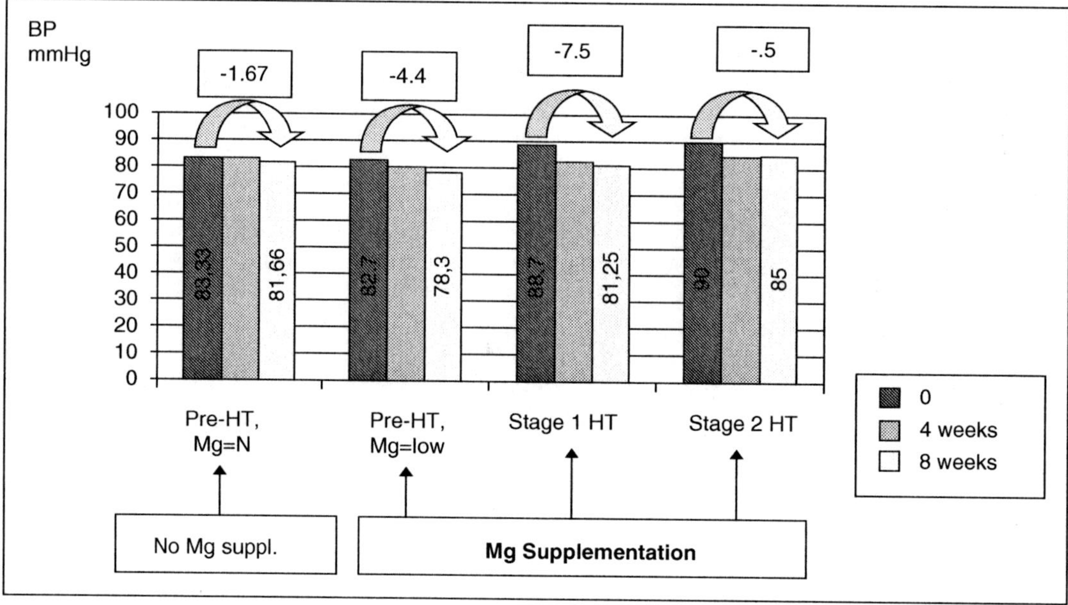

Figure 2. Mean values for DBP readings.

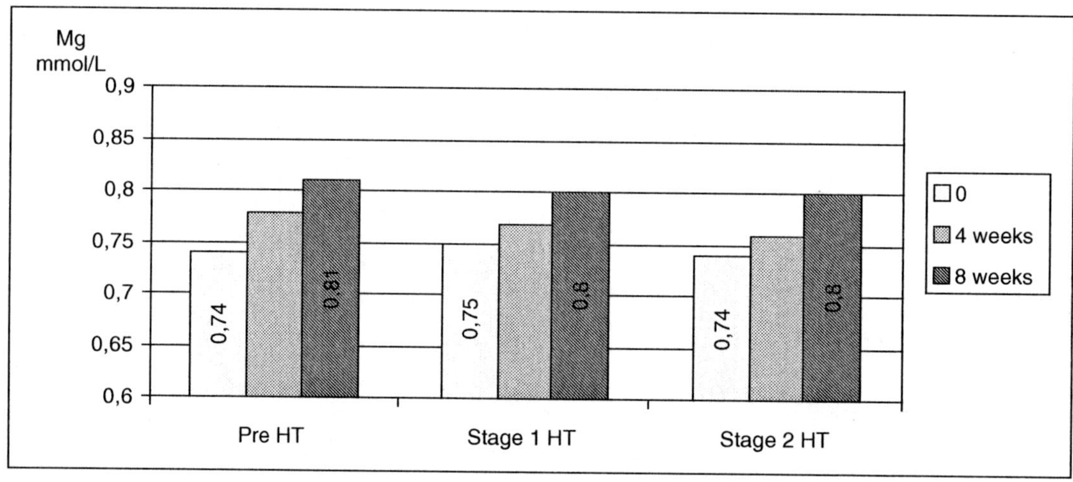

Figure 3. Mean values for serum magnesium level.

Discussions

Knowledge of the relationship between magnesium and hypertension began with the observation that the administration of parenteral magnesium sulfate reduced blood pressure in women having eclampsia or pre-eclampsia [7].

Many studies suggest that magnesium activates Na/K-ATPase, decreasing intracellular sodium concentration and lowers intracellular free calcium, relaxing the smooth muscle in blood vessels [2, 3]. Conversely, magnesium depletion inhibits sodium-pump activity, increasing intracellular sodium concentration; this, in turn, will suppress calcium's extrusion from the cell and cytosolic calcium will increase, stimulating smooth muscle contractility [2, 3].

Table IV. Recommended Dietary allowance for Magnesium.

Life stage	Age	Recommended Dietary Allowance for magnesium (mg/day)	
Infants	0-6 months	30	
	7-12 months	75	
Children	1-3 years	80	
	4-8 years	130	
	9-13 years	240	
Adolescents	14-18 years	360 – females	410 – males
Adults	19-30 years	310 – females	400 – males
	≥ 31 years	320 – females	420 – males

Adolescence is the age when the need for magnesium is maximum [6, 8]. In *Table IV*, we present the recommended dietary allowance (RDA) for magnesium, varying with age [8].
In hypertensive patients, it is important to mention that long-term use of potassium-depleting diuretics (furosemide, hydrochlorothiazide) increases the risk for renal magnesium wasting [4, 9].
The effect of magnesium on lowering blood pressure was greater for people with higher blood pressure. Those patients having the highest baseline systolic blood pressure experienced the greatest magnesium-induced fall in pressure [1].
These results support the view that oral magnesium supplementation is a useful approach to treatment of patients with essential hypertension, although it is difficult to demonstrate an independent effect of a single nutrient in such a little observational study [4, 10, 11].

Conclusions

Magnesium plays a central role in the normal regulation of blood pressure and, therefore, there is an important link between magnesium deficit and the hypertensive state. We would suggest that anyone with high blood pressure should consider increasing their intake of magnesium because oral magnesium may exert a beneficial effect on hypertension.
But monotherapy with oral magnesium cannot be considered as an efficient treatment of hypertension.
In hypertensive patients treated with drugs, sodium restriction and magnesium supplementation enhance the therapeutic effect.
The role of the health professional should be to stress maintaining a healthy weight, increasing physical activity, reducing stress and eating a healthy diet.

References

1. Dillon MJ, Ingelfinger JR. Pharmacological treatment of hypertension. In : Holliday MA, Barrat TM, Avner ED, eds. *Pediatric nephrology. 3rd Ed.* Baltimore : Williams & Wilkins, 1994 : 1165-74.

2. Ashida T, Kawano Y, Yoshimi H, Kuramochi M, Omae T. Effects of dietary salt on sodium – calcium exchange and ATP-drive calcium pump in arterial smooth muscle of Dahl rats. *J Hypertens* 1992 ; 10 : 1335-41.

3. Gilbert D'Angelo EK, Singer HA. Rembold CM. Magnesium relaxes arterial smooth muscle by decreasing intracellular Ca^{2+} without changing intracellular Mg^{2+}. *J Clin Invest* 1992 ; 89 : 1988-94.

4. Whitaker J. *Reversing hypertension-a vital new program to prevent, treat and reduce high blood pressure.* New York : Warner Books, 2000.

5. Altura BM, Altura BT. Cardiovascular risk factors and magnesium : relationships to atherosclerosis, ischemic heart disease and hypertension. *Magnes Trace Elem* 1991 ; 10 : 182-92.

6. Yetman RJ, Bonilla-Felix MA, Portman RJ. Primary hypertension in children and adolescents. In : Holliday MA, Barrat TM, Avner ED, eds. *Pediatric nephrology. 3rd Ed.* Baltimore : Williams & Wilkins, 1994 : 1117-45.

7. Kontopoulos V, Seelig MS, Dolan J. Influence of parenteral administration of magnesium sulfate to normal pregnant and to pre-eclamptic women. In : Cantin M, Seelig MM, eds. *Magnesium in health and disease. New York.* 1980 : 839-48.

8. Svetkey LP, Simons-Morton D, Vollmer WM. Effects of dietary patterns on blood pressure : subgroup analysis of the dietary approaches to stop hypertension (DASH) randomized clinical trial. *Arch Intern Med* 1997 ; 157 : 657-67.

9. Sacks FM, Brown LE, Appel L. Combinations of potassium, calcium and magnesium supplements in hypertension. *Hypertension* 1995 ; 26 : 950-6.

10. Apple LJ. A clinical trial of effects of dietary patterns and blood pressure. *N Engl J Med* 1997 ; 336 : 1117-23.

11. Dacău C, Romoşan I, Jompan A. Rolul dietei în tratamentul hipertensiunii arteriale. *Medicina familiei* 1999 ; 28 : 17-20.

Effects of magnesium sulfate and its association with mepivacaine and antihypertensives on isolated frog hearts

S.M. Popescu, F. Popescu, A.M. Popescu
UMF Craiova, Craiova, Romania

Abstract. Background. Magnesium is an important cation for the human body, and it is used as magnesium sulfate in arrhythmia and eclampsia. Mepivacaine is used often in dentistry as a proper local anesthetic for hypertensive patients, who could be under treatment with various antihypertensives. **Objectives.** The aim of this study was to observe the effects of magnesium sulfate and its associations with mepivacaine, mepivacaine-enalaprilat, mepivacaine-propranolol, mepivacaine-verapamil, mepivacaine-diltiazem on heart rate and force of contraction of the isolated frog hearts. **Methods.** The hearts were perfused with Ringer-Locke solution and heart activity has been recorded with a "Servogor" potentiometer. The substances were used in the following doses: mepivacaine 0.1 μM, enalaprilat 0.33 μM, propranolol 0.34 μM, verapamil 0.05 μM, diltiazem 0.1 μM. Magnesium sulfate was added in raising concentrations, from 16 nM to 16 mM, until cardiac arrest. Experiments were reproducible and the results were statistically interpreted and transformed into diagrams. **Results.** Magnesium sulfate had a very small effect on heart rate, but in concentrations of 16 μM it depressed inotropism with 50%. The association magnesium sulfate with mepivacaine has preserved inotropism around reference's values but has depressed heart rate with 50% at 1.6 mM magnesium sulfate. Mepivacaine-antihypertensives associations have determined a negative chronotropic effect of magnesium sulfate, which has been emphasized by concentration's growing. This action has been powerful for mepivacaine-diltiazem association, while, for mepivacaine-enalaprilat association, negative chronotropic effects have been reduced. Mepivacaine-antihypertensives associations have antagonized negative inotropic action of magnesium sulfate and have maintained ventricular inotropism around reference's values for small concentrations of magnesium sulfate. Negative inotropic effects of magnesium sulfate have been antagonized obviously even for high concentrations by mepivacaine-propranolol and mepivacaine-enalaprilat associations. Mepivacaine-verapamil association has slightly antagonized cardiodepressive effect of magnesium sulfate in high concentrations while mepivacaine-diltiazem association has determined a pronounced cardiotoxicity of magnesium sulfate in higher concentrations, with disrrhytmia and cardiac arrest.

Keywords: magnesium sulfate, mepivacaine, antihypertensives

Clinical electrophysiological effects of magnesium (Mg^{2+}) are known for more than 60 years. Mg^{2+} is a common cation in the human body and is involved in more than 300 different enzymatic reactions. Magnesium deficiency is now considered to contribute to many diseases and specific clinical conditions in which magnesium deficiency has been implicated to play a pathophysiological role are hypertension, ischemic heart disease, arrhythmia, preeclampsia, asthma and critical illness [1].
Hypomagnesemia is associated with increased all cause mortality. Altura *et al.* [2] had results that point to a role for low serum Mg^{2+} in potential development of hypertension, atherogenesis, vascular disease, and stroke. Magnesium supplementation improves myocardial metabolism, inhibits calcium accumula-

tion and myocardial cell death; it improves vascular tone, peripheral vascular resistance, after load and cardiac output, reduces cardiac arrhythmia and improves lipid metabolism. Magnesium also reduces vulnerability to oxygen-derived free radicals, improves endothelial function and inhibits platelet function, including platelet aggregation and adhesion [3].

Magnesium is implicated in L-type Ca^{2+} regulation in cardiac myocytes [4] and is known to inhibit ryanodine receptor (RyR) Ca^{2+} release channels by competing with Ca^{2+} at the cytosolic activation sites of the channel. Mg^{2+} is a critical physiological determinant of the dynamic behavior of the RyR channel, which is expected to profoundly influence the fidelity of coupling between L-type Ca^{2+} channels and RyRs in heart cells [5].

Magnesium sulfate is used in arrhythmia and eclampsia. Mepivacaine is used as a proper local anesthetic for hypertensive patients, who could be under treatment with various antihypertensives. In some cases, magnesium sulfate could interact with mepivacaine and antihypertensives.

Methods

36 Rana pipiens were killed by spinalization and hearts rapidly removed and mounted in a chamber bath system filled with Ringer-Locke solution (composition NaCl 0.98‰, KCl 10% 4.2 ml/l, Na_2CO_3 0.5 g/l, glucose 1g/l, $CaCl_2$ 1 M 1.08 ml/l, H_2O and 1000 ml) at room temperature, pH 7.4. The hearts were perfused with Ringer-Locke solution and bonded to a lever system, from which the movements were taken by a photoelectric cell and transformed into electric stimuli. With a "Servogor" potentiometer, these electrical stimuli have been transformed into mechanical stimuli, recorded on a special registration paper. This paper ruled with 60 mm on minute. In this way, atrial and ventricular contractions were recorded before (baseline) and after the substances were administered in the bath chamber. The substances were used in the following doses: mepivacaine 0.1 μM, enalaprilat 0.33 μM, propranolol 0.34 μM, verapamil 0.05 μM, diltiazem 0.1 μM. Magnesium sulfate was added in raising concentrations, from 16 nM to 16 mM, until cardiac arrest. Experiments were reproducible and the results were statistically interpreted (t paired test) and transformed into diagrams.

Results

Magnesium sulfate had a very small effect on heart rate, but in concentrations of 16 μM it depressed inotropism with 50% ($p < 0.05$). The last concentration of magnesium sulfate (16 mM) has raised inotropism to 60% from baseline ($p < 0.05$). Heart rate has been depressed slowly to 50% from baseline by the association with mepivacaine at 1.6 mM magnesium sulfate ($p < 0.05$) and then sudden to 0% by 16 mM of magnesium sulfate ($p < 0.05$) (*Table I, Figure 1*).

The association of magnesium sulfate with mepivacaine has preserved inotropism around 100% from baseline (90-110%) but the last concentration of magnesium sulfate has depressed inotropism at 0% from baseline ($p < 0.05$) (*Table II, Figure 2*).

The association of mepivacaine 0.1 μM with propranolol 0.34 μM has influenced the heart rate response to magnesium sulfate in a manner similar with mepivacaine. However, for the last concentration of magnesium sulfate, the heart rate was maintained at 17% from baseline, comparative with mepivacaine that dropped it to 0% from baseline (*Table I, Figure 1*). Also, inotropism was maintained at values similar to mepivacaine (83-106% from baseline) but for the last concentration of magnesium sulfate it reached only 31% from baseline ($p < 0.05$), comparative to 0% for mepivacaine (*Table II, Figure 2*). The presence of propranolol in bath chamber has exercised an antagonism for cardiotoxicity of mepivacaine, but only at large concentrations of magnesium sulfate.

The presence of mepivacaine 0.1 μM with enalaprilat 0.33 μM in organ bath determined a heart rate variation similar with that of mepivacaine, but at the last concentration of magnesium sulfate (16 mM) the heart rate was almost 50% from baseline ($p < 0.05$), comparative to 0% for mepivacaine (*Table I, Figure 1*). This association had an initial stimulating effect on inotropism, which reached 110% from baseline at the first concentration of magnesium sulfate and a depressing slow effect until 160 μM of magnesium sulfate, when inotropism was 60% from baseline ($p < 0.05$) (*Table II, Figure 2*).

The presence of enalaprilat antagonized the cardiotoxicity of mepivacaine and maintained heart contractility at 67% from baseline for the last concentration of magnesium sulfate ($p < 0.05$) (when mepivacaine dropped it to 0%).

The association of mepivacaine 0.1 µM with diltiazem 0.1 µM dropped heart rate to 48% ($p < 0.05$). Adding raising concentrations of magnesium sulfate continued to drop slowly heart rate until 12% from baseline ($p < 0.05$), when it remained, even at the last concentration of magnesium sulfate (*Table I, Figure 1*). Inotropism was augmented to 145% from baseline by the association of mepivacaine with diltiazem ($p < 0.05$), and then was dropped slowly by the raising concentrations of magnesium sulfate added. At the last concentration of magnesium sulfate hearts almost stopped beating (*Table II, Figure 2*). Diltiazem 0.1 µM augmented the cardiotoxicity of mepivacaine and magnesium sulfate could not antagonized this effect.

Table I. Magnesium sulfate dose-dependent effect and its associations with mepivacaine, mepivacaine-propranolol, mepivacaine-enalaprilat, mepivacaine-diltiazem, mepivacaine-verapamil on heart rate (beats/min) of frog isolated heart (mean ± SEM).

MgSO4 Molar concentrations	MgSO4 (n = 6)	Mepivacaine 0.1 µM (n = 6)	Mepivacaine + propranolol 0.34 µM (n = 6)	Mepivacaine + enalaprilat 0.33 µM (n = 6)	Mepivacaine + Diltiazem 0.1 µM (n = 6)	Mepivacaine + verapamil 0.05 µM (n = 6)
Baseline	45 ± 0.5	34.5 ± 0.5	50.5 ± 2	54 ± 2.3	53.6 ± 0.7	34.5 ± 0.7
Pretreat.		31.5 ± 0.1*	43.5 ± 2*	43 ± 1.7*	25.6 ± 2.5*	31.5 ± 1.1*
16 nM	48 ± 0.5*	29 ± 2.2*	40.5 ± 1.5*	42 ± 1.3*	16.7 ± 1.4*	29 ± 2.6
160 nM	45.3 ± 1.1	23.5 ± 1.3*	37 ± 0.1*	39.5 ± 0.6*	10.3 ± 2.4*	23.5 ± 1.9*
1.6 µM	45.3 ± 1.1	21.5 ± 1.5*	27.5 ± 0.2*	39.5 ± 0.6*	10 ± 1.7*	21.5 ± 2.7*
16 µM	41 ± 0.5*	18.5 ± 1.3*	23.5 ± 1.1*	38.5 ± 0.6*	6.7 ± 2*	18.5 ± 2.3*
160 µM	43 ± 0.5*	14 ± 2.2*	20 ± 0.4*	37.5 ± 0.6*	6.3 ± 1.7*	14 ± 2.1*
1.6 mM	42 ± 0.5*	12 ± 1.2*	17.5 ± 0.6	29.5 ± 1.1*	6.3 ± 1.6*	12 ± 1*
16 mM	41 ± 0.5*	0 ± 0*	8.5 ± 2	26 ± 1*	6.3 ± 1*	9.5 ± 1.2*

*$p < 0.05$ comparative with baseline.

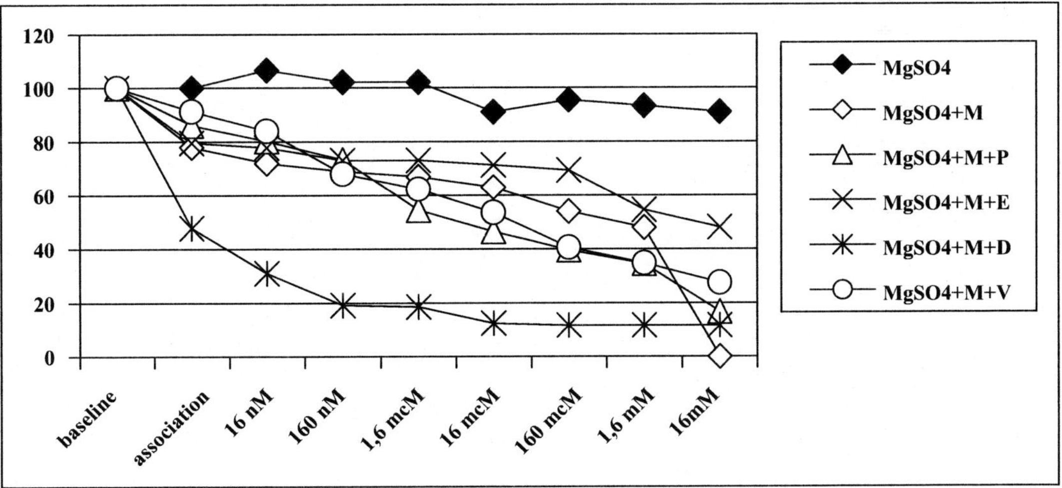

Figure 1. Magnesium sulfate effects (in raising concentrations from 16 nM to 16 mM) on heart rate (expressed as % from baseline, which is 100%) and its associations with mepivacaine, or mepivacaine and antihypertensives (propranolol, enalaprilat, diltiazem, verapamil). M = mepivacaine, P = propranolol, E = enalaprilat, D = diltiazem, V = verapamil.

Verapamil 0.05 μM associated with mepivacaine 0.1 μM determined a continuous and slow weakening of heart rate, in almost a straight line, until the last concentration of magnesium sulfate when heart rate reached 30% from baseline ($p < 0.05$) (*Table I, Figure 1*). Inotropism was maintained above 100% from baseline until 160 μM of magnesium sulfate, when it dropped to almost 50% from baseline. Comparative to mepivacaine, which dropped inotropism to 0 at the last concentration of magnesium sulfate, in the presence of verapamil heart contractility was kept to almost 50% at this concentration (*Table II, Figure 2*). In 0.05 μM concentration, verapamil antagonized cardiotoxicity of mepivacaine and maintained heart functions until the last concentration of magnesium sulfate.

Table II. Magnesium sulfate dose-dependent effect and its associations with mepivacaine, mepivacaine-propranolol, mepivacaine-enalaprilat, mepivacaine-diltiazem, mepivacaine-verapamil on ventricular contraction amplitude (mm) of frog isolated heart (mean ± SEM).

MgSO4 Molar concentrations	MgSO4 (n = 6)	Mepivacaine 0.2 μM (n = 6)	Mepivacaine + propranolol 0.34 μM (n = 6)	Mepivacaine + enalaprilat 0.33 μM (n = 6)	Mepivacaine + Diltiazem 0.1 μM (n = 6)	Mepivacaine + verapamil 0.05 μM (n = 6)
Baseline	90 ± 5	92 ± 1	102 ± 5	97.5 ± 5	103 ± 4	132.5 ± 1.1
Pretreat.		102 ± 0.8*	108 ± 3*	111 ± 2*	183 ± 1*	147.5 ± 3.3*
16 nM	67 ± 3*	101 ± 0.6*	97.5 ± 3	107.5 ± 1.1	141 ± 9	140 ± 2.2*
160 nM	60 ± 1*	100 ± 1.3*	90 ± 2	94.5 ± 1.1	78.3 ± 9	142.5 ± 1.1*
1.6 μM	50 ± 1*	99.5 ± 2	85 ± 2	83.5 ± 1.5*	73.4 ± 7	145 ± 2.2*
16 μM	42 ± 5*	90 ± 1.3	105 ± 9	70 ± 2.2*	11.7 ± 7*	137.5 ± 3.3
160 μM	40 ± 5*	85 ± 5.2	97.5 ± 8	57.5 ± 3.3*	10 ± 6*	72.5 ± 3.2
1.6 mM	42 ± 0.5*	82.5 ± 7.2	102.5 ± 7	71 ± 9	10 ± 6*	70 ± 3.1
16 mM	55 ± 2*	0 ± 0*	32.5 ± 9*	65 ± 7*	0.6 ± 0.4*	65 ± 9

*$p < 0.05$ comparative with baseline.

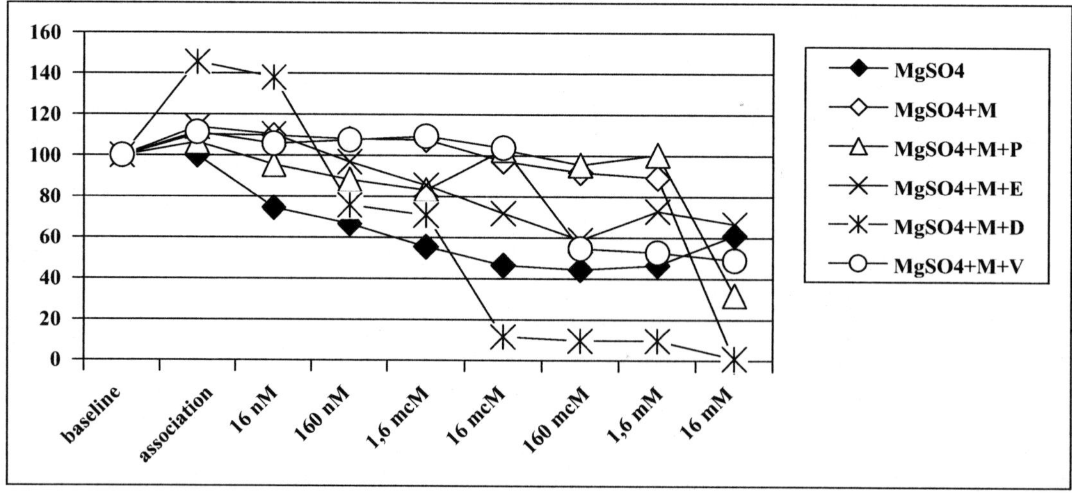

Figure 2. Magnesium sulfate effects (in raising concentrations from 16 nM to 16 mM) on ventricular amplitude of contraction (expressed as % from baseline, which is 100%) and its associations with mepivacaine, or mepivacaine and antihypertensives (propranolol, enalaprilat, diltiazem, verapamil). M = mepivacaine, P = propranolol, E = enalaprilat, D = diltiazem, V = verapamil.

Discussion

Mg^{2+} influences myocardial metabolism by its effects on contractility and electrical activity. About 12% of cardiac Mg^{2+} is found in the mitochondria and 2 to 3% in the myofibrils, a large portion being incorporated in adenosine mono-, di- and triphosphate. Mg^{2+} affects intracellular calcium by inhibiting the influx of calcium into the myocyte through sarcolemmal channels, by modulation of cyclic AMP and by competing with calcium for binding to a single high affinity site. Mg^{2+} blocks the outward current through some potassium channels resulting in an inward rectification of these channels. This suggests that internal magnesium functions as a potassium channel -blocking agent. Early afterdepolarizations are oscillations in the membrane potential and lead to triggered activity and therefore are the electrophysiological substrate of "torsade de pointes" type of ventricular flutter. Mg^{2+} is able to inhibit both early afterdepolarizations and tachyarrhythmias. Additionally Mg^{2+} interferes with the sodium-potassium-ATPase system by stabilizing the transmembrane gradient of both cations [6].

The cardiac effects of magnesium were studied by Hammad (1981) on 40 spontaneously beating rabbit atria and concentrations of 30, 60, 120 and 240 mM elicited dose dependent negative inotropic and chronotropic responses. Larger doses above 240 mM caused complete atrial arrest [7]. Our larger concentration of magnesium sulfate was 16 mM and it did not stop heart beating. Howarth *et al.* (1994) have shown that perfusion of the isolated rat heart with a physiological salt solution containing elevated extracellular Mg^{2+} $(Mg^{2+})o$ (2.4 mM-6.0 mM) resulted in a marked and progressive decrease in the amplitude of contraction compared to control $(Mg^{2+})o$ (1.2 mM) [8], results similar to us. In contrast, perfusion of hearts with low (0-0.6 mM) $(Mg^{2+})o$ caused a small transient increase in the amplitude of contraction which was often accompanied by arrhythmic activity.

Variation in the level of Mg^{2+} within physiological limits leads to mechanical changes. A decrease in the level of extracellular Mg^{2+} was accompanied by a significant increase in contractile amplitude and decrease in the velocities of contraction and relaxation. The variations in contractile parameters associated with the change in the level of Mg were statistically significant ($p < 0.01$). Variation in the contractile properties associated with change in extracellular Mg^{2+} may be effected by alteration in Ca^{2+} transients [9].

Escudero *et al.* (1987) has found that local anesthetics (lidocaine, dibucaine, procaine) markedly increased the permeability of sarcoplasmic reticulum vesicles to Mg^{2+}. The kinetic analysis of the time dependence of the light-scattering data after the osmotic shock showed that, in the absence of local anesthetics, the Mg^{2+} influx could be described as proceeding through a unique type of channel [10]. Even if the concentrations of local anesthetic used by us were much smaller, pre-treatment with mepivacaine has increased cardiotoxicity of magnesium sulfate.

Hammad (1981) concluded that the association of propranolol with magnesium could be employed in treatment of arrhythmias which are nonresponsive to propranolol [7]. In our study, magnesium sulfate was used in smaller concentrations, 16 nM-16 mM, which did not stop cardiac activity. However, the association of propranolol with mepivacaine preserved inotropism until the last concentration of magnesium sulfate, when it dropped to 31%. Mepivacaine cardiotoxicity was partial antagonized by this last concentration of magnesium sulfate, in the presence of propranolol.

Kurtzman *et al.* (1993) have shown that verapamil and nifedipine caused dose-dependent cardiac depression that was intensified by the addition of magnesium sulfate in the isolated perfused rat heart [11]. In our study, the presence of verapamil with mepivacaine augmented cardiac toxicity of magnesium sulfate, but the effect was smaller then that of mepivacaine alone.

After Quamme and Rabkin (1990), the basal $(Mg^{2+})i$ of isolated cardiomyocytes is considerably below the Mg^{2+} electrochemical equilibrium allowing passive Mg^{2+} influx. The influx pathway for Mg^{2+} is inhibited by verapamil and appears to be independent of Ca^{2+} [12].

Howarth *et al.* (1994) demonstrated that pre-treatment with a beta-adrenergic antagonist, propranolol 1 μM competitively blocked the Mg^{2+} efflux evoked by the catecholamines. Similarly, pre-treatment with a calcium (Ca^{2+}) channel blocker, verapamil 1 μM caused a significant ($P < 0.05$) decrease in net Mg^{2+} efflux evoked by isoprenaline. Their results indicated that the perturbation of $(Mg^{2+})o$ has an important influence on myocardial contractility and the mobilization of Mg^{2+} in the heart is associated with beta adrenergic stimulation possibly via an elevation in intracellular adenosine 3.5 cyclic monophosphate (cyclic AMP) [8].

Conclusions

Mepivacaine-antihypertensives associations have determined a negative chronotropic effect of magnesium sulfate, which was emphasized by concentration's growing. This action was most powerful for mepivacaine-diltiazem association, while, for mepivacaine-enalaprilat association, negative chronotropic effects were reduced. Mepivacaine-antihypertensives associations antagonized negative inotropic action of magnesium sulfate and maintained ventricular inotropism around 100% from baseline for small concentrations of magnesium sulfate. Negative inotropic effects of magnesium sulfate were antagonized obviously even at high concentrations by mepivacaine-propranolol and mepivacaine-enalaprilat associations. Mepivacaine-verapamil association slightly antagonized cardiodepressive effect of magnesium sulfate in high concentrations while mepivacaine-diltiazem association determined a pronounced cardiotoxicity of magnesium sulfate in higher concentrations, with disrrhytmia and cardiac arrest.

References

1. Touyz RM. Magnesium in clinical medicine. *Front Biosci* 2004 ; 9 : 1278-93.
2. Altura BM, Kostellow AB, Zhang A, *et al.* Expression of the nuclear factor-kappaB and proto-oncogenes c-fos and c-jun are induced by low extracellular Mg^{2+} in aortic and cerebral vascular smooth muscle cells : possible links to hypertension, atherogenesis, and stroke. *Am J Hypertens* 2003 ; 16(9 Pt 1) : 701-7.
3. Shechter M. Does magnesium have a role in the treatment of patients with coronary artery disease? *Am J Cardiovasc Drugs* 2003 ; 3(4) : 231-9.
4. Yamaoka K, Kameyama M. Regulation of L-type Ca^{2+} channels in the heart : overview of recent advances. *Mol Cell Biochem* 2003 ; 253(1-2) : 3-13.
5. Zahradnikova A, Dura M, Gyorke I, Escobar AL, Zahradnik I, Gyorke S. Regulation of dynamic behavior of cardiac ryanodine receptor by Mg2+ under simulated physiological conditions. *Am J Physiol Cell Physiol* 2003 ; 285(5) : C1059-C1070.
6. Vester EG. Clinico-electrophysiologic effects of magnesium, especially in supraventricular tachycardia. *Herz* 1997 ; 22(Suppl 1) : 40-50.
7. Hammad D, Omar SA, Varma S. Effects of magnesium on isolated rabbit atria. *Indian J Physiol Pharmacol* 1981 ; 25(2) : 123-30.
8. Howarth FC, Waring J, Hustler BI, Singh J. Effects of extracellular magnesium and beta adrenergic stimulation on contractile force and magnesium mobilization in the isolated rat heart. *Magnes Res* 1994 ; 7(3-4) : 187-97.
9. Nair P, Nair RR. Alteration in cardiomyocyte mechanics by suboptimal levels of extracellular magnesium. *Biol Trace Elem Res* 2000 ; 73(3) : 193-200.
10. Escudero B, Gutierrez-Merino C. Effects of local anesthetics on the passive permeability of sarcoplasmic reticulum vesicles to Ca^{2+} and Mg^{2+}. *Biochim Biophys Acta* 1987 ; 902(3) : 374-84.
11. Kurtzman JL, Thorp Jr. JM, Spielman FJ, Mueller RC, Cefalo RC. Do nifedipine and verapamil potentiate the cardiac toxicity of magnesium sulfate? *Am J Perinatol* 1993 ; 10(6) : 450-2.
12. Quamme GA, Rabkin SW. Cytosolic free magnesium in cardiac myocytes : identification of a Mg2+ influx pathway. *Biochem Biophys Res Commun* 1990 ; 167(3) : 1406-12.

IV. Magnesium and nutrition

Value of raising magnesium to calcium intake ratio for health

M.S. Seelig

Department of Nutrition, School of Public Health, University of North Carolina, Chapel Hill, Decatur, Georgia, USA

Calcium (Ca) and magnesium (Mg) have reciprocal effects in many functions. Their imbalance – whether caused by excess intake of one with inadequate intake of the other, or whether caused by hormones, nutrients, or drugs that enhance the absorption or cause loss of one over the other – can induce pathologic processes. The official American recommendation adopted in 1997, by the Food and Nutrition Board of the Institute of Medicine [1, 2] for high intakes of Ca, and marginal intakes of Mg, provides an intake ratio of Ca to Mg of 3-4/1, as the Dietary Reference Intake (DRI). It also establishes a new category, UL, for Tolerable Upper Limits, that results in a Ca/Mg ratio of 6-7/1. Advice as to Ca intake in Europe and the USA is based on different concepts as to requirements, leading to different interpretations of dietary adequacy [3]. Widely accepted, however, are Mg requirements that are as low as the official American recommendations, and as high as the official Ca DRI and UL. There is evidence that cessation of the Women's Health Initiative (WHI), that tested the value of estrogen repletion of postmenopausal women, was necessitated by adverse effects contributed to by high Ca and low Mg intakes. Similarly, the attempt to prevent eclampsia – that has long responded to Mg treatment – by high dose Ca supplements, was unsuccessful. There are additional data as to the health hazards of increasing Ca intake, while ignoring Mg inadequacy.

Magnesium and Calcium balance and outcome on 20th century intakes

Metabolic balance studies on customary intakes

Early extensive Mg balance studies of normal young adults have shown that at Mg intakes of less than 300 mg/day (<5 mg/kg/day), negative Mg balances are likely [4-6]. A long-term metabolic balance study of healthy young women, consuming customary diets that provided that marginal Mg intakes and Ca intakes of about 1 gram/day, resulted in Mg balances that were negative or in bare equilibrium [7]. When Mg intakes were increased to 5-6 mg/kg/day, the same or slightly lower Ca intakes were not associated with negative Mg balances, but negative Mg balances ensued on Ca intakes above 1 g/day [7]. On diets delivering 1-1.5 mg Ca, and Mg intake approximately as high as that of Ca, Mg equilibrium or positive balances were more likely than were negative Mg balances [7]. In a long-term study of the effect of Mg and Ca intakes, that were approximately those later advised by Food and Nutrition Board as the Recommended Daily Allowance (RDA) [8], persistent negative Mg balances were observed [9]. At dietary intakes of under 300 mg/day and Ca intakes of slightly over 1 g/day (Ca/Mg ratio of almost 4/1), the Mg balances of the healthy young women remained strongly negative over three consecutive 20 day balance periods, and Ca balances were strongly positive [9]. Positive Ca balance of subjects who are Mg deficient is difficult to interpret, since it may reflect either deposition in bone, or in soft tissues, such as heart, arteries and kidneys (characteristic of Mg deficiency).

Those studies were undertaken, either on self-selected diets, or on diets approximating customary diets in the mid-20th century, during which dietary surveys had shown higher Ca and lower Mg intakes than during the first third of the twentieth century, when the Ca/Mg ratio was about 2/1 [10, 11], the amounts found to maintain both Mg and Ca equilibrium in the Department of Agriculture studies [4-6].

Dietary survey and balance studies, in the last half of the century, showed further rise in Ca intake and fall in Mg intake [12-18]. This was reflected by negative Mg balances in adolescent girls, whose self-selected diets provided the recommended 1 gram of Ca/day, but as little as 193 mg of Mg [12]. In 39 day balance studies of adult men, most were in negative Mg balance on moderate to high Ca intakes [13]. As indicated above, the current DRIs for Mg and Ca intakes [1] allow for Ca/Mg intake ratios almost twice the 2/1 ratio on which both Mg and Ca were in balance [6]. More disturbing is the upper tolerable limit (UL) for Ca, and that limits Mg intakes to low amounts that result in close to a 7/1 ratio [1, 2], which greatly exceeds the 4/1 ratio that was proposed as a factor in the high cardiovascular disease morbidity and mortality rates in middle-aged men in Finland [19-21].

Sex difference in ischemic heart disease and thromboembolic events

Increase in Ischemic Heart Disease in Men in the 1960s, Associated with High Ca/Mg intake

Presented are the comparative intakes of Mg and Ca from 1901 to 1939, in the sixth decade [5-18], and today, as a result of present American official DRIs and Uls for Ca and Mg [1, 2]. A provocative comparison of deaths from ischemic heart disease (IHD), *versus* other cause of mortality, illustrated the preponderance of IHD in men as compared with women, that provoked alarm in the early 1960s. Heart disease rates in men and women were about equal in the 1920s [22], when the Mg intake had been one quarter higher than it was in the 1960s, and the Ca intake was only twice as much as that of Mg, rather than the close to four times as much, that was common by the sixth decade. Another tabulation of the sex difference in heart disease considered also the contrast associated with age in the 1960s – from which one can see that loss of estrogen secretion could explain the greater vulnerability to cardiovascular disease of women as they aged [23]. Neither of the investigators, who noted the sex difference in cardiovascular disease, attributed part of the protection in young women to better utilization of Mg, although one of them (Anderson) soon thereafter noted that higher Mg intake – to which hard (Mg rich) water contributes – protected against Mg deficiency – relating high cardiac death rates in soft water areas to low Mg intake [24, 25]. Numerous reviews – experimental, epidemiologic, and clinical – have affirmed the requirement for Mg to achieve and maintain normal arterial and cardiac structures and function, and to prevent and manage cardiovascular disease [26-35].

Increase in Thromboembolic Events, Associated with High Estrogen and Ca/Mg intake Estrogen-Containing Contraceptives and Magnesium Redistribution

Estrogen and Mg interact in ways that exert both beneficial and deleterious effects. Estrogen causes internal shift of Mg from serum to cells, including – not only those of the tissues of the reproductive system [36], but of other tissues, such as bone, heart and blood vessels, an effect that is advantageous [37-40], contributing to the lower heart disease rate in young women than in young to early middle aged men. But the lowered circulating Mg caused by high estrogen levels negates that advantage by enhancing estrogen's increase of intravascular coagulation. The rise in blood levels of estrogen, as at time of ovulation [36, 41, 42], during pregnancy as it progresses [36, 44-52], and with use of oral contraceptives [36-38, 43, 44] has been associated with lowered blood levels of Mg [35-49]. This is a major factor in the increased prevalence of venous thrombosis and complications caused by embolization to cerebral and pulmonary arteries that has occurred with complicated pregnancies, and with high dosage estrogen administration [36, 37, 42, 51, 54-61]. There have been many publication from 1967 to date, that have indicated the fall in blood Mg that occurs with rise in estrogen levels, but the relationship of that effect to increased thromboembolic disease in association with estrogen use has been largely ignored, despite Durlach's early findings [62, 63], except in reviews by Mg investigators [35, 36, 39, 49, 59]. In 1967, Durlach correlated the antithrombotic effect of Mg with phlebothrombosis in a Mg deficient woman [62], and prevented thrombotic recurrence by treating her with oral Mg. Three years later, he applied that observation to another woman with Mg deficiency, and to use of the "pill" [63]. He showed that platelet hyperaggregation induced by an oral estrogen contraceptive, that caused hypomagnesemia, could be reversed by supplementation with oral Mg during use of the estrogen [63].

Intensification of Estrogen's Thromboembolic Risk by High Ca and Low Mg

It is noteworthy that it was in the 1960s – the same period that the male predominance of ischemic heart disease attracted attention – that the increase in thromboemboli was encountered in young women. That was the era Ca intakes rose, and Mg intake fell. When, to a low or marginal supply of Mg is added estrogen's Mg lowering effect, as well as a high intake of Ca, there is elevation of the plasma Ca/Mg ratio. Since both circulating Ca and estrogen enhance coagulation by activating steps in the coagulation cascade, and Mg inhibits both the cascade and platelet aggregation, elevating estrogen levels has the potential of increasing vulnerability to thromboses and embolic events [35, 36, 39, 40, 48-53, 59] particularly in those whose Ca/Mg intake ratio is high. Estrogen administration, plus widespread use of increased amounts of Ca (to protect against osteoporosis), has been complicated by thromboembolic events in women attempting to suppress lactation [56, 64] and especially during hormonal repletion of postmenopausal women (see below).

Adverse effects caused by estrogen-induced hypercoagulation has also been reported in two categories of men: i) those at high risk of coronary artery disease who were given estrogen in hope of obtaining the protection possessed by young women, and ii) men with prostatic cancer. The attempt to lower the susceptibility of high-risk young to middle-aged men, with evidence of early-onset heart disease, by providing estrogen, was undertaken in a large therapeutic trial: "The Coronary Drug Project". It was terminated in 1973 because of development of heart attacks and strokes [65]. The limitation to only low doses of stilbestrol in men with newly diagnosed prostatic cancer, to avoid embolic accidents such as were caused by higher dose female sex hormone treatment, was indicated, first in 1970 [66], and then in a summary report of the results of three major randomized clinical trials (from 1960 to 1975) [67].

Magnesium and Estrogen Effects of Blood Lipids

Why estrogen repletion of postmenopausal women increases risk of venous thrombosis while it decreases the risk of arterial disease, was explained by Luyer *et al.* in 2001 [68] on the basis, not only of its activation of a number of pathways for coagulation, but because it lowers low density lipid-cholesterol (LDL-C) levels. Not considered in that paper was the inhibitory role of Mg in many steps of the coagulation cascade, and its activation of the enzyme, lecithin – cholesterol acyltransferase (LCAT), the enzyme that Rayssiguier *et al.* have shown (in rats) to play important roles in lipid metabolism [69-71]. It lowers levels of low density lipid cholesterol (LDL-C) and esterifies free cholesterol. LCAT activity has been found to be significantly low in atherosclerotic patients, who had significantly higher ratio of unesterified/esterified cholesterol than did a control group of normal subjects [72]. Low LCAT activity was accompanied by elevated values of phospholipids and LDL-C, a moderate increase of triglycerides, and lowered high density lipid cholesterol (HDL-C). Even in nodal subjects, Mg supplementation for four weeks produced statistically significant increases in LCAT and HDL-C [73]. Estrogen, like Mg, improves the lipoprotein profile of postmenopausal women in that it elevates the high density to low density lipoprotein cholesterol ratio. Magnesium not only raises the HDL-C/LDL-C ratio, but also lowers the triglyceride level, and esterifies free cholesterol.

Hormonal repletion of postmenopausal women

Large Scale Studies in the USA: Womens' Health Initiative (WHI) and WHIM

The National Institutes of Health (NIH) undertook a large scale study: the WHI, that was planned in 1991, to be run through 2007 [74], to test the validity of findings obtained in controlled health programs that yielded both positive and negative results during sex hormone replacement therapy (HRT), in attempts to protect against diseases that complicate the aging process in women. Explored was whether estrogen replacement therapy (ERT, using Premarin, of equine origin) alone or in combination with a synthetic progestin (HRT), was effective against increase with age of cardiovascular disease, osteoporosis, and impaired mental acuity – extending in some to dementia. A subgroup had tests of mental acuity, in the trial termed WHIM [75]. Included in the description of the planned investigations of matched groups given conjugated estrogen alone (ERT) or with a synthetic progestin (HRT), controlled by those given a placebo, was a group also supplemented with 1 gram of Ca and 400 I.U. of vitamin D daily. A 1999 preliminary paper reported that HRT did not reduce the incidence of ischemic heart disease, and that there was a three-fold increase in thromboembolisms [76]. After five years, in 2002, the data and safety monitoring board recommended terminating the HRT trial, before the target end in 2007, because the women in that group suffered more adverse effects than did controls. There were eight more strokes (to a total of 212) and peripheral emboli (to 101), and seven more coronary events (to 286) than were experienced by those in the placebo group [77]. Additionally, there were greater declines in mental scores in the group receiving sex hormones than among controls given a placebo, with forty women (of the total of 8506 receiving HRT) developing dementia. Half as many in the control group (of 8102 given placebo) developed dementia [78]. The benefit of HRT in osteoporosis was confirmed, with slightly fewer hip fractures in the HRT group than in the placebo group. There was also slightly reduced incidence of endometrial and colorectal cancers, but there were eight more invasive breast cancers than in the placebo group. Because the benefit to risk ratio was not acceptable, the HRT phase of the study was prematurely terminated. This has caused dismay among many women who feel they had benefited by the now discontinued hormone repletion, and who fear advancement of osteoporosis and other cardiovascular diseases.

European Reactions to the WHI Results

The discontinuation of the WHI study evoked responses in Europe, that were less certain than in the United States, that the NIH findings were the final word on inadvisability of hormone repletion. Gaspard *et al.*, in France, in agreement with the European Menopause Society and the International Menopause Society, are convinced that there is no alternative to HRT for menopausal symptoms and osteoporosis, and urges that physicians should continue to recommend hormones for symptomatic women [79]. They point out that the risk for cardiovascular events for women under 60 years of age is very small, and that the age of the patients in the WHI was 63 at inclusion. (Not noted in this critique was that all of the cardiovascular events were thromboembolic). They advised that those who feel good on hormones and are fully satisfied with this treatment should be made cognizant of possible harm after long-term use, but that it is important to take into account the importance of quality of life. They concluded that it is presently impossible to decide whether other estro-progestin associations, other administration routes and other molecules such as estradiol, natural progesteron or other progestins, could have an impact very different form that to the estroprogestin combination used in the WHI study. In accord with those observations, Kuttenn *et al.* [80] observed that there is need for a European study with estrogen that is not orally administered – the transdermal application being more common in Europe, an important difference, since the pro-coagulant effect is induced by the first pass liver effect only when estrogen is taken by mouth. They also point out that the synthetic progestin used in WHI inhibits estrogen's beneficial effects on the arterial wall, whereas natural progesterone does not. An unsigned report in 2003 [81], that considered the findings and limitations of the WHI study, concluded that European health authorities must organize comparative trials to assess the benefits and risk of other hormone combinations used by perimenopausal women.

Lack of Attention Given to Magnesium Intake or Status in WHI or European Comments

Information on the Mg intake and status of the participants was not secured in the WHI trial. In all likelihood, the participants were health-aware and thus probably followed advice to consume as much as 1,500 mg of Ca daily, while not concerning themselves about Mg, since the current recommendations are for Mg intakes so low that the Ca/Mg ratio is 3-4/1. A report by Karppannen *et al.* [82], has implicated a Ca/Mg ratio of 4/1 in Finland as contributory to Finnish men's very high sudden cardiac death rate, which they contrasted to the much lower rate in Japan – where the dietary Ca/Mg intake, in 1974, was less than 2/1. Six years later Karppanen [83] pointed out that in addition to the high Ca/Mg intakes, diets throughout Finland customarily contain excessive salt and saturated fat, and are also low in potassium. In southwest Finland, with just as high Ca/Mg dietary intake, but where there is three times as much Mg in water, the heart attack rate is half that in North Karelia. He noted that combined life-long Finnish mineral imbalances: too much sodium, as well as too much Ca, and both insufficient potassium and magnesium were factors in the problem of hypertension, that predisposes to the high stroke rate in Finland [84]. Public health education, to change eating habits, increasing fruit and vegetable consumption, and adding Mg to salt have since lowered the cardiovascular disease rate in Finland. In contrast, in areas in Japan, where westernized eating habits are common, cardiovascular and cerebrovascular disease rates risen and addition of Mg has been recommended [73, 85-89].

Postmenopausal osteoporosis treated with estrogen and Ca

Some diminution of hip fractures was one of the few benefits reported in the WHI. As life expectancy increases and the population ages, the problem of postmenopausal osteoporosis grows. In the mid 1980s it was estimated to afflict 15 to 20 million in the United States alone [90]. The cost for its treatment was estimated at 7-10 billion dollars annually. The nutritional approach to this serious problem has been to increase Ca and vitamin D intakes. Had it been realized that postmenopausal osteoporosis is a consequence, not only of loss of estrogen and poor Ca supply, but also of Mg insufficiency, repair of the insufficiencies of the hormone and Ca might have been accompanied by Mg supplementation. There is good reason to believe that such a practice would have averted the adverse affects caused by thromboemboli, and might have averted other disorders that commonly complicate the lives of patients with osteoporosis.

As a long ago as 1960, Daldeurp [91] called attention to the importance of Mg in formation and maintenance of bone matrix. He suggested that postmenopausal osteoporosis might be due to decreased mineralization of defective matrix, and that Mg deficiency might be contributory to inadequacy of Mg for bone apatite formation. Five years later, Miller *et al.* [92] demonstrated that a low Mg intake with excess vitamin D decreased not only the level of Mg in the femur of baby pigs, but also the Ca content, and that the bone strength and elasticity was diminished – which is a possible reflection of less matrix. In the next decade, Lai *et al.* [93] attributed the brittleness of bone of Mg- deficient rats to their decreased organic bone matrix, possibly as a result of less synthesis of protein polysaccharides and collagen. Within the next 20 years, there was additional clarification of the roles of Mg in matrix formation and bone formation. Kenney *et al.* [94] found that femurs of Mg deficient rats had less strength, expressed by fracture at lighter load stresses. Working with rat and autopsied normal human bone, *in vitro*, Makgoba and Datta [95] determined that Mg^{2+} has a critical role in cell-matrix attachment in osteoclast/osteoblast adhesion to bone during bone remodelling whereas Ca^{2+} participates in cell-matrix dissociation. Ditmar and Steidl [96] described Mg as a catalyst of bone metabolism and an important factor in formation of bone matrix and its mineralization. Since it plays a significant role in protecting against the pathogenesis of senile and postmenopausal osteoporosis, they judged that Mg should have a firm place not only in its therapy but in its prevention. In a trial of Mg supplementation, alone or with sodium fluoride, they found that best results were with drug-induced bone-wasting [97]. Stendig-Lindberg *et al.* [98] treated two groups of osteoporotic postmenopausal women, in an open controlled trial of the effect of Mg for density after one year that increased after two years. A comparable group who refused therapy, had decreases in bone density. The disparity was significant between the initial mean bone density and that one year between women taking Mg and the untreated controls.

Postmenopausal women in Italy were reported by Tranquilli *et al.* [99] to have significantly low intakes of the three major bone minerals: Ca, P and Mg, which mandated the importance of dietary evaluation in treating postmenopausal osteoporosis. Their finding, that even normal women had Mg intakes below the recommended dietary allowance (RDA), led them to advise adequate nutritional supplementation before menopause. A multiple risk survey of over 65,000 postmenopausal Swedish women, reported in 1995 by Michaelsson *et al.* [100], disclosed that – among 247 women who had sustained a first hip fracture, whose calculated daily intake of nutrients, compared with that of 893 matched women without fractures – high Ca intake did not protect against hip fractures. Their calculations disclosed that, in addition to high intakes of Mg and hormone replacement, vitamin C and iron are important in maintaining bone. Branca *et al.* [3] and Gur *et al.* [101] have more recently observed that Mg has long been known to be a component of matrix, and that other nutrients that participate in the enzymatic system involved in matrix turnover, trace minerals and vitamins, have roles in skeletal metabolism. Not considered here are details of the other minerals that participate in bone formation and maintenance. Celotti and Bignamini [102] called attention to the risk of excessive Ca intake when the Mg intake is inadequate. The dangers they cited were hypercalciuria and kidney stones. Already discussed is the risk of thromboembolism. Later will be considered the possibility that such imbalanced Ca/Mg intakes might be contributory to manifestations of the metabolic syndrome.

That estrogen loss is implicated in the bone loss of osteoporosis is widely accepted, as is the importance of estrogen repletion in arresting progression of the disease. The importance of estrogen in bone formation manifests in the increased bone mineralization in adolescent girls and young women taking oral estrogen-containing contraceptives, but not injectable progesterone contraceptives [103-105]. It is possible that some of that hormonal protection is mediated by shift of Mg to bone, as was indicated by Cohen *et al.* in 1983 [37], on observation that osteoporotic patients had subnormal bone levels of Mg, and that they retained more of parenterally administered Mg – a reflection of Mg deficiency. That estrogen could also play a role in that internal relocation of Mg in women who were not Mg deficient was evidenced by the lowered serum Mg and higher bone Mg of young women taking oral, but not depo-progesterone contraceptive agents [37, 60].

Nutritional (Mg or Ca) management of toxemias of pregnancy

Magnesium and Treatment of Eclampsia

Mg has long been considered a therapeutic pharmacologic agent in eclamptic pregnancy, and more recently also a tocolytic agent – that is useful in delaying preterm births. However, factors that increase the risk of toxemias of pregnancy, premature labor, and poor fetal outcome have in common inadequate Mg to meet maternal and fetal needs. In fact, the signs of toxemic pregnancy are of Mg deficiency. The most severe toxemia, HELLP, is characterized by abnormalities, some of which can be directly associated with Mg depletion: Hemolysis (such as can be caused by membrane instability induced by Mg inadequacy); Elevated Liver enzymes, and Low Platelets (possibly reflecting disseminated intravascular coagulation).

The value of Mg in treating toxemic pregnancy was reported in 1925 and 1926 [106, 107], and its comparative benefit over anticonvulsant and antihypertensive drugs for maternal and fetal salvage was reported by Zuspan and Ward [108] in 1965. Mg has been accepted worldwide as the preferential treatment in abnormal pregnancies, and to prevent preterm births [109-117] – apparently except for England, until recently. A survey of 1007 British obstetric consultants, reported by Hutton *et al.* [118], in 1992, disclosed that only 2% used Mg to manage eclampsia. The Eclampsia Trial Collaborative Group in Great Britain reported, in 1995, better maternal and fetal outcomes with Mg than with drugs in common use [119-121], confirming American findings of thirty years before [108].

Magnesium Prevention of Eclampsia and Fetal Damage

Important as it is to treat, with good results, women with abnormal pregnancies, preventing those life threatening disorders – of both mother and infant – would be of even greater value. A study, done with 10,000 women in 30 countries, and reported in 2002 in England [122], found that Mg halved the development of eclampsia. Nutrition and medical studies have shown that Mg intakes by pregnant mothers are commonly insufficient to meet their needs, and to prevent fetal damage [123-131]. It has been recommended that the supply over the adult female requirement should be increased by at least 150 mg [129-132]. Most prenatal vitamin-mineral supplements provide 100 mg of Mg [129]. Wynn and Wynn [133], noting the prevalence of low birth weight infants in England, and that oral supplementation with physiological amounts of Mg during pregnancy reduces pregnancy-induced-hypertension and miscarriage, intrauterine growth retardation (IUGR) and preterm birth, led them in 1988, to recommend that Mg supplement trials during pregnancy should be undertaken. In Hungary, gestational Mg deficiency has been clearly associated with poor gestational outcomes, including high rates of spontaneous abortions and preterm deliveries in regions where the water supply is very low in Mg [134]. Intervention studies with oral Mg salt supplementation of pregnant women in low Mg regions, have reduced the prevalence of spontaneous abortions and premature births [135-138]. Since Mg deficiency predisposes to complications of pregnancy that interfere with normal fetal development and growth, clinical investigators and epidemiologists should determine if Mg supplementation during pregnancy should be implemented widely.

American trial of Ca supplementation to prevent eclampsia

Based on data from countries where Ca intakes are low, and where increasing its intake has had favorable effects on pregnancy, the NIH supported a study to determine whether giving 2 grams Ca daily, in divided doses, to 2295 nulliparous American women would prevent eclampsia [139]. As noted earlier, the customary American Ca intakes are high. The supplementation was started at 13 to 21 weeks pregnancy, and results were compared with those of an equal number of nulliparous given placebo. No significant difference was commented upon in the conclusions of the report. An analysis of the tabulated data from the report shows a consistent trend towards worse maternal and infant outcomes in the Ca group than in the women given placebo.

Low cellular magnesium/calcium in the metabolic syndrome

Insulin Resistance (IR) and the Metabolic Syndrome

Insulin resistance (IR) is a common denominator of a complex of disorders which has been called the Insulin-Resistant Syndrome X and most often the Metabolic Syndrome. IR was first recognized, in 1984, by Reaven as a characteristic of Type II diabetes mellitus (non-insulin dependent diabetes mellitus: NIDDM) [140]. He noted: i) that the severity of resistance to insulin-stimulated glucose uptake increases with the magnitude of hyperglycemia; ii) that most patients with impaired glucose tolerance are hyperinsulinemic, and iii) questioned whether IR might not be the primary defect in NIDDM. The next year he ascertained that the glucose response of normal subjects to insulin was ten-fold greater than it was in patients with NIDDM- indicating the extent of the IR of NIDDM patients [141]. By 1999, Reaven commented that the ability of IR individuals to secrete large amounts of insulin is a mixed blessing, since the combination of IR and hyperinsulinemia predisposes to disorders that increase risk of coronary artery disease (CAD) [142]. He suggested that IR of muscle as a genetic change that conserved muscle mass, necessary in early times for survival, but that now is a factor in related abnormalities in diseases of Western civilization that is greatly aggravated by obesity and physical inactivity. In a 15 year follow-up of 647 Italian factory workers who were free of any disease when initially studied, he and his colleagues found that those with hyperinsulinemia initially, which was considered a surrogate determinant of IR, were at risk of later development of frank glucose intolerance, CAD and hypertension (HT) [143].

High Intracellular Calcium/Magnesium in the Metabolic Syndrome

Evidence that high intracellular Ca/Mg levels are associated with insulin resistance, and with the associated disorders that comprise the metabolic syndrome: hypertension, arteriosclerosis, hyperlipidemia, CAD, and aging, as well as obesity (particularly of central fat), was discovered and the mechanisms involved explicated by Resnick and his coworkers [144-153]. They found that intracellular Mg was significantly higher in normotensive persons than in patients with hypertension, who also showed significantly greater insulinemic response to glucose loading [144]. In that 1988 paper, they reported that even with no hyperinsulinemia, diabetics had the lowest intracellular Mg, and suggested since low Mg in cells was linked to both IR and hypertension, it might well be the common pathophysiologic basis for the clinical association of hypertension and diabetes mellitus. The next year, Resnick [145] called attention to findings of not only low cytoplasmic Mg, but high intracellular Ca of both hypertensives and diabetics, and hypothesized that hypertension, IR, and hyperinsulinemia might be different clinical manifestations of a common underlying cellular defect in divalent ion metabolism. Noting that the IR of hypertensives was closely linked to low intracellular Mg levels, Resnick and his colleagues proposed that the insulin response to glucose might be mediated by Mg [146], and that abnormal cellular handling of Ca, leading to its elevation within cells, might explain metabolic and clinical correlation of hypertension, IR, hyperinsulinemia, obesity, and NIDDM, and might be contributory to their pathophysiology and frequent clinical coexistence [147, 148]. In his 1992 paper, extending observations of abnormal divalent ion levels to therapeutic possibilities, Resnick [149] – by referring to reversal of excess free Ca accumulation and/or Mg deficit with Ca-blocking drugs, as a means of ameliorating not only hypertension but also its concurrent or excess morbidity and mortality from resultant cardiac, vascular, and metabolic effects – opened the door to the logic of therapeutic or even prophylactic use of Mg, a natural Ca-blocker. He, furthermore, showed the effects on Ca and Mg of two long-standing approaches to management of high blood pressure and its complications: salt and sugar restriction. Oral glucose loading elevates free Ca and lowers free Mg, even in normal subjects, and salt-induced hypertension is reversed by Ca-blockers [149, 150]. By 1993, he formulated the explanation of Syndrome X, on the basis of an ionic hypothesis of cardiovascular and metabolic diseases, in which a generalized defect in cell ion handling is present in all tissues, resulting in higher intracellular CA, and lower intracellular Mg and pH [151]. That syndrome comprises IR and hyperinsulinism, vasoconstriction with hypertension, ischemic heart disease and cardiac hypertrophy, and obesity. Barbagallo *et al.* [152-154] working with of the Resnick investigative group, suggested that accumulation of cytosolic Ca and depletion of cell Mg might be the missing link responsible for the frequent clinical coexistence of essential hypertension, atherosclerosis, NIDDM and other metabolic disorders of old age. They suggest that the "ionic hypothesis" of aging supposes that a sustained cellular accumulation of Ca and/or depletion of cellular Mg may provide the final common pathway for many aging-associated diseases, including HT&NIDDM.

Postmenopausal cognitive loss; roles of estrogen and magnesium

Diminished Cognition in Menopause; Protection by Estrogen

Additional to increased vulnerability to heart disease and osteoporosis, the third hazard confronted by many women in their postmenopausal years is cognitive loss, which can cause dementia, extending in some to Alzheimer's disease. Discontinuing HRT, justified by the early termination of the Women's Health Initiative [3], has created profound anxiety in postmenopausal women who were relying on hormone supplementation to maintain their health – not only physical, but mental. Those with Alzheimer's disease in their families are most apprehensive about stopping their replacement therapy. Their fear may well be justified because of the evidence that dementia occurs far less frequently in postmenopausal women receiving ERT than in those not so protected [155-161].

Estrogen has been termed "Nature's" psychoprotectant, because it favorably influences mood, mental state and memory [159]. It acts on transmitter mechanisms in the brain, increasing monoamine and neuropeptide transmitter receptors and binding sites in brain areas that control emotion, behavior, and cognition – involving verbal memory and learning capacity [159-161]. Cessation of estrogen secretion, whether induced surgically in young women, or occurring spontaneously at menopause, causes diminution of brain functional capacity, expressed by losses in learning, memory, and balance [155-161]. Data from animal models suggest that estrogen deficiency selectively increases vulnerability of estrogen-responsive neural elements: cholinergic neurons of the basal forebrain and hippocampus [155]. Increasing experimental and clinical evidence has established estrogen's role in regeneration and preservation of neurones – protecting against inflammatory reactions and apoptosis, and contributing antioxidant activity within the brain [155]. Postmenopausal estrogen deficiency has also been shown to accelerate aging of the brain and increases risk of its degenerative processes [155, 156, 158, 162-165].

Still another result of estrogen deficiency is reduced cerebral blood flow [155, 157, 161-165], although physiologic concentrations, that have been experimentally shown to cause cerebrovascular dilatation, are reversed at high concentrations. This might be contributory to the decreased clearance of beta-amyloid (seen in ovariectomized, rat models of Alzheimer's disease) [155]. There is also estrogen-related increase in blood flow during the disease, and in middle-aged postmenopausal women with early menopause [164].

Clinical and epidemiologic evidence of better cognition of estrogen-treated postmenopausal women than in those not receiving hormone replacement is now supported by experimental findings that have shown which mental functions are most improved. Determination of brain activation patterns in middle-aged and elderly postmenopausal women during performance of verbal and figural memory tasks, provide biological support for the view that estrogen protects against age-associated changes in cognition and lowers the risk of Alzheimer's disease [164-166]. ERT resulted in significantly better

performance on measures of verbal learning and memory than was seen in those who had never received hormones. Aspects of memory performance, including also encoding, retrieval, reasoning, and other higher cognitive functions were superior among women receiving estrogen.
Estrogen has profound impact on brain functioning, unlike progesterone which has been found to have minor or possibly even adverse effects [167-169]. For those who still subscribe to the menopause-is-natural philosophy this question has been posed, "why does the brain naturally have sex hormone" (*i.e.* estrogen) receptors if they are not necessary [168] ?

Magnesium Enhancement of Estrogen's Central Nervous System Protection

Mg has anti-spasmotic effects on arterial smooth muscle, which gives rise to vasodilatation of blood vessels, allowing for increased regional blood supply, including the brain [171-173]. It exerts this effect through its Ca blocking activity [170-176], and by increasing prostacycline secretion [174] as well as by exerting anticonstrictor effects against vascular mediators [175]. It also exerts direct neuroprotective effects – which are being applied in stroke victims [175, 176], and have been demonstrated in animals with brain trauma [177-187].
Human brain trauma is associated with Mg loss [186-189]. The rationale for limiting brain damage after stress [190], and trauma is based on Mg's multiple roles for functioning of key enzymes involved in energy metabolism, protein synthesis and nucleic acid metabolism, in maintaining the stability and normal function of cell membranes of excitable tissues, and in blocking Ca, which enters damaged nerve tissue [182]. The cellular disruption of brain trauma initiates complex, secondary neurochemical changes which include free radical production, membrane phospholipid breakdown, altered neurotransmitter release and energy failure – in all of which Mg is protective [182, 191].
Considering the widely recommended high Ca intake in postmenopausal women (see above) the Ca-blocking effect of Mg is useful – with and without estrogen repletion. In view of the evidence that Mg levels tend to be lower and Ca levels higher in postmenopausal women [192] the effect of Mg administration can both increase cerebral blood flow and block Ca. Additionally, ischemia – resulting from degrees of blood flow impairment, also results in uptake of Ca, and is protected against by Mg. There is also evidence that Mg-low cerebral tissues, as reflected by low cerebrospinal levels [193] are more vulnerable to ischemia than are Mg replete tissue [186]. This brings us back to brain damage of estrogen deficiency, that is associated with impaired cerebral circulation (which Mg can help to correct), and with resultant cognitive loss. Indirect links to the putative role of Mg in protecting against the processes leading to Alzheimer's disease, are its use in recovery (measured by improvement in learning tests) of brain traumatized animals, and its favorable effect on up regulation of beta amyloid precursor protein in brain injured patients [188]. Direct association of loss of cognition with low magnesium levels is the evidence that cognitive impairment was likelier in hypertensive patients with low serum magnesium than normomagnesemic hypertensives [194]. And patients with chronic fatigue syndrome, associated with latent tetany of magnesium deficiency, commonly have cognitive impairment, that is responsive to magnesium repletion [195].

Concluding comments

The DRI for Ca having been set at 1,500 mg/day, without regard for the inadequacy of the usual diet to provide sufficient Mg to maintain Mg equilibrium, the risk of inducing Mg deficiency has been increased. Considered in this report, are major problems that developed by the 1960s, by which time the Ca/Mg ratios of the American diet had increased to between 3-4/1. It was then that the male preponderance in ischemic heart disease (IHD) caused alarm. It was also then that young women, whose natural estrogen levels protected them from IHD that afflicted men, began to suffer strokes and heart attacks-that were blamed on high estrogen oral contraceptive agents (OCAs). Changing the formulations of the OCAs-reducing estrogen content and adding progesterone reduced that problem. The recent early termination of the WHI and WHIM studies of the NIH, in which (equine) estrogen alone, or combined with progestin in postmenopausal women, was discontinued because of thromboembolic events, recalls the comparable events with administration of estrogen-rich OCA to young women. In Europe, where transdermal estrogen is more common, and considered less likely to cause intravascular coagulation, new investigations are being urged to test repletion with natural estrogen and progesterone, not orally administered. It is recommended that it should also be determined whether supplementing with Mg would further lower risk of thromboembolisms, and even reduce cognitive loss. Now that American postmenopausal women are increasingly being deprived of their estrogen repletion, and are thus at increased risk of bone loss, we must be cognizant of the harm that results from high Ca and low Mg intakes, as well as of the benefit of adequate Mg in bone maintenance. Will a new epidemic of IHD develop in postmenopausal women without estrogen repletion, and subjected to excess Ca, such as was first seen in men in the 1960s? Still another question should be addressed. Might high Ca intake, in association with low dietary Mg, contribute to the high intracellular Ca/Mg implicated as a common denominator in the different manifestations of the metabolic syndrome? During growth and development, pregnancy, heavy exercise and other stresses, which are anabolic states in which there is increased need for Mg for new tissue formation, the likelihood of Mg deficiency is intensified.

References

1. Food and Nutrition Board. *Institute of Medicine: Panel on Calcium and Related Nutrients*. Washington DC: National Academy Press, 1997.
2. Yates AA, Schlicker SA, Suitor CW. Dietary Reference Intakes: the new basis for recommendations for calcium and related nutrients, B vitamins, and choline. *J Am Diet Assoc* 1998; 98: 699-706.
3. Branca F, Valtuena S, Valtuena S. Calcium, physical activity and bone health-building bones for a stronger future. *Public Health Nutr* 2001; 4: 117-23.
4. Hathaway M. In: *Magnesium in human nutrition. Home Econ Research Report 19*. Agric Washington DC: Res Serv US Dept Agriculture, 1962: 1-94.
5. Leverton R, Leichsenring JM, Linkswiler H, Fox H, Meyer FL. In: *The metabolic response of young women to a standardized diet. Home Economics Research Report 16*. Washington DC: US Dept of Agriculture, 1962: 1-45.
6. Seelig MS. The requirement of magnesium by the normal adult. *Am J Clin Nutr* 1964; 14: 342-90.
7. Scoular FI, Pace JK, Davis AM. The calcium, phosphorus and magnesium balances of young college women consuming self-selected diets. *J Nutr* 1957; 62: 489-501.
8. Commission on Life Sciences. *National Research Council: Recommended Dietary. Food and Nutrition Board. Allowances, 10th Ed.* Washington DC: National Academy Press, 1989.
9. Irwin M, Feeley R. Frequency and size of meals and serum lipids, nitrogen and mineral retention, fat digestibility and urinary thiamine and riboflavin in young women. *Am J Clin Nutr* 1967; 20: 816-24.
10. Friend B. Nutrients in United States food supply. A review of trends. *Am J Clin Nutr* 1967; 20: 907-14.
11. Walker MA, Page L. Nutritive content of college meals. III. Mineral Elements. *J Am Diet Assoc* 1977; 70: 260-6.
12. Greger JL, Baligar P, Abernathy RP, Bennett OA, Peterson T. Calcium, magnesium, phosphorus, copper, and manganese balance in adolescent females. *Am J Clin Nutr* 1978; 31(1): 117-21.
13. Greger JL, Smith SA, Snedeker SM. Effect of dietary calcium and phosphorus level on the utilization of calcium, phosphorus, magnesium, manganese, and selenium by adult males. *Nutr Res* 1981; 1: 315-25.
14. Bogert L, McKittreck E. Interrelations between calcium and magnesium metabolism. *J Biol Chem* 1922; 54: 363-74.
15. Pao EM, Mickle SJ. Problem nutrients in the United States. *Food Technol* 1981; 35: 58-69.
16. Lakshmanan FL, Rao RB, Kim WW, Kelsay JL. Magnesium intakes, balances, and blood levels of adults consuming self-selected diets. *Am J Clin Nutr* 1984; 40: 1380-9.
17. Morgan KJ, Stampley GL, Zabik ME, Fischer DR. Magnesium and calcium dietary intakes of the U.S. population. *J Am Coll Nutr* 1985; 4: 195-206.
18. Morgan KJ, Stampley GL. Dietary intake levels and food sources of magnesium and calcium for selected segments of the US population. *Magnesium* 1988; 7: 225-33.
19. Karppanen H, Neuvonen PJ. (Letter to Editor). Ischaemic heart-disease and soil magnesium in Finland. *Lancet* 1973; 2: 1390.
20. Karppanen H, Pennanen R, Passinen L. Minerals, coronary heart disease and sudden coronary death. *Adv Cardiol* 1978; 25: 9-24.
21. Karppannen H. Epidemiologic evidence for considering magnesium deficiency as a risk factor for cardiovascular diseases. *Magnes Bull* 1990; 12: 80-6.
22. Anderson TW. The changing pattern of ischemic heart disease. *Canad Med Assoc J* 1973; 108: 1500-4.
23. Stamler JS. The relationship of sex and gonadal hormones to atherosclerosis. In: Sandler M, Boume GH, eds. *Atherosclerosis and Its Origin*. New York: Academic Press, 1963: 231-62.

24. Anderson TW, Neri LC, Schreiber GB, Talbot FDF, Zdrojewski A. Ischemic heart disease, water hardness and myocardial magnesium. *Canad Med Assc J* 1975; 113: 199-203.

25. Anderson TW, Leriche WH, Hewitt D, Neri LC. Magnesium, water hardness, and heart disease. In: Cantin M, Seelig MS, eds. *Magnesium in Health & Disease Spectrum. New York.* 1980: 565-71; (2nd Intl Sympos on Mg, Quebec, 1976).

26. Seelig MS, Heggtveit HA. Magnesium interrelationships in ischemic heart disease: A review. *Am J Clin Nutr* 1974; 27: 59-79.

27. Seelig MS, Haddy FJ. Magnesium and the Arteries: I. Effects of magnesium deficiency on arteries and on retention of sodium, potassium, and calcium. In: Cantin M, Seelig MS, eds. *Magnesium in Health & Disease.* New York: Publ Spectrum Press, 1980: 605-38.

28. Haddy FJ, Seelig MS. In: *Magnesium and the arteries: II. Physiologic effects of electrolyte abnormalities on arterial resistance.* Ibid, 1980: 639-57.

29. Sheehan JP, Seelig MS. Interactions of magnesium and potassium in the pathogenesis of cardiovascular disease. *Magnesium* 1984; 3: 301-14.

30. Altura BM, Altura BT. New perspectives on the role of magnesium in the pathophysiology of the cardiovascular system. I. Clinical aspects. *Magnesium* 1985; 4: 226-44.

31. Altura BM, Altura BT. Biochemistry and pathophysiology of congestive heart failure. Is there a role for magnesium? *Magnesium* 1986; 5: 134-43.

32. Altura BT, Altura BM. Cardiovascular actions of magnesium: Importance in etiology and treatment of high blood pressure. *Magnesium Bul* 1987; 9: 6-21.

33. Seelig MS. Cardiovascular consequences of magnesium deficiency and loss; pathogenesis, prevalence and manifestations - magnesium and chloride loss in refractory potassium repletion. *Am J Cardiol* 1989; 63: 4G-24G.

34. Altura BM, Altura BT. Cardiovascular risk factors and magnesium: relationships to atherosclerosis, ischemic heart disease and hypertension. *Magnes Trace Elem* 1991; 10: 182-92.

35. Seelig MS. Epidemiologic data on magnesium deficiency -associated cardiovascular disease and osteoporosis; consideration of risks of current recommendations for high calcium intake. In: Rayssiguier Y, ed. *Advances in Magnesium research: Nutrition and Health.* Vichy: Proc of 9th International Symposium on Magnesium, Chapt 27, 2001: 177-90.

36. Goldsmith N. Physiologic relationship between magnesium and female reproductive apparatus. In: Durlach J, ed. *Serum magnesium variations in normal subjects: the effect of estrogen on magnesium distribution.* Vittel: Proc Intl Sympos on Magnesium, 1971/1973: 439-58.

37. Cohen L, Laor A, Kitzes R. Bone magnesium, crystallinity index and state of body magnesium in subjects with senile osteoporosis, maturity-onset diabetes and women treated with contraceptive preparations. *Magnesium* 1983; 2: 70-5.

38. Classen HG, Baier S, Schimatscheck HF, Classen CU. Clinically relevant interactions between hormones and magnesium metabolism. *Magnesium Bul* 1995; 17: 96-103.

39. Seelig MS. Increased magnesium need with use of combined estrogen and calcium for osteoporosis. *Magnesium Res* 1990; 3: 197-215.

40. Seelig MS. Interrelationship of magnesium and estrogen in cardiovascular and bone disorders, eclampsia, migraine and premenstrual syndrome. *J Am Coll Nutr* 1993; 12: 442-58.

41. Muneyvirci-Delale O, Nacharaju VL, Altura BM, Altura BT. Sex steroid hormones modulate serum ionized magnesium and calcium levels throughout the menstrual cycle in women. *Fertil Steril* 1998; 69: 958-62.

42. O'Shaughnessy A, Muneyyirci-Delale O, Nacharaju VL, Dalloul M, Altura BM, Altura BT. Circulating divalent cations in asymptomatic ovarian hyperstimulation and in vitro fertilization patients. *Gynecol Obstet Invest* 2001; 52: 237-42.

43. DeJorge FB, Canato C, Medici C, Delascio D. Effects of the progestin-estrogen oral contraceptives on blood serum copper, copper oxidase, magnesium and sulphur concentrations. *Matern Infanc (Sao Paulo)* 1967; 26: 261-7.

44. Muller P, Dellenbach P, Frenot M. Magnesium deficiency in obstetrics and gynecology. In: Durlach J, ed. *1st Intl Sympos on Magnesium Deficit in Human Pathology.* Vittel, 1971: 459-80.

45. Ryzen E, Greenspoon JS, Diesfield P. Rude. Blood mononuclear cell magnesium in normal pregnancy and preeclampsia. *J Am Coll Nutr* 1987; 6: 121-4.

46. Ajayi G. Serum magnesium concentration in premenopausal, menopausal women, during normal and EPH-gestosis pregnancy and the effect of diuretic therapy in EPH-gestosis. *Magnesium Bul* 1988; 10: 72-6.

47. Borella P, Szilagyi A, Than G, Csaba I, Giardino A, Facchinetti F. Maternal plasma concentrations of magnesium, calcium, zinc and copper in normal and pathological pregnancies. *Sci Total Environ* 1990; 99: 67-76.

48. Kurzel RB. Serum magnesium levels in pregnancy and preterm labor. *Am J Perinatol* 1991; 8: 119-27.

49. Handwerker SM, Altura BT, Altura BM. Ionized serum magnesium and potassium levels in pregnant women with preeclampsia and eclampsia. *J Reprod Med* 1995; 40(3): 201-8.

50. Lidegaard O. Oral contraceptives, pregnancy and the risk of cerebral thromboembolism: the influence of diabetes, hypertension, migraine and previous thrombotic disease. *Br J Obstet Gynaecol* 1995; 102: 153-9.

51. Standley CA, Whitty JE, Mason BA, Cotton DB. Serum ionized magnesium levels in normal and preeclamptic gestation. *Obstet Gynecol* 1997; 89(1): 24-7.

52. Arikan GM, Panzitt T, Gucer F, *et al.* Course of maternal serum magnesium levels in low-risk gestations and in preterm labor and delivery. *Fetal Diagn Ther* 1999; 14(6): 332-6.

53. Kisters K, Barenbrock M, Louwen F, Hausberg M, Rahn KH, Kosch M. Membrane, intracellular, and plasma magnesium and calcium concentrations in preeclampsia. *Am J Hypertens* 2000; 13: 765-9.

54. Nilsson IM, Kullander S. Coagulation and fibrinolytic studies during use of gestagens. *Acta Obst Gynec Scand* 1967; 46: 286-303.

55. Cole M. Strokes in young women using oral contraceptives. *Arch Intern Med* 1967; 120: 551-5.

56. Kalman SM. Effects of oral contraceptives. *Annu Rev Pharmacol* 1969; 9: 363-78.

57. Anderson TW. Oral contraceptives and female mortality trends. *Canad Med Assoc J* 1970; 102: 1156-60.

58. Bottiger LE, Westerholm B. Oral contraceptives and thromboembolic disease. *Acta Med Scand* 1971; 190: 455-63.

59. Goldsmith NF, Johnston JO. Magnesium-estrogen hypothesis: thromboembolic and mineralization ratios. In: Cantin M, Seelig MS, eds. *Magnesium in Health & Disease.* New York: Spectrum, 1980: 313-23; (2nd Intl Mg Sympos, Quebec, 1976).

60. Meade TW. Oral contraceptives, clotting factors, and thrombosis. *Am J Obst Gynecol* 1982; 142: 758-61.

61. Bonnar J. Coagulation effects of oral contraception. *Am J Obstet Gynecol* 1987; 157: 1042-8.

62. Durlach J. Physiologic antithrombotic role of magnesium, pertinent to phlebothrombotic disease due to magnesium deficiency. *Coeur Med Interne* 1967; 6: 213-32.

63. Durlach J. The pill and thrombosis (platelets, estrogens and magnesium). *Rev Franc Endocrinol Clin* 1970; 11: 45-54.

64. Jeffcoate TNA, Jeffcoate TNA, Miller J, Roos RF, Tindall VR. Puerperal thromboembolism in relation to the inhibition of lactation by oestrogen therapy. *BMJ* 1968; 4: 19-25.

65. Unsigned. The Coronary Drug Project. Findings leading to discontinuation of the 2.5 mg day estrogen group. *JAMA* 1973; 226: 652-7.

66. Blackard CE, Doe RP, Mellinger GT, Byar DP. Incidence of cardiovascular disease and death in patients receiving diethylstilbestrol for carcinoma of the prostate. *Cancer* 1970; 26: 249-56.

67. Byar DP, Corle DK. Hormone therapy for prostate cancer: results of the Veterans Administration Cooperative Urological Research Group studies. *NCI Monogr* 1988; 7: 165-70.

68. Luyer MD, Khosla S, Owen WG, Miller VM. Prospective randomized study of effects of unopposed estrogen replacement therapy on markers of coagulation and inflammation in postmenopausal women. *J Clin Endocrinol Metab* 2001; 86(8): 3629-34.

69. Gueux E, Rayssiguier Y, Piot M-C, Alcindor L. Reduction of plasma lecithin-cholesterol-acyltransferase activity by acute magnesium deficiency in the rat. *J Nutr* 1984; 114: 1479-83.

70. Rayssiguier Y. Magnesium, lipids and vascular diseases. Experimental evidence in animal models. *Magnesium* 5: 182-190.

71. Rayssiguier Y, Mazur A, Cardot P, Gueux E. Effects of magnesium on lipid metabolism and cardiovascular disease. In: Itokawa Y, Durlach J, eds. *Magnesium in Health and Disease.* Publ John Libbey&Co Ltd, 1989: 199-207; (5th Intl Mg Sympos, Kyoto, Japan 1988).

72. Solajic-Bozicevic N, Stavljenic A, Sesto M. Lecithin: cholesterol acyltransferase activity in patients with acute myocardial infarction and coronary heart disease. *Artery* 1991; 18: 326-40.

73. Itoh K, Kawasaka T, Nakamura M. The effects of high oral magnesium supplementation on blood pressure, serum lipids and related variables in apparently healthy Japanese subjects. *Br J Nutr* 1997; 78: 737-50.

74. Unsigned. The Women's Health Initiative Study Group. Design of the Women's Health Initiative Clinical Trial and Observational Study. *Control Clin Trials* 1998; 19: 61-109.

75. Shumaker SA, Reboussin BA, Espeland MA, *et al.* The Women's Health Initiative Memory Study (WHIMS): A trial of the effect of estrogen therapy in preventing and slowing the progression of dementia. *Control Clin Trials* 1998; 19: 604-21.

76. Rossouw JE. Hormone replacement therapy and cardiovascular disease. *Curr Opin Lipidol* 1999; 10: 429-34.

77. Rossouw JE, Anderson GL, Prentice RL, *et al.* Writing Group for the Women's Health Initiative Investigators. Risks and benefits of estrogen plus progestin in healthy postmenopausal women: principal results From the Women's Health Initiative randomized controlled trial. *JAMA* 2002; 288: 321-33.

78. Rapp SR, Espeland MA, Shumaker SA, *et al.* WHIMS Investigators. Effect of estrogen plus progestin on global cognitive function in post menopausal women: the Women's Health Initiative Memory Study: a randomized controlled trial. *JAMA* 2003; 289: 2663-72.

79. Gaspard U, van den Brule F, Pintiaux A, Foidart JM. Clinical study of the month. Benefit/risk balance of postmenopausal estrogen-progestin treatment in

peril in the Women's Health Initiative study: practical attitude of the clinician. *Rev Med Liege* 2002; 57(8): 556-62.

80. Kuttenn F, Gerson M, de Ligneres B. Effects of hormone replacement therapy in menopause on cardiovascular risk. Need for a European study. *Presse Med* 2002; 31(10): 468-75.
81. Unsigned. Postmenopausal hormone therapy: cardiovascular risks. *Prescrire Int* 2000; 12: 65-9.
82. Karppanen H, Pennanen R, Passinen L. Minerals, coronary heart disease and sudden coronary death. *Adv Cardiol* 1978; 25: 9-24.
83. Karppanen H. Ischaemic heart disease. An epidemiological perspective with special reference to electrolytes. *Drugs* 1984; 28(Suppl 1): 17-27.
84. Karppanen H. Minerals and blood pressure. *Ann Med* 1991; 23: 299-305.
85. Goto S. Magnesium intake and its balance in Japanese. *Magnesium-Bul* 1987; 9: 51.
86. Omura T, Hisamatsu S, Takizawa Y, Minowa M, Yanagawa H, Shigematsu I. Geographical distribution of cerebrovascular disease mortality and food intakes in Japan. *Soc Sci Med* 1987; 24(5): 401-7.
87. Sei M, Nakamura H, Miyoshi T. Nutritional epidemiological study on mineral intake and mortality from cardiovascular disease. *Tokushima J Exp Med* 1993; 40(3-4): 199-207.
88. Ogihara T, Hiwada K, Matsuoka H, *et al.* Guidelines on treatment of hypertension in the elderly. 1995- A tentative plan for comprehensive research projects on aging and health- Members of the Research Group for "Guidelines on Treatment of Hypertension in the Elderly", Comprehensive Research Projects on Aging and Health, the Ministry of Health and Welfare of Japan. *Nippon Ronen Igakkai Zasshi* 1996; 33(12): 945-75.
89. Imakita M, Yutani C, Sakurai I, *et al.* The second nationwide study of atherosclerosis in infants, children, and young adults in Japan. Comparison with the first study carried out 13 years ago. *Ann N Y Acad Sci* 2000; 902: 364-8.
90. Volpe SL, Taper LJ, Meacham S. The relationship between boron and magnesium status and bone mineral density in the human: a review. *Magnes Res* 1993; 6: 291-6.
91. Dalderup LM. The role of magnesium in osteoporosis and idiopathic hypercalcemia. *Voeding* 1960; 21: 424-35.
92. Miller ER, Ullrey DE, Zutaut CL, *et al.* Magnesium requirements of the baby pig. *J Nutr* 1965; 85: 13-20.
93. Lai CC, Singer L, Armstrong W. Bone composition and phosphatase activity in magnesium deficiency in rats. *J Bone Joint Surg* 1975; 57: 516-22.
94. Kenney MA, McCoy H, Williams L. Effects of magnesium deficiency on strength, mass and composition of rat femur. *Calcif Tissue Int* 1994; 1: 44-9.
95. Makgoba MW, Datta HK. The critical role of magnesium ions in osteoclast-matrix interaction: implications for divalent cations in the study of osteoclastic adhesion molecules and bone resorption. *Eur J Clin Invest* 1992; 22: 692-6.
96. Ditmar R, Steidl L. The significance of magnesium in orthopedics. V. Magnesium in osteoporosis. *Acta Chir Orthop Traumatol Czech* 1989; 56: 143-59.
97. Steidl L, Ditmar R. Osteoporosis treated with magnesium lactate. *Acta Univ Palacki Olomuc Fac Med* 1991; 129: 99-106.
98. Stendig-Lindberg G, Tepper R, Leichter I. Trabecular bone density in a two year controlled trial f peroral magnesium in osteoporosis. *Magnes Res* 1993; 6: 155-63.
99. Tranquilli AL, Lucino E, Garzetti GG, Romanini C. Calcium, phosphorus and magnesium intakes correlate with bone mineral content in postmenopausal women. *Gynecol Endocrinol* 1994; 8: 55-8.
100. Michaelsson K, Holmberg L, Mallmin H, *et al.* Diet and hip fracture risk: a case-control study. Study Group of the Multiple Risk Survey on Swedish Women for Eating Assessment. *Int J Epidemiol* 1995; 24: 771-82.
101. Gur A, Colpan L, Nas K, *et al.* The role of trace minerals in the pathogenesis of postmenopausal osteoporosis and a new effect of calcitonin. *J Bone Miner Metab* 2002; 20: 39-43.
102. Celotti F, Bignamini A. Dietary calcium and mineral/vitamin supplementation: a controversial problem. *J Int Med Res* 1999; 27: 1-14.
103. Goldsmith NF, Johnston JO. Bone mineral: effects of oral contraceptives, pregnancy, and lactation. *J Bone Joint Surg* 1975; 57-A(5): 657-68.
104. Sultana S, Choudhury S, Choudhury SA. Serum alkaline phosphatase and bone mineral density: to assess bone loss in oral contraceptive pill user. *Mymensingh Med J* 2002; 1(2): 107-9.
105. Cromer BA. Bone mineral density in adolescent and young adult women on injectable or oral contraception. *Curr Opin Obstet Gynecol* 2003; 15(5): 353-7.
106. Lazard EM. A preliminary report on the intravenous use of magnesium sulphate in puerperal eclampsia. *Am J Obst Gynec* 1925; 9: 178-88.
107. McNeile LG, Vruwink J. Magnesium sulphate intravenously in the care and treatment of preeclampsia and eclampsia. *JAMA* 1926; 87: 236-40.
108. Zuspan GP, Ward MC. Improved fetal salvage in eclampsia. *Obst Gynec* 1965; 26: 893-7.
109. Conradt A, Weidinger H, Algayer H. Magnesium therapy decreased the rate of intrauterine fetal retardation, premature rupture of membranes and premature delivery in risk pregnancies treated with betamimetics. *Magnesium* 1985; 4: 20-8.

110. Spaetling L, Spaetling G. Magnesium supplementation in pregnancy. A double-blind study. *Br J Obstet Gynaecol* 1988; 95: 120-5.

111. Dommisse J. Phenytoin sodium and magnesium sulphate in the management of eclampsia. *Br J Obstet Gynaecol* 1990; 97: 104-19.

112. Kyank H. MgSO4 therapy in severe pre-eclampsia/eclampsia. *Zentralbl Gynakol* 1990; 112: 5-10.

113. Rudnicki M, Frolich A, Rasmussen WF, McNair P. The effect of magnesium on maternal blood pressure in pregnancy-induced hypertension. A randomized double-blind placebo-controlled trial. *Acta Obstet Gynecol Scand* 1991; 70: 445-50.

114. Editorial Comment. A safer and more effective treatment regimen for eclampsia. *Aust NZ Obstet Gynecol* 1994; 34: 144-5.

115. Anthony J, Johanson RB, Duley L. Role of magnesium sulfate in seizure prevention in patients with eclampsia and pre-eclampsia. *Drug Saf* 1996; 15: 188-99.

116. Jerie P. Hypertension and its treatment in pregnancy. *Cas Lek Cesk* 1998; 137: 467-72.

117. Rey E, Lelorier J, Burgess E, Lange IR, Leduc L. Report of the Canadian Hypertension Society Consensus Conference 3. Pharmacologic Treatment of Hypertensive Disorders in Pregnancy. *Canad Med Assoc J* 1997; 57: 1245-54.

118. Hutton JD, James DK, Stirrat GM, *et al.* Management of severe pre-eclampsia and eclampsia by UK consultants. *Br J Obstet Gynaecol* 1992; 99: 554-6.

119. The Eclampsia Trial Collaborative Group. Which anticonvulsant for women with eclampsia? Evidence from the collaborative eclampsia trial. *Lancet* 1995; 345: 1455-63.

120. Saunders N, Hammersley B. Magnesium for eclampsia. *Lancet* 1995; 34: 788-9.

121. Robson SC. Magnesium sulphate: the time of reckoning. *Br J Obstet Gynaecol* 1996; 103: 99-102.

122. Magpie Trial Collaboration Group. Do women with pre-eclampsia, and their babies, benefit from magnesium sulphate? The Magpie Trial: a randomised placebo-controlled trial. *Lancet* 2002; 359: 1877-90.

123. Seelig MS. Human requirements of magnesium; factors that increase needs. In: Durlach J, ed. *Proc. 1st Intl Sympos on Magnesium Deficit in Human Path.* France: Vittel, 1971: 11-38.

124. Ashe JR, Schofield FA, Gram MR. The retention of calcium, iron, phosphorus, and magnesium during pregnancy: the adequacy of prenatal diets with and without supplementation. *Am J Clin Nutr* 1979; 32: 286-91.

125. Seelig MS. *Magnesium in the Pathogenesis of Disease: Early Roots of Cardiovascular, Skeletal and Renal Abnormalities.* New York: Plenum Press, 1980.

126. Johnson NE, Philipps C. Magnesium content of diets of pregnant women. In: Cantin M, Seelig MS, eds. *Magnesium in Health & Disease.* New York: Spectrum, 1980: 827-31; (2nd Int'l. Magnesium Sympos. Quebec, Canada, 1976).

127. Kontopoulos V, Seelig MS, Dolan J, Berger AR, Ross RS. *Influence of parenteral administration of magnesium sulfate to normal pregnant and to pre-eclamptic women.* Ibid, 839-848.

128. Weaver K. A *possible anticoagulant effect of magnesium in pre-eclampsia.* Ibid, 833-8.

129. Franz KB. Magnesium deficiency during pregnancy. *Magnesium* 1987; 6: 18-27.

130. Weaver K. Magnesium and fetal growth. *Tr Subst Environm Hlth* 1988; 23: 136-42.

131. McGarvey ST, Zinner SH, Willett WC, Rosner B. Maternal prenatal dietary potassium, calcium, magnesium, and infant blood pressure. *Hypertension* 1991; 17: 218-24.

132. Lichton IJ. Dietary intake levels and requirements of Mg and Ca for different segments of the U.S. population. *Magnesium* 1989; 8: 117-23.

133. Wynn A, Wynn M. Magnesium and other nutrient deficiencies as possible causes of hypertension and low birth weight. *Nutr Health* 1988; 6: 69-88.

134. Losonczy J, Adorjan G, Novak M, Toth MO. Correlation between the incidence of preterm delivery and the concentration of magnesium in drinking water in Szabolcs-Szatmar County, Northeast Hungary. *Magnesium Res* 1989; 2: 229-30.

135. Kuti V, Balazs M, Morvay F, Varenka Z, Szekly A, Szucs M. Effect of maternal magnesium supply on spontaneous abortion and premature birth and on intrauterine foetal development: experimental epidemiological study. *Magnesium Bul* 1981; 3: 73-9.

136. Kovacs L, Molnar BG, Huhn E, Bodis L. Magnesium substitution in pregnancy. A prospective, randomized double-blind study. *Geburtshilfe Frauenheilkd* 1988; 48: 595-600.

137. Molnar BG, Kovacs L. Experiences with magnesium substitution during pregnancy. *Magnesium Res* 1989; 2: 30-1.

138. Novak M, Losonczy J, Kornafeld J, Toth MO. Experiences on giving magnesium citrate to pregnant mothers. *Magnesium Res* 1989; 2: 230.

139. Levine RJ, Hauth JC, Curet LB, *et al.* Trial of calcium to prevent preeclampsia. *N Engl J Med* 1997; 337: 68-76.

140. Reaven GM. Insulin secretion and insulin action in non-insulin-dependent diabetes mellitus: which defect is primary? *Diabetes Care* 1984; 7(Suppl 1): 17-24.

141. Reaven GM, Chen YD, Donner CC, *et al.* How insulin resistant are patients with non-insulin-dependent diabetes mellitus? *J Clin Endocrinol Metab* 1985; 61(1): 32-6.

142. Reaven GM. Insulin resistance: a chicken that has come to roost. *Ann N Y Acad Sci* 1999; 892: 45-57.

143. Zavaroni I, Bonini L, Gasparini P, *et al.* Hyperinsulinemia in a normal population as a predictor of non-insulin-dependent diabetes mellitus, hypertension, and coronary heart disease: the Barilla factory revisited. *Metabolism* 1999; 48: 989-94.

144. Resnick LM, Gupta RK, Gruenspan H, Laragh JH. Intracellular free magnesium in hypertension: Relation to peripheral insulin resistance and obesity. *J Hypertens* 1988(Suppl 4): S199-S201.

145. Resnick LM. Hypertension and abnormal glucose homeostasis. Possible role of divalent ion metabolism. *Am J Med* 1989; 87(6A): 17S-22S.

146. Resnick LM, Gupta RK, Gruenspan H, Alderman MH, Laragh JH. Hypertension and peripheral insulin resistance. Possible mediating role of intracellular free magnesium. *Am J Hypertens* 1990; 3(5 Pt 1): 373-9.

147. Resnick LM. Calcium metabolism in hypertension and allied metabolic disorders. *Diabetes Care* 1991; 14: 505-20.

148. Resnick LM, Gupta RK, Bhargava KK, Gruenspan H, Alderman MH, Laragh JH. Cellular ions in hypertension, diabetes, and obesity. A nuclear magnetic resonance spectroscopic study. *Hypertension* 1991; 17(6 Pt 2): 951-7.

149. Resnick LM. Cellular calcium and magnesium metabolism in the pathophysiology and treatment of hypertension and related metabolic disorders. *Am J Med* 1992; 93(2A): 11S-20S.

150. Resnick LM. Cellular ions in hypertension, insulin resistance, obesity, and diabetes: a unifying theme. *J Am Soc Nephrol* 1992; 3(4 Suppl): S78-S85.

151. Resnick LM. Ionic basis of hypertension, insulin resistance, vascular disease, and related disorders. The mechanism of "syndrome X". *Am J Hypertens* 1993; 6: 123S-134S.

152. Barbagallo M, Resnick LM, Dominguez LJ, Licata G. Diabetes mellitus, hypertension and ageing: the ionic hypothesis of ageing and cardiovascular-metabolic diseases. *Diabetes Metab* 1997; 23: 281-94.

153. Barbagallo M, Dominguez LJ, Licata G, Resnick LM. Effects of aging on serum ionized and cytosolic free calcium: relation to hypertension and diabetes. *Hypertension* 1999; 34(4 Pt 2): 9026.

154. Barbagallo M, Gupta RK, Dominguez LJ, Resnick LM. Cellular ionic alterations with age: relation to hypertension and diabetes. *J Am Geriatr Soc* 2000; 48: 1111-6.

155. Birge SJ. Is there a role for estrogen replacement therapy in the prevention and treatment of dementia? *J Am Geriatr Soc* 1996; 44: 865-70.

156. Greene RA. Estrogen and cerebral blood flow: a mechanism to explain the impact of estrogen on the incidence and treatment of Alzheimer's disease. *Int J Fertil Womens Med* 2000; 45: 253-7.

157. Compton J, van Amelsvoort T, Murphy D. Mood, cognition and Alzheimer's disease. *Best Pract Res Clin Obstet Gynaecol* 2002; 16: 357-70.

158. Cholerton B, Gleason CE, Baker LD, Asthana S. Estrogen and Alzheimer's disease: the story so far. *Drugs Aging* 2002; 19: 405-27.

159. Fnk G, Sumner BE, Rosie R, Grace O, Quinn JP. Estrogen control of central neurotransmission: effect on mood, mental state, and memory. *Cell Mol Neurobiol* 1996; 16: 325-44.

160. Birge SJ, McEwen BS, Wise PM. Effects of estrogen deficiency on brain function. Implications for the treatment of postmenopausal women. *Postgrad Med Spec* 2001: 11-6.

161. Brinton RD. Cellular and molecular mechanisms of estrogen regulation of memory function and neuroprotection against Alzheimer's disease: recent insights and remaining challenges. *Learn Mem* 2001; 8: 121-33.

162. Genazzani AR, Spinetti A, Gallo R, Bernardi F. Menopause and the central nervous system: intervention options. *Maturitas* 1999; 31: 103-10.

163. Li W, Zheng T, Altura BM, Altura BT. Sex steroid hormones exert biphasic effects on cytosolic magnesium ions in cerebral vascular smooth muscle cells: possible relationships to migraine frequency in premenstrual syndromes and stroke incidence. *Brain Res Bull* 2001; 54: 83-9.

164. Maki PM, Resnick SM. Effects of estrogen on patterns of brain activity at rest and during cognitive activity: a review of neuroimaging studies. *Neuroimage* 2001; 14: 789-801.

165. Resnick SM, Maki PM. Effects of hormone replacement therapy on cognitive and brain aging. *Ann N Y Acad Sci* 2001; 949: 203-14.

166. Maki P, Zonderman A, Resnick S. Enhanced verbal memory in non-demented elderly women receiving hormone-replacement therapy. *Am J Psychiatry* 2001; 158: 227-33.

167. Rice MM, Graves AB, McCurry SM, Larson EB. Estrogen replacement therapy and cognitive function in postmenopausal women without dementia. *Am J Med* 1997; 103: 26S-35S.

168. Greene RA, Dixon W. The role of reproductive hormones in maintaining cognition. *Obstet Gynecol Clin North Am* 2002; 29: 437-53.

169. Maki PM, Rich JB, Rosenbaum RS. Implicit memory varies across the menstrual cycle: estrogen effects in young women. *Neuropsychologia* 2002; 40: 518-29.

170. Altura BM, Altura B. Microcirculatory actions and uses of naturally-occurring (magnesium) and novel synthetic calcium channel blockers. *Microcirc Endothelium Lymphatics* 1984; 1: 185-220.

171. Altura BM, Altura BT. New perspectives on the role of magnesium in the pathophysiology of the cardiovascular system. II. Experimental aspects. *Magnesium* 1985; 4: 245-71.

172. Chi OZ, Pollak P, Weiss HR. Effects of magnesium sulfate and nifedipine on regional cerebral blood flow during middle cerebral artery ligation in the rat. *Arch Int Pharmacodyn Ther* 1990; 304: 196-205.

173. Iseri LT, French JH. Magnesium: nature's physiologic calcium blocker. *Am Heart J* 1984; 108: 188-93.

174. Nadler JL, Goodson S, Rude RK. Evidence that prostacyclin mediates the vascular action of magnesium in humans. *Hypertension* 1987; 9: 379-83.

175. Muir KW. New experimental and clinical data on the efficacy of pharmacological magnesium infusions in cerebral infarcts. *Magnes Res* 1998; 11: 43-56.

176. Muir KW. Magnesium in stroke treatment. *Postgrad Med J* 2002; 78: 641-5.

177. Heath DL, Vink R. Blood-free magnesium concentration declines following graded experimental traumatic brain injury. *Scand J Clin Lab Invest* 1998; 58: 161-6.

178. McIntosh TK, Vink R, Yamakami I, Faden AI. Magnesium protects against neurological deficit after brain injury. *Brain Res* 1989; 482: 252-60.

179. Shapira Y, Yadid G, Cotev S, Shohami E. Accumulation of calcium in the brain following head trauma. *Neurol Res* 1989; 11: 169-72.

180. Vink R, Heath DL, McIntosh TK. Acute and prolonged alterations in brain free magnesium following fluid percussion_induced brain trauma in rats. *J Neurochem* 1996; 66: 2477-83.

181. Feldman Z, Gurevitch B, Artru AA, *et al.* Effect of magnesium given 1 hour after head trauma on brain edema and neurological outcome. *J Neurosurg* 1996; 85: 131-7.

182. Cernak I, Vink R. Magnesium as a regulatory cation in direct and indirect traumatic brain injury. *Magnesium Res* 1999; 12: 223-4.

183. Bareyre FM, Saatman KE, Raghupathi R, McIntosh TK. Postinjury treatment with magnesium chloride attenuates cortical damage after traumatic brain injury in rats. *J Neurotrauma* 2000; 17: 1029-39.

184. Cernak I, O'Connor C, Vink R. Activation of cyclo-oxygenase-2 contributes to motor and cognitive dysfunction following diffuse traumatic brain injury in rats. *Clin Exp Pharmacol Physiol* 2001; 28: 922-5.

185. Vink R, O'Connor CA, Nimmo AJ, Heath DL. Magnesium attenuates persistent functional deficits following diffuse traumatic brain injury in rats. *Neurosci Lett* 2003; 336: 41-4.

186. Altura BM, Altura BT. Association of alcohol in brain injury, headaches, and stroke with brain tissue; serum levels of ionized magnesium: a review of recent findings and mechanisms of action. *Alcohol* 1999; 19: 119-30.

187. Cernak I, Savic VJ, Kotur J, Prokic V, *et al.* Characterization of plasma magnesium concentration and oxidative stress following graded traumatic brain injury in humans. *J Neurotrauma* 2000; 17: 53-68.

188. Reilly PL. Brain injury: the pathophysiology of the first hours.'Talk and Die revisited'. *J Clin Neurosci* 2001; 8: 398-403.

189. Polderman KH, Bloemers FW, Peerdeman SM, Girbes AR. Hypomagnesemia and hypophosphatemia at admission in patients with severe head injury. *Crit Care Med* 2000; 28: 2022-5.

190. Cernak I, Savic V, Kotur J, *et al.* Alterations in magnesium and oxidative status during chronic emotional stress. *Magnes Res* 2000; 13: 29-36.

191. Vink R, Nimmo AJ, Cernak I. An overview of new and novel pharmacotherapies for use in traumatic brain injury. *Clin Exp Pharmacol Physiol* 2001; 28: 919-21.

192. Muneyyirci-Delale O, Nacharaju VL, Dalloul M, Altura BM, Altura BT. Serum ionized magnesium and calcium in women after menopause: inverse relation of estrogen with ionized magnesium. *Fertil Steril* 1999; 71: 869-72.

193. Lampl Y, Geva D, Gilad R, Eshel Y, Ronen L. Cerebrospinal fluid magnesium level as a prognostic factor in ischaemic stroke. *J Neurol* 1998; 245: 584-8.

194. Corsonello A, Pedone C, Pahor M, *et al.* Gruppo Italiano di Farmacovigilanza Anziano (GIFA): Serum magnesium levels and cognitive impairment in hospitalized hypertensive patients. *Magnes Res* 2001; 14: 273-82.

195. Seelig MS. Review and hypothesis: might patients with the chronic fatigue syndrome have latent tetany of magnesium deficiency. *J Chron Fatigue Syndr* 1998; 4: 77-108.

Plasma magnesium content in patients with chronic renal failure treated with a low-protein diet

E. Planells[1], C. Sánchez[1], M. Larrubia[3], P. Aranda[1], A. Pérez de la Cruz[2], C. Asensio[2], P. Galindo[2], J. Mataix[1], J. Llopis[1]

1. Department of Physiology, School of Pharmacy and Institute of Nutrition and Food Technology, University of Granada ; 2. Virgen de las Nieves Hospital SAS; 3. I+D Sanaví, S.A., Granada, (Spain)

Keywords: chronic renal failure, hypo-protean diet, magnesium

Treatment of patients with chronic renal failure (CRF) habitually includes nutritional therapy low in protein, potassium and phosphorus [1-6]. This dietary prescription implies some practical problems, such as a limitation in the consumption of many foods, which can result in an inadequate caloric and micro-nutritional (mineral and vitamin) contribution. Hence, the abnormal metabolism of macrominerals and trace metals may be one of the factors influencing clinical disorders in pre-dialysis patients [7, 8]. Mg is an important but commonly neglected electrolyte in the clinical setting [9]. Human magnesium homeostasis mainly occurs via renal function. Kidney functions altered by chronic renal failure [10] and are found to be related to plasma magnesium [11, 12], include maintenance of the volume and ionic composition of body fluids, regulation of systemic blood pressure, and degradation of peptide hormones (PTH).

In the present study, patients with CRF included in the experimental group received a nutritional therapy consisting of a balanced hypo-protein diet adjusted to their needs based on common commercial food (G2) or prepared from manufactured low-protein dietary products (LPDP) (G3). Both groups were compared with a control group (G1). The effects of hypo-protein diet intake on plasma magnesium content in pre-dialysis patients with CRF were evaluated after 10 months.

Patients and methods

Patients and diets

Experimental groups of patients were given nutritional therapy consisting of a balanced diet adjusted to their needs (300-400 mg Mg/day) prepared from manufactured (Sanavi®) low-protein dietary products (LPDP) of similar appearance and texture to usually consumed foods (milk substitute, flours, bread, confectionery, four types of Italian pasta, biscuits, and fast desserts), offering sufficient calories in a varied diet designed to be more attractive and agreeable to the patients subject to protein restriction.

Forty-six patients with CRF were included in the study on the basis of the following criteria: blood creatinine > 3 mg/dL, creatinine clearance < 20 ml/min, and a stable clinical condition.

The patients were divided into three groups: group 1 (G1) patients (n = 16) consumed their regular diet, group 2 (G2, n = 15) received a balanced diet adjusted to their needs, and group 3 (G3, n = 15) a balanced diet consisting of LPDP.

The diets were designed on the basis of the results of a nutritional questionnaire that included 24 hour recall (recorded on three different days, one of them a nonworking day) and consumption frequency. Diets were prepared following CRF recommended intakes [13, 14]: Protein: 0.6-0.8 g/kg/day (7% Kcal); Calcium: 500-600 mg/day; Potassium: 1800-2600 mg/day; Sodium: 1-3 g/day; Phosphorus: 0.5-0.7 g/day. The Mg intake was that recommended for healthy subjects (300-400 mg/day). At five-monthly follow-up visits, blood samples were obtained and centrifuged. The plasma was used for Mg measurements.

Analytical techniques

Plasma Mg concentrations were measured by atomic absorption spectrometry (AAS) (Perkin Elmer Aanalyst 300), in wet-mineralized samples (NO_3 H/ClO $_4$H, 10: 1).
Human Serum (Certified Reference Material Seronorm™ 201405, Asker-Norway, UK) was used for quality control assays.

Statistical analyses

The data for three groups of patients were compared with Student's *t* test. All analyses were done with the SPSS software package. Differences were considered significant at the 5% probability level.

Results and discussion

At the beginning of the study, the Mg intake of the CRF patients was approximately 30% below recommended intakes for healthy people, but their plasma Mg levels were within the normal range of values (1.9 – 2.5 mg Mg/dL). Mg intakes and plasma concentrations are shown in *Table I.*
After 10 months of balanced nutritional therapy, plasma Mg concentrations reached normal values and Mg levels were similar among Groups 1, 2 and 3, although in previous studies we had observed a significant tendency for Mg to increase in patients after 6 months on an LPDP diet. The present study showed plasma Mg concentrations within the normal range in the three groups at 10 months of treatment, possibly due to the homeostatic role of intestinal Mg absorption and an improved bioavailability of LPDP.

Conclusion

Plasma Mg levels in patients with CRF who consumed the adjusted diet based on usually consumed foods or LPDP tended to remain within normal values for a long period, regardless of the type of low-protein nutritional therapy. Given the clinical disorders that alteration of Mg homeostasis may produce in CRF patients, control of the Mg status should be part of routine clinical practice.

Table I. Initial and Final (10 mo) Magnesium concentrations in plasma and Mg intake in CRF patients.

	Initial pl Mg	**Final pl Mg**	**Initial intake**	**Final intake**
GROUP 1 (G1) n = 16	2.38 ± 0.16	2.09 ± 0.170	230.3 ± 26.6	230.0 ± 20.0
GROUP 2 (G2) n = 15	2.37 ± 0.15	2.30 ± 0.097	240.3 ± 17.7	347.5 ± 20.0*°
GROUP 3 (G3) n = 15	2.33 ± 0.24	2.52 ± 0.200	244.0 ± 17.5	354.0 ± 20.0*°

Mean values ± EEM; Mg intake: *G2,3 vs G1, $p < 0.05$; G initial vs G final, $p < 0.05$. Units: Mg Plasma: mg Mg/dL; Mg intake: mg/day.

References

1. Kopple D, Berg R, Houser H, Steinman TI, Teschan P. Nutritional status of patients with different levels of CRI. *Kidney Int* 1989 ; 36(suppl. 27) : 184-94.
2. Kopple J. Nitrogen metabolism. In : Massry S, Sellers A, eds. *Clinical aspects of uremia and dialysis.* Springfield : Charles C Tomas, 1976 : 241-73.
3. Wong ET, Rude RK, Singer FR. A high prevalence of hypomagnesemia and hypermagnesemia in hospitalized patients. *Am J Clin Pathol* 1983 ; 79 : 348-52.
4. Brener B, Meyer T, Hostetter T. Dietary protein intake and the progressive nature of pathogenesis of progressive glomerular sclerosis in aging, renal ablation and intrinsic renal disease. *N Engl J Med* 1982 ; 307 : 652-9.
5. May RC, Kelly RA, Mitch WE. Pathophysiology of uremia. In : Brenner BM, Rector FC, eds. *The kidney.* Philadelphia : WB Saunders Company, 1991 : 1.997-2.108.
6. Harter HR. Review of significant findings from the National Cooperative Dialysis study and recommendations. *Kidney Int* 1983 ; 23(suppl 13) : 107-12.
7. Maschio G, Oldrizi L, Tessitore N, *et al.* Effects of dietary and phosphorus restrictions on the progression of chronic renal failure. *Kidney Int* 1982 ; 22 : 371-6.
8. Pietrzak I, Bladek K, Bulikowski W. Comparison of magnesium and zinc levels in blood in end stage renal disease patients treated by hemodialysis or peritoneal dialysis. *Magnes Res* 2002 ; 15(3-4) : 229-36.
9. Al-Ghamdi SMG, Cameron EC, Sutton RAL. Magnesium deficiency : pathophysiologic and clinical overview. *Am J Kidney Dis* 1994 ; 24 : 737-52.
10. Mitch WE, Klahr S. *Handbook of Nutrition and the Kidney.* Philadelphia, New York : Lippincott-Raven Publishers, 1998.
11. Resnick LM. Cellular ions in hypertension, insulin resistance, obesity and diabetes : a unifying theme. *J Am Soc Nephrol* 1992 ; 3(suppl) : 78-85.
12. Planells E, Llopis J, Perán F, Aranda P. Changes in tissue calcium and phosphorus content and plasma concentration of parathyroid hormone and calcitonin after long-term magnesium deficiency in rats. *J Am Coll Nutr* 1995 ; 14 : 292-8.
13. Wiggins K. *Guidelines for Nutrition Care of Renal Patients.* Edit. American Dietetic Association, 2002 ; (150).
14. Lorenzo V, De Bonis E, Torres A. La dieta en el paciente con IRC. Jano 22-27/9/1995. Vol XLIX. N° 1.136 pag. 93-98.

Comparative study about the blood levels of calcium and magnesium at two groups of elderly persons related to their nutritional intake

V. Gavăt, F.D. Petrariu, D.C. Labă, A. Albu, M. Luca
Hygiene Discipline, Faculty of Medicine, "Gr.T.Popa" University of Medicine and Pharmacy Iasi, Romania

Magnesium is a macro element required by every cell of our body. About one half our body's magnesium stores are inside of cells tissues and organs, and the other half is combined with calcium and phosphorus in bones. Only 1% of our body magnesium is found in blood. It is a difficult task to keep blood levels of magnesium constant.

Magnesium is needed for more than 300 biochemical reactions in the body. It helps to maintain normal muscle and nerve function, bones strong, keeps heart rhythm steady and it is involved in protein synthesis and in energetic metabolism.

Sources of magnesium. Green vegetables such as spinach provide magnesium because the center of the chlorophyll molecule contains this macro element. Nuts, seeds, and some whole grains are also valuable sources of magnesium [1]. Magnesium is present in many foods, but usually occurs in small quantities in: wheat germs, almonds, cereals, oats, pumpkin seeds, bran flakes, potatoes, soybeans, peanuts, peanut butter, chocolate, green beans, bananas, avocados.

Daily needs for magnesium cannot be met from a single kind of food. Eating a large variety of foods – including the formula of five servings of fruits and vegetables daily – and a plenty of whole grains, helps to ensure an adequate intake of magnesium.

The magnesium content of refined foods is usually low [1]. For example whole-wheat bread, contains twice as much magnesium as white bread, because the magnesium-rich germ and bran are removed when white flour is industrially processed.

Signs of magnesium deficiency include confusion, disorientation, loss of appetite, depression, cramps and muscle contractions, tingling, abnormal heart rhythms, coronary spasm and seizures.

When magnesium deficiency does occur, it is usually due to excessive loss of magnesium in urine, gastrointestinal system troubles that cause a loss of magnesium and limit magnesium absorption, or due a chronically low intake of magnesium.

The loss of magnesium through diarrhea and fat malabsorption usually occurs after intestinal surgery or infection, but it can occur with chronic malabsorptive problems such as Crohn's disease, gluten sensitive enteropathy, and regional enteritis.

Thiazide diuretics can increase loss of magnesium in urine, medicines that are widely used to treat cancer and antibiotics can cause a hyperexcretion of magnesium. Physicians routinely monitor magnesium levels of individuals who take these drugs and prescribe magnesium supplements if indicated [2].

Also uncontrolled diabetes increases loss of magnesium in urine and may stimulate an individual's need for magnesium. In this situation is useful to determine the need for extra magnesium. Routine supplementation with magnesium is not recommended for individuals with well-controlled diabetes [3].

A person who drinks a lot of alcohol is exposed at high risk for magnesium deficiency, because alcohol stimulates urinary excretion of magnesium. Low blood levels of magnesium occur in 30-60% of alcoholics, and in almost 90% of patients experiencing alcohol withdrawal. Alcoholics, who substitute alcohol for food, will usually have lower magnesium intakes [4].
Individuals with chronically low blood levels of potassium and calcium may have an underlying problem with magnesium deficiency. Including magnesium supplements in their diets may make potassium and calcium supplementation more effective for them [5].

Objective

To establish the correlation between nutritional intake of magnesium and calcium and the blood levels of these macro elements at elderly people.

Material and method

We have followed the daily nutritional intake of magnesium and calcium for each person, using the nutritional inquiry method, for the food provided at two closed groups (A and B) of elderly persons, with a different habitational and nutritional status. In the same time we have analyzed the blood levels of magnesium and calcium in these two groups. The first group was composed by 112 persons (72 women and 40 men) and the second group consisted in 100 persons (55 women and 45 men). Magnesium Recommended Dietary Allowances varies for males from 420-350 mg/day and for females from 320-280 mg/day [6]. Calcium Recommended Dietary Allowance is for males 800 mg/day and for females varies from 800-1,000mg/day [7]. We've performed four nutritional inquiries in both locations (2 in spring and 2 in autumn). If we identify the lower nutritional intake of calcium and magnesium as risk factors, we can assess the correlations between the exposure and the biological effects. The biochemical analyses were performed by colorimetric method, on a Blood Chemistry Analyzer HITACHI 705®.

Results and discussions

Group analysis. Our study groups were identified in two Elderly Care Centers from Iasi. The research period was settled between 2002 and 2003.
The first group (A) is larger and it has an F/M ratio of 1.8, comparative with the second one (B), which is 1.12 times smaller, and it has an F/M ratio of 1.22.
The distribution by age groups resembles in both locations, the main extended age group is between 65 and 85 years (in group A represents 73%, and in group B 74%) (*Figure 1*).
These elderly persons receive different food alimony, in the second group this value is with 12% higher than in the first group.
Spring nutritional inquiries showed in the first location (A) an average calcium intake/person of 415 mg and in the second location (B) an average calcium intake 1,083 mg.
The same inquiries showed for magnesium in the first location (A) an average magnesium intake/person of 263 mg and in the second location (B) an average magnesium intake of 317 mg.
Autumn nutritional inquiries showed in the first location (A) an average calcium intake/person of 671 mg and in the second location (B) an average calcium intake 662 mg.
The same inquiries showed for magnesium in the first location (A) an average magnesium intake/person of 351 mg and in the second location (B) an average magnesium intake of 312 mg.
Analyzing the nutrition of these two groups, we can see that the nutritional intake for macro elements varies from spring to autumn, especially for calcium and less for magnesium.
Magnesium average intake/person/day was in the first location 307 mg and in the second location 314.5 mg. Men are disadvantaged by these intakes because their magnesium RDA is 1.31 times higher than women's RDA (*Figure 2*).

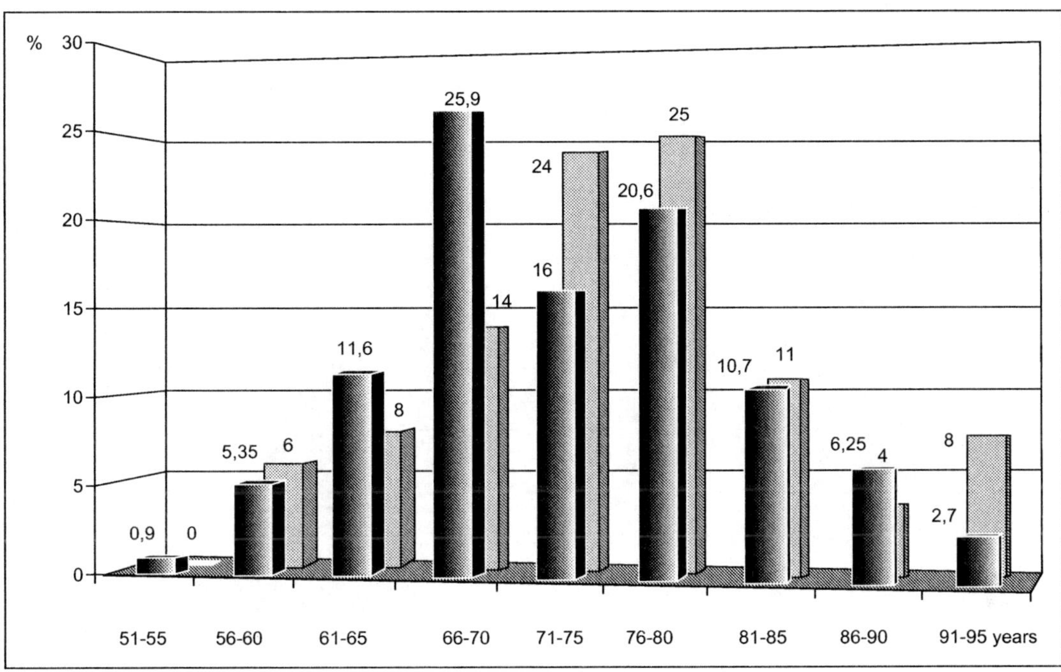

Figure 1. Distribution by age groups of our patients.

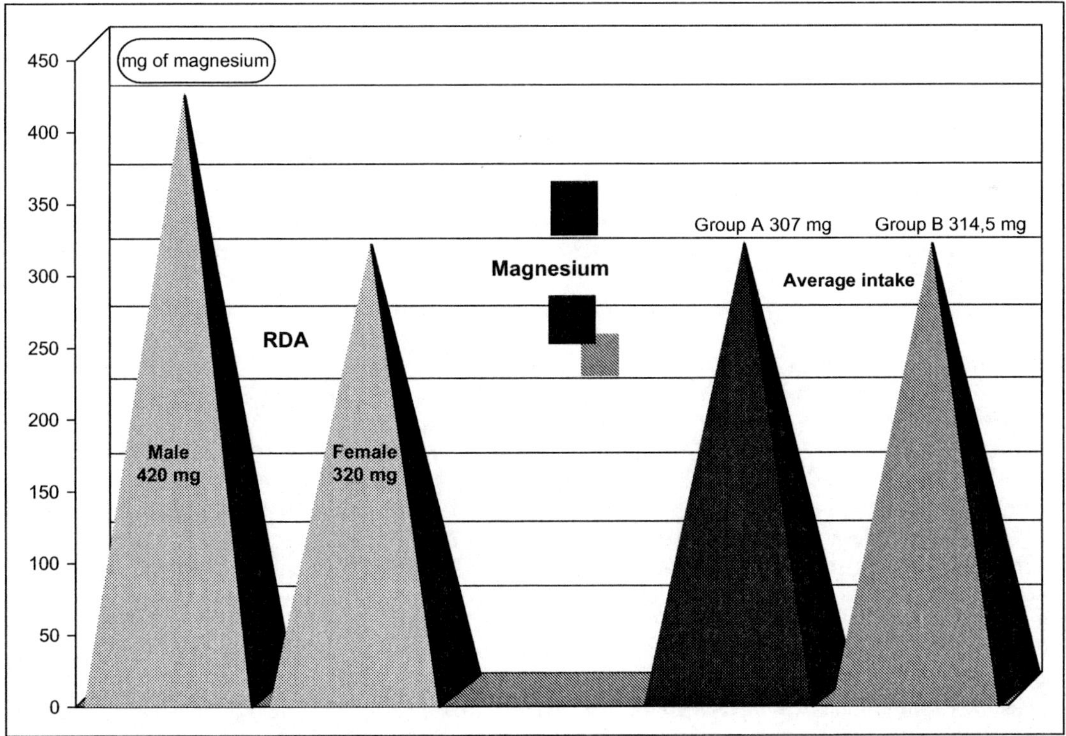

Figure 2. RDA and magnesium average intake.

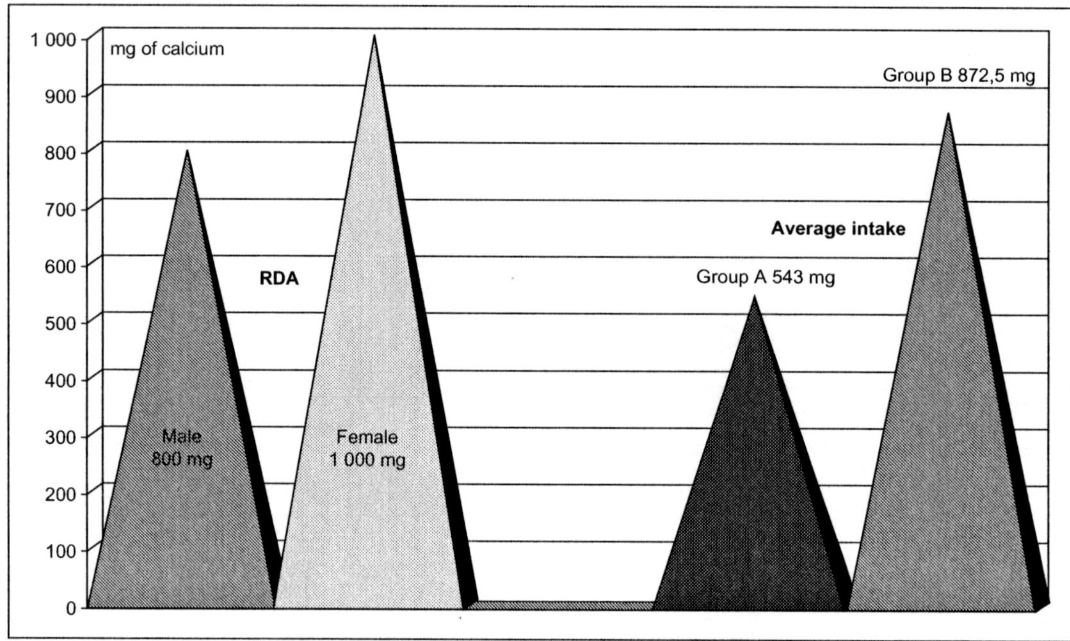

Figure 3. RDA and average calcium intake.

Calcium average intake/person/day was in the first location 543 mg and in the second location 872.5 mg. In this case women are disadvantaged by these intakes because their magnesium RDA is 1.25 times higher than men's RDA (*Figure 3*).

Drinking water can provide magnesium and calcium, but the amount varies according to the water supply. Regular dietary surveys don't assess magnesium intake from water, which may generate an underestimating total magnesium intake [8]. In our study, the first ECC is supplied from *Prut River* source and the second one from *Timisesti* source. Every elderly person drinks approximately 1.5 liter of water/day from any source, so for group A the average intake of calcium is 99 mg/day and 28.5 mg of magnesium and for group B the average intake of calcium is 60 mg/day and 33 mg of magnesium (*Figure 4*).

Summarizing, the total daily intake of calcium is for the first group 642 mg and for the second 932.5 mg and of magnesium 335.5 mg in-group A, respective 347.5 mg in-group B.

The second part of our analysis focused on the biochemical status of all the assisted persons, as a possible consequence of their nutrition. We've performed, with the courtesy of a clinical laboratory, the blood chemistry tests and the results are presented in the next graphs.

For magnesium blood levels we've registered values below 1.9 mg% in 32.1% from all the cases (group A) and in 30% of all the cases from the second group (B) (*Figure 5*).

For calcium blood levels we've encountered in-group A 86.6% of the values below 9 mg% and in group B 70% of the values lower than the same limit (*Figure 6*).

After we have obtained these results, the significance test χ^2 was applied and it shows that (at $p = 0.01$) there are significant differences for calcium nutritional intake and the frequency of normal calcemia between analyzed groups. For magnesium we haven't established a correlation between an average intake (which has almost the same value in both locations) and the number of affected persons (in this case with hypomagnesemia).

A possible explanation is taking in account that magnesium is present in many foods, but unfortunately in small amounts. As with most nutrients, daily needs for magnesium cannot be met from a single food group.

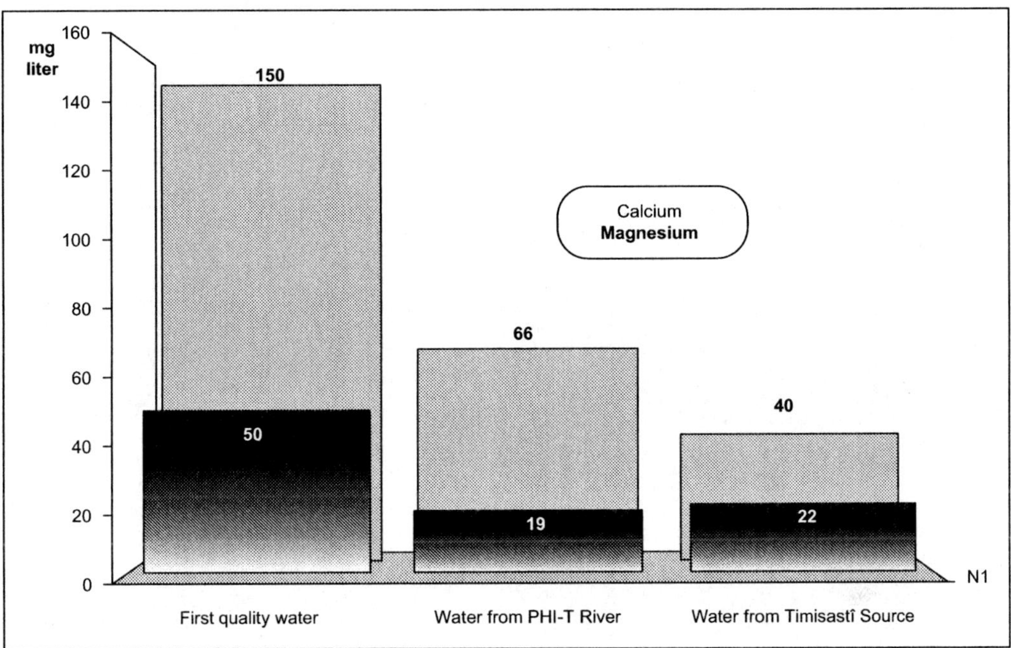

Figure 4. Assessment of water quality.

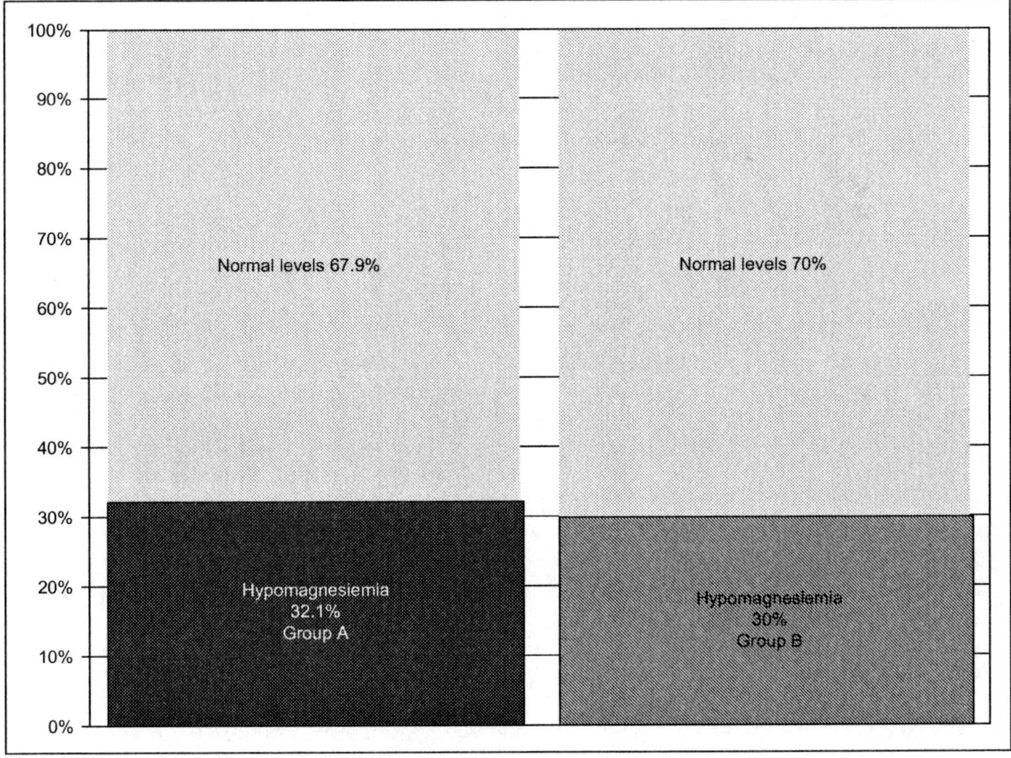

Figure 5. Distributions of mgnesium blood levels in the study groups.

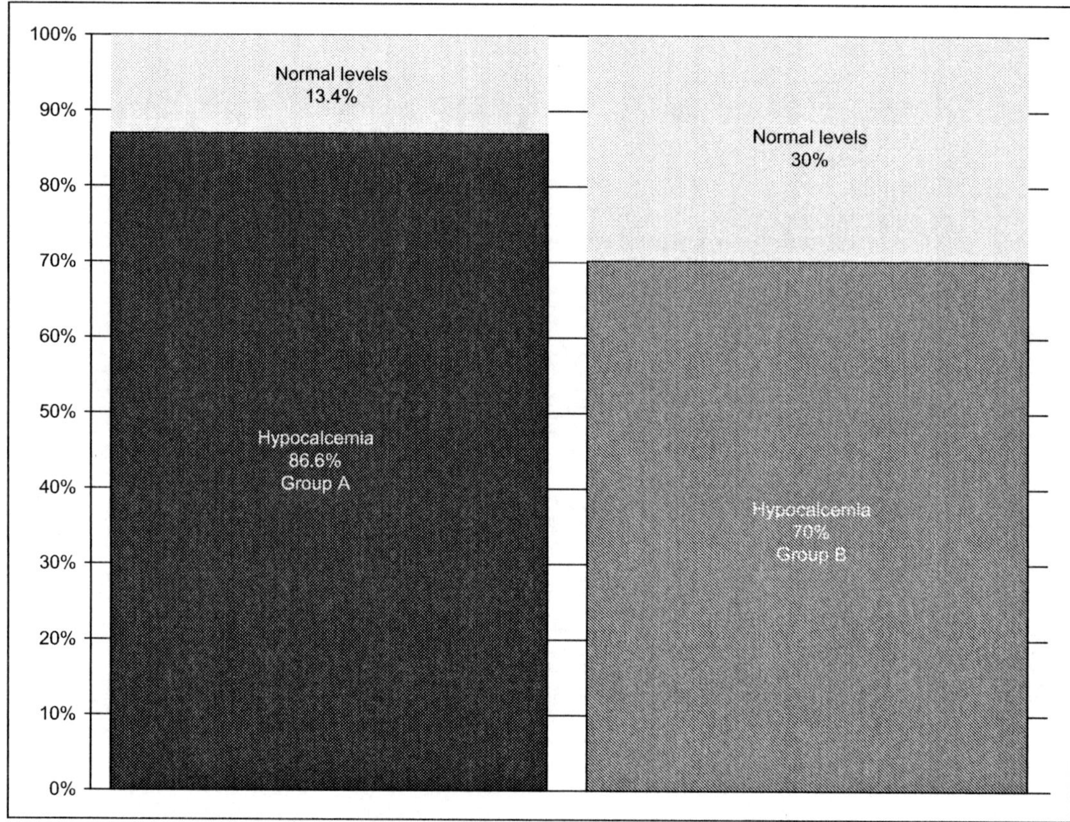

Figure 6. Distributions of calcium blood levels in the study groups.

Conclusions

– The study groups consisted mainly (3/4) by elderly persons with ages between 65-85 years.

– The total average intake of calcium/person was of 642 mg in the first group and of 932.5 mg in the second group.

– The total average intake of magnesium/person was of 335.5 mg in the first group and of 347.5 mg in the second group.

– We have established a positive correlation between the nutritional intake of calcium and the number of persons with normal calcemia ($p = 0.01$)

– For magnesium the nutritional intake was similar in both study groups (magnesium is frequent encountered in food groups) so the blood levels of magnesium also resembled.

– As a preventive measure, we can decrease the cardiovascular disease incidence in these two elderly groups, by supplementing the magnesium and calcium intake from bottled mineral water because the tap water is not hard enough.

– Adjusting the nutritional intake for these macro minerals should be focused on consuming natural food like: green vegetables (spinach) nuts, seeds, and whole grains and not on refined foods.

References

1. Institute of Medicine. *Food and Nutrition Board. Dietary Reference Intakes: calcium, Phosphorus, Magnesium, Vitamin D, and Fluoride.* Washington DC: National Academy Press, 1999.

2. Kelepouris E, Agus ZS. *Hypomagnesemia: renal magnesium handling.* Semin Nephrol, 1998.

3. Tosiello L. *Hypomagnesemia and diabetes mellitus. A review of clinical implications.* Arch Intern Med, 1996.

4. Elisaf M, Bairaktari E, Kalaitzidis R, Siamopoulos K. Hypomagnesemia in alcoholic patients. *Alcohol Clin Exp Res* 1998: 22.

5. Elisaf M, Milionis H, Siamopoulos K. *Hypomagnesemic hypokalemia and hypocalcemia: Clinical and laboratory characteristics.* Mineral Electrolyte Metabolism, 1997.

6. Whitney Noss E, Rolfes Rady S. *Understanding Nutrition.* USA: West Publishing Company, 1996.

7. Florea M. *Explorări clinice şi morfofuncţionale în medicină. Cluj-Napoca: Editura Tipomur.* 1993.

8. Berca C. *Apa şi sănătatea.* Bucureşti: Editura Ceres, 1994.

Importance of magnesium dose in the treatment of hypertension

A. Rosanoff
13-1255 Malama St., Pahoa, HI 96778 USA

Intervention studies with oral magnesium therapy for hypertensive patients have led to conflicting results: Several have shown a significant drop in blood pressure with daily magnesium supplements [1-17] while several others have not [18-32]. Perhaps a comparison of these studies with respect to daily magnesium dose can lend some understanding to this seeming conflict.

Methods

A Medline search was conducted for clinical studies testing oral magnesium supplements' effect on blood pressure. More studies were found from these studies' references. Data from all abstracts were sorted in ascending order of daily magnesium dose and placed in a summarizing table. Full manuscripts for several of the studies were retrieved and surveyed for any circumstances of the experimental design, form of magnesium, or subject population that might impact results, and to ascertain whether antihypertensive medications were in use by subjects (treated) or not (untreated).

Results and discussion

Thirty-two human studies were found that used oral magnesium supplements of a stated daily dose and had change in blood pressure as a measured outcome. All were included in the comparison (*Tables IA, B*).

Part a - Less than or equal to 15 mmol per day Mg supplements

Half of the studies were conducted at or below a daily dose of 15 mmol magnesium. Eleven of these studies showed no significant drop in blood pressure while five did. Of the 11 studies resulting in "no change" in blood pressure, ten were on subjects untreated for hypertension. Four of the five studies showing a drop in blood pressure were on treated subjects. It is tempting to generalize that when orally supplemented with levels of magnesium at or below 15 mmol/day, hypertensive subjects treated with diuretics, ace inhibitors, beta blockers and/or calcium channel blockers appear to show a significant drop in blood pressure while untreated subjects do not. But studies #4 and #5, which do not fit this pattern, must be explained before any conclusion can be made.

Study #5, Henderson, *et al.* [20], on treated subjects showed no blood pressure change. At baseline, the placebo group's diastolic blood pressure was significantly higher than that of the magnesium test group's, so the results for this study are not statistically valid. One wonders, if the placebo group had gotten the magnesium, would this study have shown a blood pressure drop? Unfortunately, we cannot include this study in our conclusions. This leaves all four studies on treated subjects receiving 10-15 mmol oral magnesium per day showing a blood pressure drop.

Study #4, Sebekova, *et al.* [2] presents the one case where untreated hypertensives showed a drop in blood pressure at a daily dose at or below 15 mmol magnesium. These subjects were off antihypertensive medications during the study and for at least two weeks before the study began, but were possibly,

Table IA. Summary of human studies testing magnesium supplements' effect on blood pressure; arranged from smallest to largest daily magnesium dose
Part a. Magnesium supplement dose =/< 15 mmol/day.

Study[f]	Citation [Reference]	Mg dose mmol per day	Form of Mg	N[a]	Results on bp	Anti-HT Medications[b]
1	Nowson & Morgan, 1989 [18]	10	Aspartate	12/25	No change	Untreated
2[cd]	Doyle *et al.*, 1999[19]	10.3	$Mg(OH)_2$	26	No change	Untreated
3	Shafique & Misbah-ul-Ain Ashraf, 1993 [1]	10	$MgCl_2$	22/42	**Drop**	Treated
4	Sebekova *et al.*, 1992 [2]	10.5 liq.	Dichloro-aspartate-hydrochloride	12	Drop	Untreated
5[e]	Henderson *et al.*, 1986 [20]	12.5	MgO	20/40	No change	Treated
6	Michon P, 2002 [3]	13??	Slow-Mag B_6	92	**Drop**	Treated
7[c]	Sacks *et al.*, 1998 [21]	14	Lactate	50/153	No change	Untreated
8[c]	Sibai *et al.*, 1989 [22]	15	Aspartate hydrochloride	185/374	No change	Untreated Pregnancy
9[d]	Cappuccio *et al.* 1985 [23]	15	Aspartate	17	No change	Untreated
10[c]	TOPH, 1992 [24a] Yamamoto *et al.*, 1995 [24b] Whelton *et al.*, 1997 [24c]	15	Diglycine	227/461	No change	Untreated
11	Lind *et al.*, 1991 [25]	15	Lactate & citrate	71	No change	Untreated
12	Ferrara, *et al.*, 1992 [26]	15	Pidolate	n/14	No change (from placebo)	Untreated
13	Sacks *et al.*, 1995 [27]	15	Diglycine chelate (given with K or Ca)	65/96	No change (from placebo)	Untreated
14[d]	Plum-Wirell *et al.*, 1994 [28]	15	Aspartate	39	No change	Untreated
15[d]	Wirell *et al.* 1994 [4]	15	Aspartate	39	**Drop** when Mg followed placebo	Treated (beta blockers)
16	Dyckner & Wester, 1983 [5]	15	Aspartate hydrochloride	20	**Drop**	Treated

[a] N = number of subjects in Mg test group/Mg + placebo or number of subjects tested w/ Mg in crossover or non-placebo study.
[b] "Treated" = antihypertensive medications; "Untreated" = no antihypertensive medications.
[c] Study performed on normotensive population.
[d] Cross-over design: can minimize change in blood pressure.
[e] Significant difference between placebo and test groups at baseline.
[f] Low sodium diet.

even probably "treated" before that given their high baseline blood pressure mean of 162/96 and a mean duration of hypertension reported at 9.4 years. We *could* consider these subjects at least affected by treatment. If we accept this single case assumption, we can generalize that oral Mg supplements at or below 15 mmol per day will only lower blood pressure in individuals treated with antihypertensive medications. There is added evidence for this generalization in studies #14, Plum-Wirell, *et al.* [28] and #15, Wirell, *et al.* [4]: These two studies, conducted by the same research group, using the same length of study, same dose and same form of magnesium, differ only in that study #15 was on treated subjects while study #14 was on untreated subjects. They clearly show that magnesium aspartate, administered as an oral supplement at 15 mmol per day will cause a blood pressure drop in subjects treated with beta blockers but no such blood pressure drop in untreated subjects.

Table IB. Part b. Magnesium supplement dose > 15 mmol/day.

Study[f]	Citation [reference]	Mg dose mmol per day	Form of Mg	N[a]	Results on bp	Andi-HT Medications[b]
17[d]	Widman *et al.*, 1993 [6]	15	$Mg(OH)_2$	17	No change	Untreated
		30			Drop (n.s.)	
		40			Drop	
18	Reyes *et al.*, 1984 [7]	15.8	$MgCl_2$	13/21	Drop SBP	Treated
19[f]	Olhaberry *et al.*, 1987 [29]	15.8	$MgCl_2$	7/14	No change	Untreated
20[d]	Purvis *et al.*, 1994 [8]	16	$MgCl_2$	28	Drop SBP	Unknown
21	Cohen *et al.*, 1984 [30]	18.75	MgO	8	No change	Untreated
22	Itoh *et al.*, 1997 [9]	17 – 24	$Mg(OH)_2$	23/33	Drop	Some treated
23[d]	Kawano *et al.*, 1998 [10]	20	MgO	60	Drop	Most treated
24	Witteman *et al.*, 1994 [11]	20	Aspartate HCl	47/91	Drop	Untreated
25[d]	Patki *et al.*, 1990 [12]	20	$MgCl_2$ (with K)	37	Drop	Untreated
26	Daly *et al.*, 1990 [13]	20.6	MgO	20/40	Drop	Untreated
27[c]	Kisters *et al.*, 1993 [31]	20.8	Mg hydrogen aspartate	37/69	No change	Untreated
28[d]	Sanjuliani *et al.*, 1996 [14]	25	MgO	15	Drop	Untreated??
29[e]	Haga, 1992 [15]	25	MgO	17/25	Drop	Untreated HT
		25	MgO	8/25	No change	Normotensive
30	Motoyama *et al.*, 1989 [16]	25	MgO	21	Drop	Untreated??
31[d]	Saito *et al.*, 1988 [17a] Hattori *et al.*, 1988 [17b]	25	MgO	20	Drop	Treated
32[g,e]	Zemel *et al.*, 1990 [32]	40	Aspartate	7/13	No Change	Untreated

[a] N = number of subjects in Mg test group/Mg + placebo or number of subjects tested w/ Mg in crossover or non-placebo study.
[b] "Treated" = antihypertensive medications; "Untreated" = no antihypertensive medications.
[c] Study performed on normotensive population.
[d] Cross-over design: can minimize change in blood pressure.
[e] Significant difference between placebo and test groups at baseline.
[f] Low sodium diet.
[g] Presumed Mg repleted.

Part b - Greater than 15 mmol per day Mg supplements

Of these 16 studies, most (75%) showed a drop in blood pressure. Before we conclude that magnesium supplements greater than 15 mmol per day do lower a high blood pressure, whether or not subjects are treated with antihypertensive medications, we must look closely at the four studies above 15 mmol magnesium per day that showed no drop in blood pressure.

Study #19, Olhaberry, *et al.* [29]. In this study, subjects were placed on a low sodium diet along with a placebo or magnesium supplement just above 15 mmol (15.8) per day. Morris, *et al.* [33] reported that a low sodium diet usually restricts foods high in Ca, Fe, Mg and vitamin B_6. It is safe to assume that these subjects lowered their dietary Mg intake during the study, probably enough to make their magnesium supplement more like a 15 mmol daily dose or below, a dose we know cannot lower a high blood pressure. There's direct evidence for this assumption in the paper: the placebo group (also on the low sodium diet) showed a significant drop in serum Mg by the end of the study.

Study #21, Cohen, *et al.* [30]. In this small study (n = 8), subjects were generally younger than in the other studies and were selected for grade II vasoconstriction in the retinal artery. The study found that retinal vasoconstriction and hypomagnesemia were corrected with magnesium supplements. But this magnesium dose of 18.75 mmol per day as MgO did not correct the high blood pressure of these subjects. The magnesium supplement was given in 3 packets per day. If in the three month study there was 80% compliance, which is pretty good, the magnesium dose would be only 15 mmol per day, a dose that we know will not lower high blood pressure in untreated subjects.

Study #27, Kisters *et al.* [31]. In this study, magnesium supplements just above 20 mmol per day did not lower blood pressure. Why not? This was primarily a cholesterol study that measured change in blood pressure. Subjects were all hyperlipidemic, but their blood pressure mean at baseline was 130/83, normotensive according to the World Health Organization criteria of 140/85. Three other studies conducted on normotensive subjects, studies #7, #8 and #10, were on daily oral magnesium doses too low (14-15 mmol/day) to draw conclusions about magnesium's ability to lower blood pressure in the normal range. But, Haga, *et al.* [15] provides a direct comparison of normotensives and hypertensives at a daily magnesium dose high enough to lower high blood pressure in study #29: Magnesium supplements of 25 mmol per day as MgO lowered blood pressure in hypertensive subjects but not in normotensive subjects. Together, Haga, *et al.* [15] and Kisters, *et al.* [31] provide evidence for the conclusion that oral magnesium supplements do not lower blood pressure in normotensive subjects. Should we expect oral magnesium supplements to lower a normal blood pressure as medications do? As an essential nutrient, magnesium might be better expected to normalize high blood pressure in those with a magnesium deficit, not lower the normal-range blood pressure of magnesium-sufficient individuals.

Study #32, Zemel, *et al.* [32]. In both their article and abstract, these authors conclude that Mg supplements as high as 40 mmol per day will not lower high blood pressure in "**repleted**" hypertensive individuals. What do they mean by "repleted?" In this study, total and HDL cholesterol values were determined at baseline and throughout the study. At baseline, the total cholesterol of the placebo group was significantly higher than that of the Mg test group, but this significance was not found throughout the study. However, the placebo group had much higher mean total cholesterol values than the Mg test group throughout the study. Compared with the values deemed safe by the American Heart Assn Standard for total cholesterol, the placebo group's values were either "too high" or 16% above "desirable" levels while the Mg test group's values were at the "desirable" level or very near (within 5%) throughout the study. The authors did not calculate the Total to HDL cholesterol ratios, but the data are available. When compared with the American Heart Assn Standard for Cholesterol: HDL Ratio, this ratio for the Mg test group was "optimal" or within 5% of optimal throughout the study while the placebo group's ratios remained above the "safe" cut-off value except for one reading which was 7% above "optimal". Aware that magnesium deficit can impact cholesterol values, these authors concluded that the differences found between placebo and test groups signified that the Mg test group was "Mg repleted" at baseline. We learn from this study that magnesium-replete individuals can have high blood pressure, and that giving such individuals magnesium supplements, even at high doses, will not correct or normalize the blood pressure. In such cases, something other than magnesium deficit is causing the high blood pressure.

Titrating and cross-over design. Study #17, Widman, *et al.* [6]

Widman and co-workers give us the only study on multiple doses of magnesium supplement. This is closer to the clinical studies for antihypertensive medications which, to minimize side-effects, often try several doses of a drug to ascertain the lowest possible effective dose. They begin subjects on one dose and if that doesn't lower the blood pressure, they raise the dose, sometimes more than once. It is called "titrating". We can see from the Widman, *et al.* study that daily magnesium dose is important in hypertension therapy: 15 mmol was ineffective in lowering high blood pressure, as the other studies have already demonstrated, but 30 mmol showed an insignificant drop, and 40 mmol a significant drop in blood pressure. We might wonder if the insignificant drop at 30 mmol magnesium negates the other studies showing a significant drop in blood pressure at doses of 20 + mmol/day magnesium. This study was a crossover design which can falsely minimize any blood pressure drop in studies of essential nutrients. Perhaps the Widman, et al insignificant drop on 30 mmol Mg with the crossover design might have been a significant drop with a parallel design.

The crossover design points to some basic differences between essential nutrients and pharmaceutical medications. Medications act on specific functions and then wash out of the body, some slowly, some rapidly. One has to keep taking the medication to keep the action working. A drug can, presumably, be fully washed out of a body. Sometimes a larger dose of a particular medication is required to maintain the desired effect. Essential nutrients are different. When a subject with a nutritional deficiency takes a nutritional supplement, they will first correct any deficiency that exists and then build up a body store of that nutrient. A lower intake is then needed to maintain the sufficiency. If that nutrient's supply is then cut off, as in a placebo period, the body will conserve that store by minimizing the nutrient's loss from the body. Ideally, no one ever completely washes out any essential nutrient, for this would mean sickness or even death. These basic differences between medications and essential nutrients can present problems with the crossover design. When testing antihypertensive medications, the crossover design makes total sense. When testing magnesium, an essential nutrient, for its effect on high blood pressure, the crossover design can falsely minimize a drop in blood pressure. Subjects given magnesium first, followed by the placebo after the crossover, may get enough magnesium during the first half of the study to correct a borderline deficiency and even build up a small store. Even after a wash-out period, these subjects cannot be considered the "same" as at baseline with regards to magnesium status. Statistically, these subjects do not fulfil the "independence" requirement of many statistical tests. When these people go on placebo, such stores, however small, may see them through part or all of the zero magnesium supplement period, making their placebo blood pressure value artificially low. Such people "tilt" the results in blood pressure studies to minimize the effect of magnesium on blood pressure. This is what could be at work in study #15 where only those subjects given magnesium AFTER the placebo period showed a significant drop in blood pressure.
Magnesium supplements at or under 15 mmol per day may not be enough to normalize a high blood pressure, but they have been shown to correct other aspects of magnesium deficits (*e.g.* See Doyle, *et al.* [19], and Cohen, *et al.* [30]). Subjects given magnesium after a placebo period, without a crossover, may be a better test of supplemental magnesium's effect on hypertension in studies of this duration, *i.e.* weeks to months.

Conclusion

These studies show that oral magnesium supplements of 25 mmol/day do not lower blood pressure in normotensive subjects, that supplements at or below 15 mmol per day will lower a high blood pressure only in hypertensive subjects treated with antihypertensive medications, while supplements above 20 mmol per day will lower a high blood pressure in subjects, treated or untreated, that are not magnesium repleted.

References

1. Shafique M, Misbah-ul-Ain Ashraf M. Role of magnesium in the management of hypertension. *J Pak Med Assoc* 1993 ; 43 : 77-8.

2. Sebekova K, Revusova V, Polakovicova D, Drahosova J, Zverkova D, Dzurik R. Anti-hypertensive treatment with magnesium-aspartate-dichloride and its influence on peripheral serotonin metabolism in man : a subacute study. *Cor Vasa* 1992 ; 34 : 390-401.

3. Michon P. Poziom frakcji calkowitej i zjonizaowanej. magnezu na podstawie analizy biochemicznej krwi i wlosow oraz wplyw suplementacji magnezem (Slow Mag B6) na wybrane parametry w chorobie nadcisnieniowej u pacjentow leczonych roznymi grupami lekow. *Ann Acad Med Stetin* 2002 ; 48 : 85-97.

4. Wirell MP, Wester PO, Stegmayr BG. Nutritional dose of magnesium in hypertensive patients on beta blockers lowers systolic blood pressure : a double-blind, cross-over study. *J Intern Med* 1994 ; 236 : 189-95.

5. Dyckner T, Wester PO. Effect of magnesium on blood pressure. *Br Med J (Clin Res Ed)* 1983 ; 286 : 1847-9.

6. Widman L, Wester PO, Stegmayr BK, Wirell M. The dose-dependent reduction in blood pressure through administration of magnesium. A double blind placebo controlled cross-over study. *Am J Hypertens* 1993 ; 6 : 41-5.

7. Reyes AJ, Leary WP, Acosta-Barrios TN, Davis WH. Magnesium supplementation in hypertension treated with hydrochlorothiazide. *Curr Therap Res* 1984 ; 36 : 332-40.

8. Purvis JR, Cummings DM, Landsman P, *et al.* Effect of oral magnesium supplementation on selected cardiovascular risk factors in non-insulin-dependent diabetics. *Arch Fam Med* 1994 ; 3 : 503-8.

9. Itoh K, Kawasaka T, Nakamura M. The effects of high oral magnesium supplementation on blood pressure, serum lipids and related variables in apparently healthy Japanese subjects. *Br J Nutr* 1997 ; 78 : 737-50.

10. Kawano Y, Matsuoka H, Takishita S, Omae T. Effects of magnesium supplementation in hypertensive patients : assessment by office, home, and ambulatory blood pressures. *Hypertension* 1998 ; 32 : 260-5.

11. Witteman JC, Grobbee DE, Derkx FH, Bouillon R, de Bruijn AM, Hofman A. Reduction of blood pressure with oral magnesium supplementation in women with mild to moderate hypertension. *Am J Clin Nutr* 1994 ; 60 : 129-35.

12. Patki PS, Singh J, Gokhale SV, Bulakh PM, Shrotri DS, Patwardhan B. Efficacy of potassium and magnesium in essential hypertension : a double-blind, placebo controlled, crossover study. *BMJ* 1990 ; 301 : 521-3.

13. Daly NM, Allen KGD, Harris M. Magnesium Supplementation and Blood Pressure in Borderline Hypertensive Subjects : A Double Blind Study. *Mag-Bul* 1990 ; 12 : 149-54.

14. Sanjuliani AF, de Abreu Fagundes VG, Francischetti EA. Effects of magnesium on blood pressure and intracellular ion levels of Brazilian hypertensive patients. *Int J Cardiol* 1996 ; 56 : 177-83.

15. Haga H. Effects of dietary magnesium supplementation on diurnal variations of blood pressure and plasma Na+, K(+)-ATPase activity in essential hypertension. *Jpn Heart J* 1992 ; 33 : 785-800.

16. Motoyama T, Sano H, Fukuzaki H. Oral magnesium supplementation in patients with essential hypertension. *Hypertension* 1989 ; 13 : 227-32.

17. (a) Saito K, Hattori K, Omatsu T, Hirouchi H, Sano H, Fukuzaki H. Effects of oral magnesium on blood pressure and red cell sodium transport in patients receiving long-term thiazide diuretics for hypertension. *Am. J. Hypertens.* 1988 ;1 :71S–4S ; (b) Hattori K, Saito K, Sano H, Fukuzaki H. Intracellular magnesium deficiency and effect of oral magnesium on blood pressure and red cell sodium transport in diuretic-treated hypertensive patients. *Jpn. Circ. J.* 1988 ;52 :1249–56.

18. Nowson CA, Morgan TO. Magnesium supplementation in mild hypertensive patients on a moderately low sodium diet. *Clin Exp Pharmacol Physiol* 1989 ; 16 : 299-302.

19. Doyle L, Flynn A, Cashman K. The effect of magnesium supplementation on biochemical markers of bone metabolism or blood pressure in healthy young adult females. *Eur J Clin Nutr* 1999 ; 53 : 255-61.

20. Henderson DG, Schierup J, Schodt T. Effect of magnesium supplementation on blood pressure and electrolyte concentrations in hypertensive patients receiving long term diuretic treatment. *Br Med J (Clin Res Ed)* 1986 ; 293 : 664-5.

21. Sacks FM, Willett WC, Smith A, Brown LE, Rosner B, Moore TJ. Effect on blood pressure of potassium, calcium, and magnesium in women with low habitual intake. *Hypertension* 1998 ; 31 : 131-8.

22. Sibai BM, Villar MA, Bray E. Magnesium supplementation during pregnancy : a double-blind randomized controlled clinical trial. *Am J Obstet Gynecol* 1989 ; 161 : 115-9.

23. Cappuccio FP, Markandu ND, Beynon GW, Shore AC, Sampson B, MacGregor GA. Lack of effect of oral magnesium on high blood pressure : a double blind study. *Br Med J (Clin Res Ed)* 1985 ; 291 : 235-8.

24. (a) The effects of nonpharmacologic interventions on blood pressure of persons with high normal levels. Results of the Trials of Hypertension Prevention, Phase I. *JAMA* 1992 ;267 :1213–20 ; (b) Yamamoto ME, Applegate WB, Klag MJ, et al. Lack of blood pressure effect with calcium and magnesium supplementation in adults with high-normal blood pressure. Results from Phase I of the Trials of Hypertension Prevention (TOHP). Trials of Hypertension Prevention (TOHP) Collaborative Research Group. *Ann. Epidemiol.* 1995 ;5 :96–107 ; (c) Whelton PK, Kumanyika SK, Cook NR, et al. Efficacy of nonpharmacologic interventions in adults with high-normal blood pressure : results from phase 1 of the Trials of Hypertension Prevention. Trials of Hypertension Prevention Collaborative Research Group. *Am. J. Clin. Nutr.* 1997 ;65 :652S–60S.

25. Lind L, Lithell H, Pollare T, Ljunghall S. Blood pressure response during long-term treatment with magnesium is dependent on magnesium status. A double-blind, placebo-controlled study in essential hypertension and in subjects with high-normal blood pressure. *Am J Hypertens* 1991 ; 4 : 674-9.

26. Ferrara LA, Iannuzzi R, Castaldo A, Iannuzzi A, Dello Russo A, Mancini M. Long-term magnesium supplementation in essential hypertension. *Cardiology* 1992 ; 81 : 25-33.

27. Sacks FM, Brown LE, Appel L, Borhani NO, Evans D, Whelton P. Combinations of potassium, calcium, and magnesium supplements in hypertension. *Hypertension* 1995 ; 26 : 950-6.

28. Plum-Wirell M, Stegmayr BG, Wester PO. Nutritional magnesium supplementation does not change blood pressure nor serum or muscle potassium and magnesium in untreated hypertension. A double-blind crossover study. *Magnes Res* 1994 ; 7 : 277-83.

29. Olhaberry JV, Reyes AJ, Acosta-Barrios TN, Leary WP, Queiruga G. Pilot evaluation of the putative antihypertensive effect of magnesium. *Mag-Bul* 1987 ; 9 : 181-4.

30. Cohen L, Laor A, Kitzes R. Reversible retinal vasospasm in magnesium-treated hypertension despite no significant change in blood pressure. *Magnesium* 1984 ; 3 : 159-63.

31. Kisters K, Spieker C, Tepel M, Zidek W. New data about the effects of oral physiological magnesium supplementation on several cardiovascular risk factors (lipids and blood pressure). *Magnes Res* 1993 ; 6 : 355-60.

32. Zemel PC, Zemel MB, Urberg M, Douglas FL, Geiser R, Sowers JR. Metabolic and hemodynamic effects of magnesium supplementation in patients with essential hypertension. *Am J Clin Nutr* 1990 ; 51 : 665-9.

33. Morris CD. Effect of dietary sodium restriction on overall nutrient intake. *Am J Clin Nutr* 1997 ; 65 : 687S-691S.

V. Magnesium and pediatrics

The 30-years experience of the Cluj-Napoca 2nd Pediatrics Clinic concerning the children's deficiency of magnesium

N. Miu, M. Mărgescu, L. Slăvescu, M. Andreica, M. Marc, G. Sur, A. Bizo, D. Şerban, S. Fritea, C. Mărgescu, C. Blag, C. Aldea, D. Deleanu, T.L. Pop

University of Medicine and Pharmacy Cluj-Napoca, 2nd Pediatrics Clinic, Romania

Importance of magnesium in physiology

Magnesium represents the second intracellular cation with major importance in body's physiology. It is needed for anatomical and functional integrity of different cellular **organites** and it is involved in all important metabolic pathways: glucidic, proteic and lipidic metabolism. Magnesium has the role of biologic catalyzator of energy producing systems that is needed for cell reactions. Magnesium is involved in ionic changes and keeps potassium inside cells.

Needed for cell integrity and present through the body, magnesium is involved in body's growing and development from intrauterine life and then in neuromuscular physiology, osseous system, cardiovascular system and blood cells, in renal, digestive, hepatic, endocrine physiology or in sensitive system.

Magnesium plays a role in different defense processes: antistress, homeostatic of temperature, antihypoxic, regulatory of acid-base balance, antianaphylaxis, regulatory of oxidative phosphorilation, anti-inflammatory system, antitoxins and also in unspecific or specific immunity [1-5].

This work aims to demonstrate the complex and important physiologic actions of magnesium by emphasizing the clinical and biological manifestations of magnesium deficiency present in pediatric activity and the corrections of this manifestations after treatment with magnesium.

Neurologic and muscular manifestations of magnesium deficiency

We have studied a number of 1500 patients admitted in our clinic over 30 years that presented different forms of neuromuscular hyperexcitability due to magnesium deficiency. In these patients were tested the levels of serum and erythrocyte magnesium, serum calcium, serum phosphate and selectively electrocardiograms, electroencephalograms and electromyograms were performed.

Magnesium deficiency was grouped in primary deficiency (658 patients) and secondary deficiency (842 patients) (*Figure 1*).

As primary deficiency we encountered the following entities:

1. Congenital chronic hypomagnesemia, with familial characteristic in 530 patients. Neonatal onset was present in 34 patients, the rest of them was with disease onset during infancy.
2. Transitory idiopathic neonatal hypomagnesemia in 126 patients. These patients are different that the first group by the rapid disappearance of manifestations after magnesium treatment.
3. Primary selective magnesium **malabsorbtion** in 2 patients. This is an autosomal recessive disease and parenteral magnesium in repeated administration is needed.

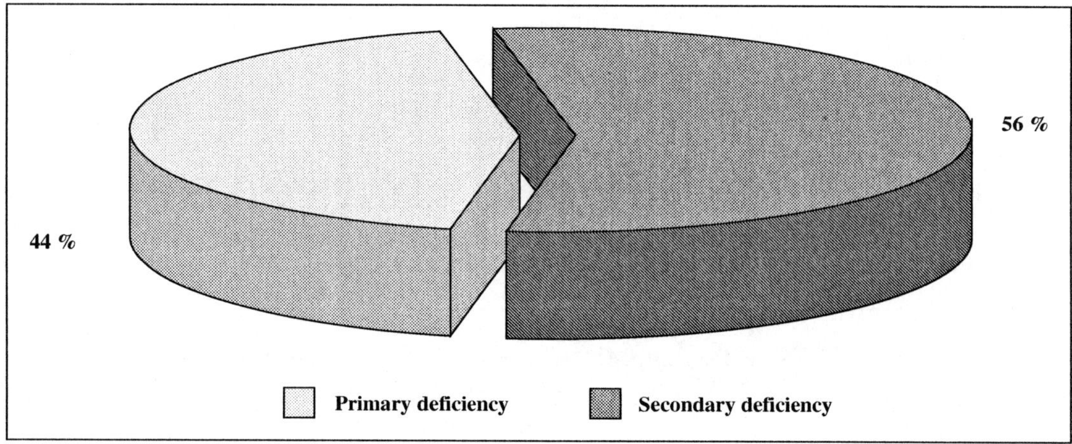

Figure 1. Primary and secondary deficiency.

Primary deficiency of magnesium was diagnosed in 658 patients, the mothers of 149 patients presented clinical and biological symptoms that suggested hypoparathyroidism. In these cases neuromuscular hyperexcitability was rapidly manifested, sometimes by neonatal convulsions:

- Congenital hypoparathyroidism with confirmed familial type: 52 cases,
- Neonatal hyperparathyroidism with mother's hyperparathyroidism: 7 cases,
- DiGeorge syndrome (thymus and parathyroids aplasia): 3 cases,
- Prematurity: 59 cases,
- Dysmaturity: 34 cases,
- Toxemia, eclampsia or pre-eclampsia: 92 mother with neonates with hypomagnesaemia.

Secondary deficiency of magnesium had the following causes:

1. Alimentary intake deficiency, malabsorption (chronic diarrhea with malabsorption syndrome and malnutrition): 515 cases
2. Endocrine, metabolic or nervous regulation disorders after neonatal period (211 cases): hypoparathyroidism, pseudohypoparathyroidism, hyperthyroidism, hyperaldosteronism, Schwartz-Bartter syndrome, juvenile diabetes mellitus, stress;
3. Urinary hypersecretion of magnesium: some nephrotic syndromes, tubular acidosis, chronic renal failure during **poliuric** phase, prolonged treatment with Furosemide and Prednisone, overdose of vitamin D: 116 cases.

Neuromuscular manifestation of magnesium deficiency

1. Neonates (160 cases): convulsions, extreme agitation, intestinal spasm with/without vomiting, sleep disturbances, spasm of larynx with stridor;
2. Infant (538 cases): recurrent convulsions, successfully treated with magnesium, agitation without motivation, irritability, sleep disturbances with sleep-awake rhythm changes, muscle fibrillations, intestinal spasms (colic) treated with magnesium, crying spasm, laryngospasm;
3. Children (802 cases): positive Chvostek sign, fasciculations, fibrillations, muscular cramps, tetany, biliary dysfunctions, eso-gastro-intestinal dysfunctions, bronchi's or cardiovascular dysfunctions, tremor, asthenia, irritability, insomnia, headaches (sometimes with migraine characteristic), lower capacity of mental concentration, attention or memory, anxious hyperemotivity, syncope, vertigo, enuresis.

Attention Deficit Hyperactivity Disorder (ADHD) Syndrome

Neuromuscular hyperexcitability syndrome had a latent evolution with periods of unbalance due to precipitating factors (stress, infections) [6-8]. We could not demonstrate a direct link between serum or erythrocyte level of magnesium and the type of clinical manifestations, but we observed in cases with magnesium level under 1 mEq/l (in serum) or under 2.5 mEq/l (in erythrocyte) clinical manifestations with severity related with the importance of deficiency.

In 34 patients the epilepsy was also diagnosed (neurologic exam, electroencephalogram and **electromiograms** were performed in order to differentiate the tetany and epilepsy). In these patients the association of therapy with magnesium had improved the neurologic manifestations in most patients (73%).

Biochemical changes encountered in these patients were:

1. Isolated hypomagnesemia: 608 cases:
   a. Low level of serum magnesium and low level of erythrocyte magnesium: 498 cases
   b. Normal level of serum magnesium and low level of erythrocyte magnesium: 106 cases
2. Hypomagnesemia and hyperphosphatemia: 591 cases
3. Hypomagnesemia, hyperphosphatemia and hypocalcemia: 125 cases
4. Hypomagnesemia and hypocalcemia: 120 cases

Therapy with magnesium was administered as attack dose of 10 mg/kg/day, for 3-4 weeks, then 5 mg/kg/day for a certain period of time. Vitamin B6, vitamin D, vitamin E and orotic acid were also administered as magnesium fixating agents. Magne B6 and Magnerot were the drugs used in these patients. Neuromuscular hyperexcitability manifestations had disappeared or clearly improved in most cases as effect of magnesium association. That fact confirms the main role of magnesium in the pathogenesis of these manifestations (*Figure 2*).

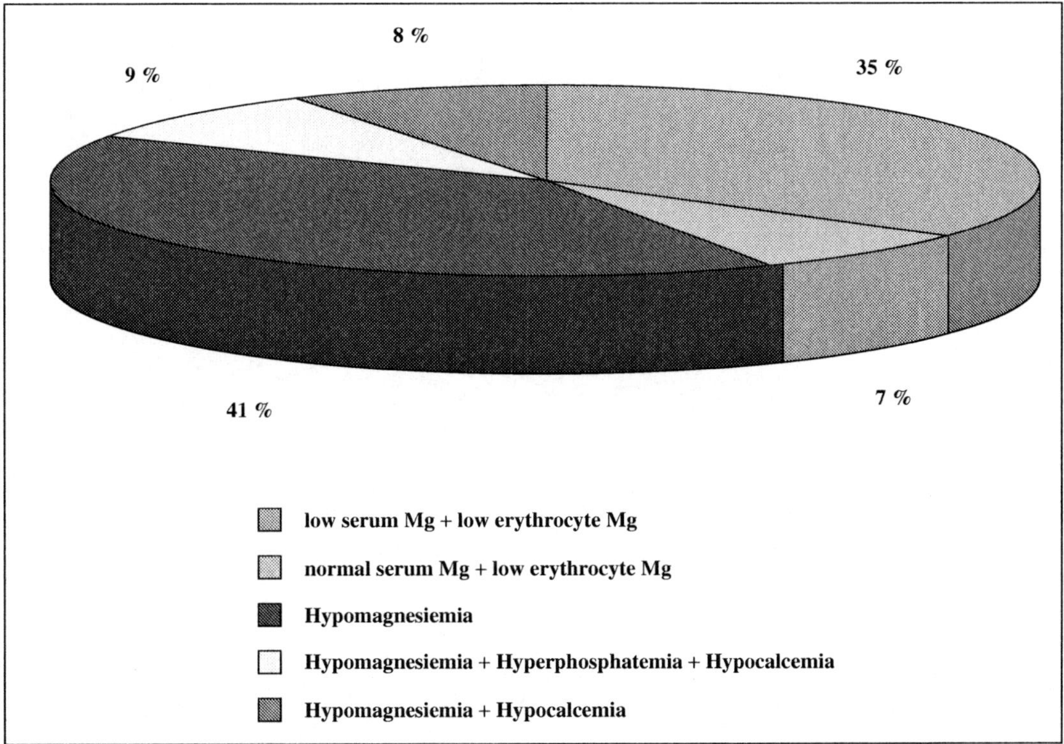

Figure 2. Biochemical changes.

Osteopenia in magnesium deficiency

A study was performed in 91 patients with age between 3 months and 2 years with clinical signs of rickets aiming to demonstrate the correlation between magnesium deficiency and the degree of disorder in the metabolism of proteins and of osseous collagen amino acids, knowing the role of magnesium in ribosomal activity. The semiquantitative chromatography of urinary amino acids performed in 24 patients showed the presence of hydroxiproline and hydroxilysine in urine of 71% of these patients. These are certified markers of osseous matrix organic degradation.

In order to verify if there is a hypercatabolic state with osseous collagen degradation in other 22 patients, the loading test with Methionine S35 intravenous was performed. In 12 infants the level of urinary elimination of methionine metabolites was increased (17-35%), demonstrating the increasing of transmethylation processes. In these cases the levels of erythrocyte and serum magnesium were decreased, a deficiency of enzyme activation was present, knowing the role of magnesium in metabolic pathway of methionine. The short-term deficiency of magnesium induce a secondary hyperparathyroidism and a long-term deficiency induce hypoparathyroidism with osseous magnesium deficiency leading to ATP-ase inactivation, with insufficient organic matrix and decreased alkaline phosphatase elimination (in rickets with magnesium deficiency *the level of alkaline phosphatase is frequently decreased, not increased).*

Association of magnesium supplements in therapy have determined the clinical, biological and radiological improvement. If there is a overdose of vitamin D hypermagnesiuria is the first biological sign [9] (*Figure 3*).

Magnesium deficiency is involved in physiologic processes of teeth development, inducing teeth dystrophies.

Cyclosporine and azathyoprine used in immunosuppression after kidney transplantation is associated with low level of magnesium in serum or urine by increased excretion in over 2/3 of transplanted patients and this fact justify magnesium supplementation in therapy. A clear correlation between the serum level of cyclosporine and urinary excretion of magnesium it was not demonstrated.

Most of diuretics drugs (excepted spironolactone, amiloride and acetazolamide) determines the appearance of hypermagnesiuria and hyperkaliemia and the use of these drugs in patients with kidney or cardiac diseases should be associated with the evaluation of magnesium and potassium levels [10].

Long-term treatment with corticosteroids has negative effects on osseous metabolism (during growing and mineralization), sometimes deficiency of magnesium is associated.

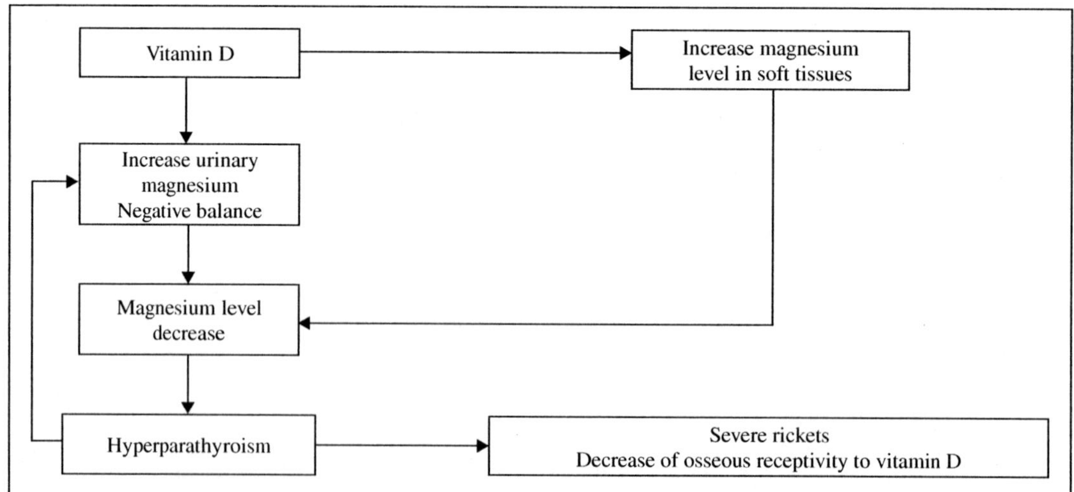

Figure 3. High doses of vitamin D in magnesium metabolism disorders.

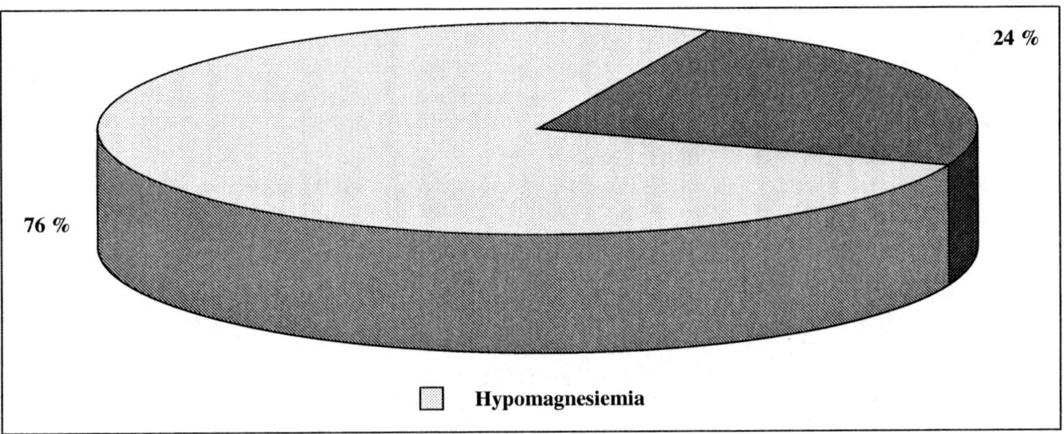

Figure 4. Hypomagnesemia in cardiac disease.

Aminoglicosides use in toxic doses could be associated with hypermagnesiuria and magnesium deficiency by lesions in proximal contorts tubes.

Magnesium deficiency in cardiovascular disease in children

Prolapse of mitral valve was diagnosed (clinical, echographic and by electrocardiogram) in 79 patients and it was the most frequent valvular disease in children and adolescents. In 60 patients (76%) hypomagnesemia was present with a hyperkinetic activity of valvular system, without the presence of anatomic changes. This fact confirmed the important role of magnesium deficiency in pathogeny of this disease in children and adolescents [11].

Clinical manifestations were pre-cordial pain, palpitations, asthenia, dizziness, dyspnea associated with other manifestations of magnesium deficiency, including Raynaud syndrome. Another proof of magnesium deficiency importance in the prolapse of mitral valve pathogeny was the disappearance of clinical manifestations, of ultrasound or EKG changes after magnesium therapy associated with orotic acid or vitamin B6 (sometimes associated with β-blocking drugs) for 3 to 6 months (*Figure 4*).

In patients with cardiac shock (58 patients) there were two situations:

1) Magnesium deficiency (14 cases) that determined a more severe shock and increased and longer ischemic manifestations. In these patients glycemia and serum electrolytes were difficult to normalized.

2) The appearance of magnesium deficiency in 19 of the 24 patients with **prologed** shock over 24 hours. It is important to notice the risk of hypomagnesiemia in patients with cardiac failure treated with digital medication and with diuretics from thyazides class without magnesium therapy. The risk of arrythmia and sudden death is present and there is a known correlation between stress and magnesium deficiency.

Magnesium in asthma attack

Magnesium is an antagonist of calcium in smooth muscle, decreasing the calcium level and relaxing of the muscle. Magnesium has more actions: inhibition of acetylcholine at neuromuscular junction level, inhibition of histamine, direct inhibition of smooth muscle contraction, activation of adenilat-cyclase with product of AMPc with **bronchodilatatore** role, induction of sedation with the decrease of respiratory effort [12-15].

These positive effects were demonstrated in 74 patients with asthma attack of mild or severe form, patients that have received intravenous MgSO4 (20-75 mg/kg in 20 minutes) associated with specific medication, with the serum magnesium level monitoring. We have observed the reduced duration and severity of the attack, more rapid improvement of ventilation test results and the decrease in days of hospitalization.

The long-term treatment with oral magnesium (in order to maintain the magnesium level at 2 mEq/l) associated with specific medication determined the improvement of disease evolution in 1/4 of the patients.

Serum magnesium level changes in acute enterocolites with acute dehydration syndrome

The low level of serum magnesium is frequent (serum magnesium of 1.27 ± 0.32 mEq/l) and is associated with hypomagnesiuria (2.46 ± 1.48 mEq/l), demonstrating the role of renal tubule in the regulation of magnesium metabolism. Frequently hypokaliemia and hypernatriemia (an aldosteronism) is associated. In patients with therapy with intravenous magnesium, the inhibition of excessive aldosterone production leads to more rapid normalization of electrolyte disturbances. This fact was demonstrated comparing those patients with the patients without magnesium therapy.

Magnesium deficiency favors eso-gastro-duodenal diskynesia (regurgitations, after meal abdominal distention) and biliary diskynesia with vomiting and diarrhea with aggravation of clinical and biological condition.

The quickest correction of magnesium deficiency after severe diarrhea, chronic diarrhea is needed in order to prevent the hypomagnesemia to become chronic with all possible complications.

Hypomagnesemia was frequent in hepatic cirrhosis cases. There are many causes of the low level of magnesium: insufficient magnesium in diet, decreased intestinal absorption and increased intestinal loses by complex mechanism, increased renal excretion. The signs and symptoms of hypomagnesemia are **overposed** with those of hepatic failure and sometimes are difficult to recognize, the magnesium balance and electrophysiological examinations (electromyogram, electroencephalogram, electrocardiogram) are needed.

Magnesium and malnutrition

Magnesium deficiency has complex causes, in many cases these are associated: insufficient diet intake, unilateral diet, malabsorbtion, chronic diarrhea (infectious or food intolerance). Multi-centric studies have demonstrated the presence of a direct link between malnutrition (from intrauterine life) – magnesium deficiency – deficiency of proteic synthesis – mental retardation. Level of magnesium in serum and cephalorachidian liquid are decreased in children with protein caloric malnutrition, with changes in cerebral metabolism, suggesting a direct link between magnesium deficiency and mental retardation.

The signs of severe extracellular deficiency could appear also during the weight increase and protein synthesis, due to the entrance of cations into cells and due to the increased need of magnesium in protein synthesis.

The study performed in dystrophic children including 24 infants with mild or severe malnutrition without electrolytes changes at the beginning of endovenous or enteral nutrition have demonstrated the appearance of hypomagnesemia (13 patients) and hypokaliemia (12 patients) form the third day of nutrition. The severity of magnesium deficiency was in direct correlation with the severity of malnutrition. An intake of 20 mg/kg/day magnesium is needed. The feeding of infants with malnutrition with cow-milk or with adapted formula has demonstrated a more important retention and a decreased elimination of magnesium and calcium for the age and weight.

Serum magnesium in evolution of type 1 diabetes mellitus

We have studied 25 patients with type 1 diabetes mellitus with medium age of 12± 6 years, 6 patients at the onset of diabetes and 19 patients with known diabetes (medium duration of evolution 5.8 ± 3.2 years). The following parameters were tested: serum magnesium, using blue **xylidyn** color reaction (normal value 1.65-2.2 mg/l), glycosilated hemoglobin (HbA1c) using minicolumns chromatography test (normal value 4.5-6.5%). The medium values of glycosilated hemoglobin and serum magnesium level were compared at the onset and during evolution, using **nule** hypothesis and Student's test to find the p value.

It is known that the glycosilated hemoglobine level is a control parameter for the previous 2-3 months of evolution. Increased level of glycosilated hemoglobin at the onset of diabetes (HbA1c > 13 ± 2.2%) (*Figure 5*) is correlated with increased glycemia for long time until the clinic diagnostic is possible. The same change appear if the control of diabetes is unsatisfactory.

The level of serum magnesium is decreased in patients with diabetes mellitus at onset and when the control is deficient (demonstrated by increased HbA1c) due to osmotic diuresis that increased magnesium excretion.

All patients with diabetes mellitus had long term treatment with Magne B6, aspacardin, Magnerot. In 1999 in 6 patients with diabetes mellitus the levels of magnesium and HbA1c was followed during Magnerot treatment for 2 months periods.

The follow-up during the indicated period and followed strictly by the patients with diabetes mellitus have shown a favorable clinic evolution with serum magnesium levels of 1.7 ± 0.2 mEq/l and HbA1c level of 7.8 ± 1.5%.

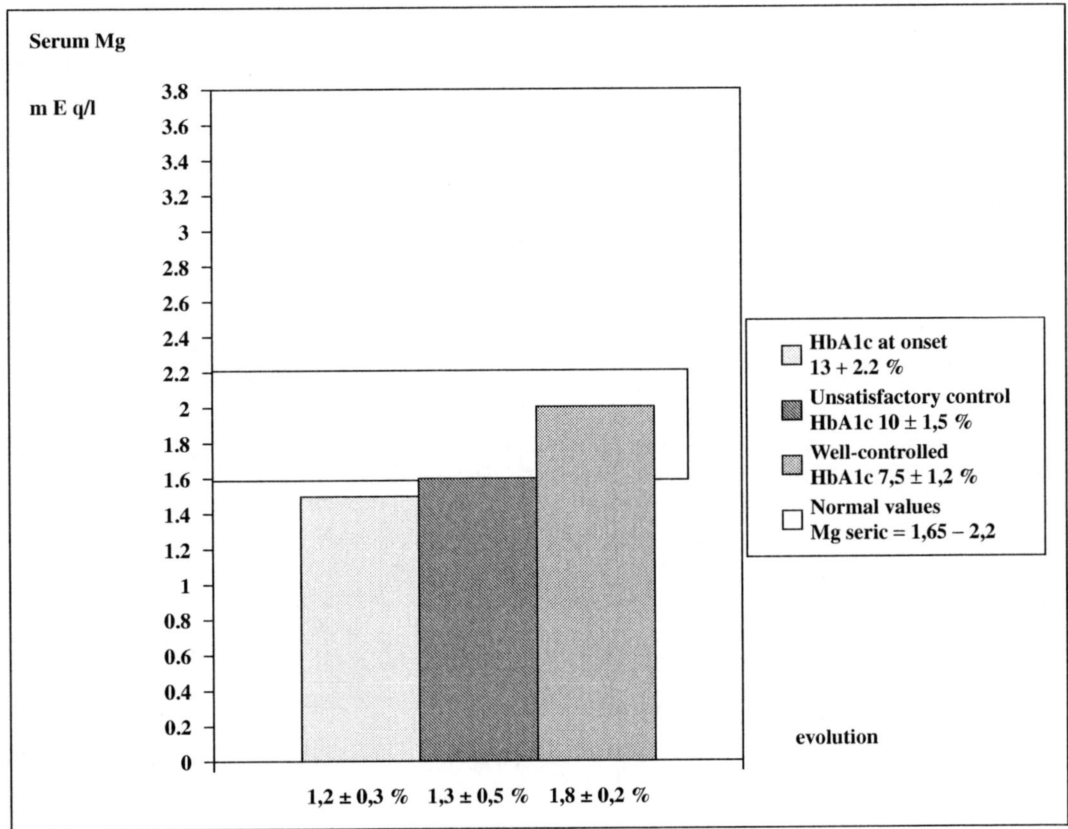

Figure 5. Level of magnesium in diabetes mellitus.

Conclusions. Magnesium deficiency is the most frequent disorder of mineral metabolism in juvenile diabetes mellitus. The level of serum magnesium depends of the control level of diabetes, demonstrated by HBA1c tested in same time with magnesium level: the well controlled diabetes mellitus is, the near normal magnesium level is. The serum level of magnesium depends also on the supplementary magnesium intake that should be followed strictly by the patient. Magnesium deficiency favors the appearance of vascular complications of diabetes [16-18].

Treatment of magnesium deficiency

In order to increase the magnesium intake by physiologic doses we could improve the diet, taking into consideration also the hydric magnesium, or we could prescribe magnesium salts. The treatment consist of a daily magnesium intake of 5 mg/kg/day as salts easy absorbable. Doses should be greater in children, especially during growing periods, as more as 8-10 mg/kg/day. The doses of 5 mg/kg/day could be increased in pregnancy, lactation, intense physical effort and exercises.

The magnesium prescription should be followed for minimum three weeks, ideally should be four weeks of treatment. After this period of time a clinical, biochemical and electric evaluation should be done, mainly in spasmophilic, tetany forms and in those with neurological functional complications (central forms of magnesium deficiency). After this first evaluation the results could be clinically very good, but sometimes without normal level of magnesium. Then the treatment should be continued for more 3-4 weeks and after this time a new evaluation should be done, including phosphor-calcium evaluation.

Loading with magnesium could determine a lose of calcium. In cases with magnesium deficiency with tetany or spasmophilia a complete investigation, including serum, erythrocyte, urinary magnesium level, calcium level, phosphor level, pH, sodium, chloride, hormone levels, should be done.

If using the above doses clinical, biochemical and electric results are not obtained the following possibilities should be taken into consideration:

- Magnesium deficiency causes that decrease intestinal absorption;
- Real malabsorbtion (rarely only magnesium selective);
- Magnesium loses (tubulopathy, intestinal hypersecretion, excessive lactation, intense physical exercises, intense sweating);
- Stress with excessive lose of magnesium;
- Endocrine disease with disorder of magnesium metabolism;
- Deficiency of magnesium fixating agents;
- Need of association of "saving" magnesium therapy;
- Iatrogenic factors (intense treatment with cortisone, digitalis and furosemide treatment, laxative drugs).

In these situations etiopathogenetic treatment should be administered.

We emphasized that in simple forms of magnesium deficiency, after the first month of follow-up the results are good and the complex examination should be done just in special situations. The tolerance of physiologic doses is excellent, increase of intestinal transit to diarrhea was observed in those with other associations (ex. Colitis) or when the magnesium was overdosed.

Parenteral treatment with magnesium is indicated in three cases:

- Acute forms of deficiency – emergency cases;
- Impossibility of oral administration;
- Negative results with oral therapy (malabsorption) [19].

References

1. Dacey MJ. Hypomagnesemic disorders. *Crit Care Clin* 2001; 17(1): 155-73.
2. Dreosti. Ivor E. Magnesium status and health. *Nutr Rev* 1995; 53(9): S23-S27.
3. Fox C, Ramsoomair D, Carter C. Magnesium: its proven and potential clinical significance. *South Med J* 2001; 94(12): 1195-201; (Review).
4. Johnson S. The multifaceted and widespread pathology of magnesium deficiency. *Med Hypotheses* 2001; 56(2): 163-70.
5. Knochel JP. Disorders of magnesium metabolism. In: Fauci AS, Braunwald E, Isselbacher KJ, eds. *Harrison's Principles of Internal Medicine. 14th Ed.* New York: McGraw-Hill Companies, Inc, 1998: 2263.
6. Baumgaertel A. Alternative and controversial treatments for attention-deficit/hyperactivity disorder. *Pediatr Clin North Am* 1999; 46(5): 977-92.
7. Hornyak M, Voderholzer U, Hohagen F, Berger M, Riemann D. Magnesium therapy for periodic leg movements-related insomnia and restless legs syndrome: an open pilot study. *Sleep* 1998; 21(5): 501-5.
8. Kozielec T, Starobrat-Hermelin B. Assessment of magnesium levels in children with attention deficit hyperactivity disorder. *Magnes Res* 1997; 10(2): 143-8.
9. Dimai HP, *et al.* Daily oral magnesium supplementation suppresses bone turnover in young adult males. *J Clin Endocrinol Metab* 1998; 83(8): 2742-8.
10. Dorup I, Skajaa K, Thybo NK. Oral magnesium supplementation restores the concentrations of magnesium, potassium and sodium-potassium pumps in skeletal muscle of patients receiving diuretic treatment. *J Intern Med* 1993; 233(2): 117-23.
11. Lichodziejewska B, *et al.* Clinical symptoms of mitral valve prolapse are related to hypomagnesemia and attenuated by magnesium supplementation. *Am J Cardiol* 1997; 79(3): 768-72.
12. Britton J, *et al.* Dietary magnesium, lung function, wheezing, and airway hyper-reactivity in a random adult population sample. *Lancet* 1994; 344(8): 357-62.
13. Ciarallo L, Brousseau D, Reinert S. Higher-dose intravenous magnesium therapy for children with moderate to severe acute asthma. *Arch Ped Adol Med* 2000; 154(10): 979-83.
14. Gilliland FD, Berhane KT, Li YF, Kim DH, Margolis HG. Dietary magnesium, potassium, sodium, and children's lung function. *Am J Epidemiol* 2002; 155(2): 125-31.
15. Rowe BH, Edmonds ML, Spooner CH, Camargo CA. Evidence-based treatments for acute asthma. *Respir Care* 2001; 46(12): 1380-90; (Review).
16. American Diabetes Association. Magnesium supplementation in the treatment of diabetes. *Diabetes Care* 1992; 15: 1065-7.
17. De Valk HW. Magnesium in diabetes mellitus. *Neth J Med* 1999; 54(4): 139-46.
18. Tosiello L. Hypomagnesemia and diabetes mellitus. *Arch Intern Med* 1996; 156(10): 1143-8.
19. Durlach J, Durlach V, Bac P, Bara M, Guiet-Bara A. Magnesium and therapeutics. *Magnes Res* 1994; 7(3-4): 313-28.

Magnesium homeostasis in early neonatal period

L. Blaga[1], G. Panga[2], M. Hedesiu[3], I. Lupea[1], D. Duma[1], V. Popa[1]

1. Neonatology Departement, University of Medicine "Iuliu Hatieganu" Cluj Napoca, Romania ; 2. Paediatric Departement, University of Medicine "Iuliu Hatieganu" Cluj Napoca, Romania ; 3. Departement of Radiology, University of Medicine "Iuliu Hatieganu" Cluj Napoca, Romania

Abstract. Early neonatal period is characterized by rapid changes in mineral homeostasis. Magnesium transfer across placenta increase with gestational age, heaving a peack in the 8th month of pregnancy, thus magnesium levels in cord blood are same in term and preterm newborns. Aricetta *et al.* demonstrate that total magnesium levels are higher in preterm than term newborns. Magnesium level fall in the first two days after birth, than it has a gradual increase during the first month of life. Hypomagnesemia, defined as magnesium values less than 1.6 mg/dl after Loughead studies, is often meet in early neonatal in premature newborns and in small for gestational age newborns, but not so often as hypocalcemia.

In this paper the authors want to establish the inferior physiological lower limit of magnesium in term and preterm newborns and the incidence of hypomagnesemia at 48 hours after birth in low birth weight infants. We have studied 121 newborns (20 appropriate for gestational age term newborns, 50 small for gestational age term newborns, 26 premature newborns and 25 small for gestational age preterm newborns). Magnesium was determinated at 48 hours after birth by colorimetric method. Magnesium levels are significantly higher in premature newborns than in term newborns ($p < 0{,}005$) and insignificantly higher than small for gestational age term and preterm newborns. The incidence of hypomagnesemia was 19.2% in preterm newborns with gestational age less than 30 weeks and only 6% in small for gestational age term newborns. No symptoms have been found. We concluded that magnesium levels are significantly higher in preterm newborns than in appropriate for gestational age term infants. Incidence of hypomagnesemia is higher in premature newborns than in small for age term and preterm infants.

The early neonatal period is characterized by rapid changes in mineral homeostasis. Since the transplacentar transfer of Mg reaches its peak from the second quarter of pregnancy and then remains constant, similar Mg concentrations are to be found in the umbilical cord of both the 8 month premature baby and term new-born baby (between 1.43-2.45 mg/dl) [1]. Klaus and Fanaroff [2] have found magnesemia in full-term babies between 1.5-2.8 mg/dl, in close connection with maternal magnesemia. As shown by Aricetta *et al*, but also by others [3-7], there is an inverse ratio between the serous Mg values and the gestational age. However Aricetta shows that this ratio referring to total amount of magnesium, while the amount of ultrafiltrable magnesium is similar in premature and term newborns. Once the umbilical cord is severed and thus, the maternal contribution in minerals, Mg values drop to 1.87 mg/dl during the first 48 postnatal hours [1] and then rice progressively all through the first week of autonomous life. Authors differ in their opinions on Mg concentrations during the first 48 hours as well as in their appreciation of the minimal level. Loughead [1] and other authors [3-6] have found the physiological lower limit of Mg to be 1.6 mg/dl; Koo and Tsang [8, 9] believe it to be 1.5 mg/dl. It also varies narrowly due to the direct influence of calcium concentrations and to the indirect influence of phosphatemia [1, 2]. Although much more rarer than hypocalcemia, lower than 1.6 mg/dl hypomagnesemia is often associated with maternal diabetes mellitus, prematurity and intrauterine growth retardation [7, 10, 11].

Table I. Description of the four groups which have been studied.

Diagnostic	Nr. cases	Gestational age (weeks)	Birth weight (g)	Ponderal index	Male	Female
AGA term newborn	20	38-41 39.3 ± 0,9	2900-4200 3430 ± 365	2.0-2.2 2.1 ± 0.07	45%	55%
SGA term newborn	50	38-43 39.7 ± 1	1200-2500 2250 ± 219	1.2-2.0 1.6 ± 0.23	44%	56%
AGA preterm	26	28-36 31.7 ± 2	700-2000 1523 ± 352	1.9-2.4 2.06 ± 0.2	69%	31%
SGA preterm	25	30-36 33.9 ± 2	700-1900 1354 ± 340	1.3-1.9 1.6 ± 0.12	36%	64%

About 30-50% of the babies born of diabetic mothers develop hypomagnesemia during the first 48 postnatal hours [12]. Miu and colleagues [9] further linked neonatal hypomagnesemia with intrauterine growth restriction in 14.4% of the cases and with prematurity in 37% of the cases. However, there is no consensus as to the occurrence of hypomagnesemia in intrauterine growth restriction newborn and to the way in which the factors that induced intrauterine growth retardation influence mineral homeostasis. The present study intends to establish the normal values and the physiological lower limit of magnesemia during the first 48 postnatal hours in both the full term baby and the "small for age" babies, in close connection with our geographical area. We also intend to establish the occurrence of early hypomagnesemia in premature newborn and in those with intrauterine growth retardation.

Material and method

The study includes 20 full term babies free of perinatal pathology (the witnesses' lot), aged 38-41 weeks (39.3 ± 0.9), weighing between 2900-4200 g at birth (3430 g ± 365) and with a ponderal index between 2.0-2.2 (2.1 ± 0.07); the sex-ratio being 9 to 11 (45% males *vs.* 55% females).

The second lot was made up of 101 underweight babies, 50 of which are full term babies with fetal hypotrophy, aged 38-43 weeks (39.7 ± 1), weighing between 1200-2500 g at birth (2250 ± 219) and with a ponderal index of 1.2-2.0 (1.6 ± 0.23); the sex-ratio was 22 to 28 (44% males *vs.* 56% females). Another 26 are premature babies whose weight was appropriate to their gestational age. They are 28-36 weeks old (31.7 ± 2), weigh between 700 and 2000 g (1523 ± 352) and have a ponderal index of 1.9-2.4 (2.06 ± 0.2), their sex-ratio being 18 to 8 (69% males *vs.* 31% females).

The last 25 are premature babies with intrauterine growth retard, aged 30-36 weeks (33.9 ± 2), weighing 700- 1900 g (1354 g ± 340) and with a ponderal index of 1.3-1.9 (1.6 ± 0,12); their sex-ratio being 9 to 16 (36% males *vs.* 64% females) (*Table I*).

Magnesemia was determined at 48 hours after birth by the colorimetric method, from a sample of 2 ml of blood collected a jeun in the morning. The gestational age was established from anamnestic and echographic data whereas after birth by the Dubovitz-Ballard score. Babies qualified as hypotrophic if their weight was below the 10 percentage mark on the intrauterine growth curves.

Results

In our study, magnesemia of full term babies appropriate for gestational age (AGA), free of perinatal pathology, oscillated between 1.89-2.56 mg/dl, (2.24 ± 0.18). For small for age (SGA) term newborns magnesemia oscillated between 1.17-2.69 mg/dl (2.23 ± 0.38). For premature babies with an appropriate age/weight balance, magnesemia varied between 1.32-2.92 mg/dl (2.34 ± 0.4). Lastly, for small for age premature newborns magnesemia varied between 1.3-2.67 mg/dl (2.22 ± 0.4) (*Table II*).

Table II. Magnesium levels in the four groups.

Diagnostic	AGA term newborn	SGA term newborn	AGA preterm	SGA preterm
Magnesium values (mg/dl)	1.89-2.56 2.24 ± 0.18	1.17-2.69 2.23 ± 0.38	1.32-2.92 2.34 ± 0.4	1.30-2.67 2.22 ± 0.4

Discussions

Appropriate for gestational age full term babies' magnesemia is irrelevantly higher than that of small for age term newborns (kw = 0.0919) and significantly lower (kw = 0.0113) than that of premature babies with an appropriate age-weight ratio. The values of magnesemia in premature babies with fetal hypotrophy are irrelevantly lower as compared to premature babies with an appropriate age-weight ratio (p = 0.6397). Lastly, magnesemia concentrations are irrelevantly higher when we compare full term hypotrophic babies to premature hypotrophic babies (p = 0.9393).

Our study used the following values as the lower limit (the average 2 DS) of magnesemia: 1.88 mg/dl for AGA full term babies; 1.47 mg/dl for SGA full term babies; 1.54 mg/dl for premature babies with an appropriate age-weight ratio and 1.42 mg/dl for SGA premature babies.

3 out of 50 cases (6%) of full term babies with intrauterine growth restriction exhibited precocious hypomagnesemia in our study, 2 of which also exhibited hypocalcemia, the sex ratio being 1 to 2. Hypomagnesemia was also detected in 5 out of 26 (19.2%) cases of premature babies with an appropriate age-weight ratio, none of them having associated hypocalcemia; their sex ratio was 3 to 2. Hypomagnesemia in SGA premature babies was to be found in 4 out of 25 cases (16%), one case exhibiting associated hypocalcemia. The sex ratio in this case was 2 to 2.

The only clinical symptoms were registered in the case of associated hypomagnesemia and hypocalcemia. The positive result of the therapeutic calcium test proved that hypocalcemia and hypomagnesemia were not interrelated.

As for the risk factors, our study pointed out the presence of hypomagnesemia in intrauterine growth restricted newborns and in babies with a low ponderal index [1, 2]. Premature babies with an appropriate age-weight ratio developed hypomagnesia at gestational age between 28-34 weeks (4 out of 5 cases were aged below 30 weeks). SGA premature babies that had hypomagnesemia were aged 31-36 weeks and had a ponderal index of 1.5-1.8 (*Figure 1*).

The mothers of hypotrophic hypomagnesemic full term babies were affected by pregnancy toxemia (2 cases) and by post-term pregnancy (1 case). Hypotrophic hypomagnesemic premature babies came from hypertensive pregnancy mothers (2), twin pregnancy (1) and chronic placental insufficiency (1) (*Figure 2*).

SGA full term babies that had associated hypomagnesemia and hypocalcemia came from post-term pregnancy (1 case) and from hypertensive pregnancy (1 case). The case of one preterm new born with both fetal hypotrophy and associated hypocalcemia and hypomagnesemia came from an arterial hypertensive pregnancy.

Conclusions

1. Magnesemia in AGA full term babies was irrelevantly higher than in SGA full term babies, and significantly lower than in premature babies with an appropriate age-weight ratio. SGA full term babies' magnesemia was irrelevantly higher than that of SGA premature babies. Premature babies with fetal malnutrition had irrelevantly lower magnesium values than premature babies with an appropriate age-weight ratio.

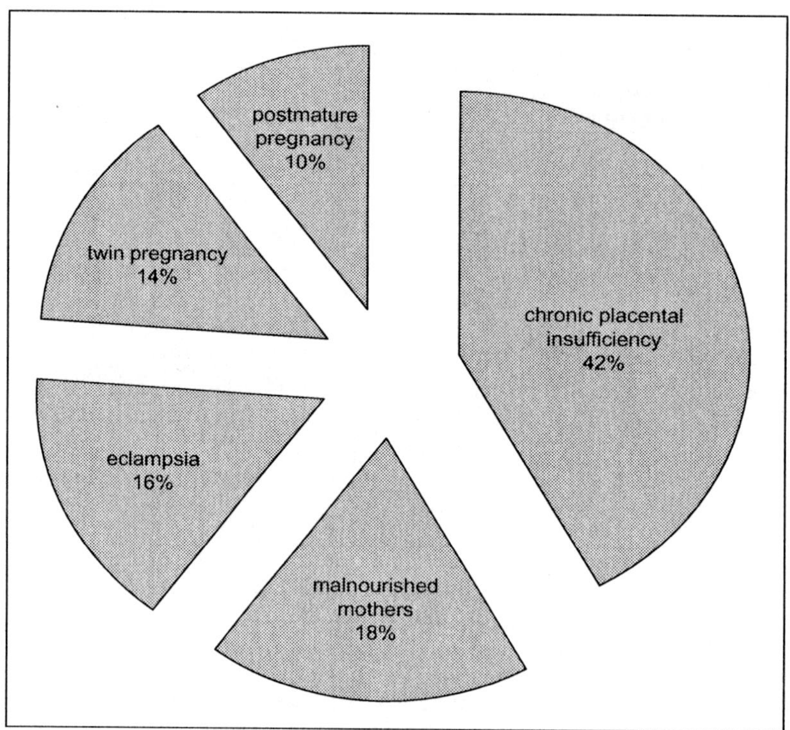

Figure 1. Risk factors of IUGR in SGA term newborns.

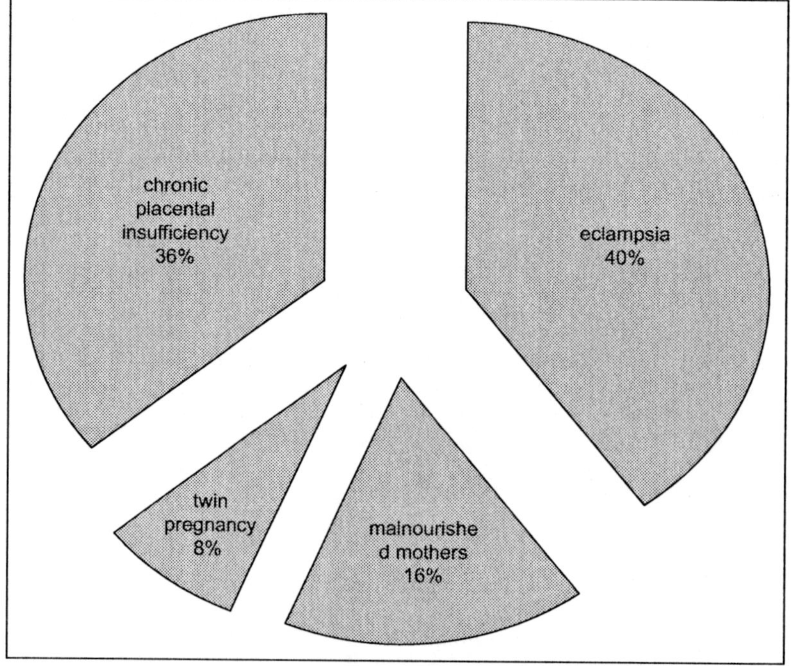

Figure 2. Risk factors of IUGR in preterm newborns.

2. Statistically, the occurrence of precocious hypomagnesemia registered the following values: 19.2% in premature babies with an appropriate age-weight ratio, 16% in premature babies with intrauterine underdevelopment and 6% in full term babies with fetal malnutrition.

3. Precocious hypomagnesemia was asymptomatic irrespective of its type: either isolated or associated to hypocalcemia.

4. The risk factors in precocious hypomagnesemia were prematurity and intrauterine growth restriction. The gestational age below 30 weeks in premature newborns and the lower ponderal index (below 1.2 in the case of small for age term newborns and below 1.8 in small for age premature newborns) are the more frequent occurrence of hypomagnesemia.

References

1. Loughead JL, Tsang RC. In: *Neonatal mineral metabolism*. New York: Principles of Perinatal-Neonatal Metabolism, 1991: 472-99.
2. Klaus F. In: *Care of the high risk neonate. Fifth Ed.* 2001: 301-23.
3. Ariceta G, Vallo A, Rodriguez-Soriano J. Calcium and magnesium homeostasis in term and preterm infant. *Pediatr Nephrol* 1993; 7: C13.
4. Ariceta G, Soriano JR, Valla A. Magnesium homeostasis in premature and full term neonates. *Pediatr Nephrol* 1995; 9: 423-7.
5. Cook LA, Mimouni FB. Whole blood ionized magnesium in the healthy neonate. *J Am Coll Nutr* 1997; 16(2): 181-3.
6. Marcus JC, Valencia GB, Altura BT, *et al.* Serum ionized magnesium in premature and term infants. *Pediatr Neurol* 1998; 18(4): 311-4.
7. Tsang RC, Oh W. Serum magnesium levels in low birth weight infants. *Am J Dis Child* 1970; 1120: 44-8.
8. Koo WWK, Tsang RC. Calcium and magnesium homeostasis. In: *Neonatology, Pathophysiology and Management of the new-born. Philadelphia.* 1994: 575-603.
9. Miu N, de magneziu D. In: *Magneziul in biologia si patologia umana.* 2000: 73-4.
10. Tsang RC. Neonatal magnesium disturbance. Am J Dis Child 124: 282-93.
11. Stigson L, Kjellmer I. Serum levels of magnesium at birth related to complications of immaturity. *Acta Paediatr* 1997; 86(9): 991-4.
12. Surcel V, Neamtu A. Metabolismul magneziului in sarcina. In: *Magneziul in biologia si patologia umana.* 2000: 153-65.

VI. Magnesium, gastroenterology, metabolism and endocrinology

Digestive manifestations of magnesium deficiency

P.J. Porr

3rd Medical Clinic, Str. Croitorilor 21, RO-400162 Cluj-Napoca, Romania

Magnesium deficiency (MD) is an ubiquitous phenomenon in the human pathology – from the most developed to the third world countries, which may occur at all ages, more marked during growth, pregnancy and breast feeding [1]. MD appears with increased frequency after stress, under all its forms – physical, biological (various diseases) or mental, which determines higher magnesiuria and consecutive MD by excessive Mg loss [1-3]. It clearly affects women more than men, who seem to have a protection mechanism against magnesiuria [4].

The clinical manifestations are characterized by a great diversity, which in turn is due to the various physiological and pathophysiological implications of the Mg^{2+} ion (for example Mg is a catalyzer in over 300 enzymatic reactions of the protein, lipid and sugar metabolism) [1, 3]. This variety of the clinical manifestations can determine an underdiagnosis of MD, so that, unfortunately, there may be situations, in which the patient is sent from one specialist to another, from the gynecologist to the internist, from the cardiologist to the endocrinologist or psychiatrist [3]. "It is highly regrettable that the deficiency of such an inexpensive, low-toxicity nutrient result in diseases that cause incalculable suffering and expense throughout the world" [5].

The classical manifestations of MD are neuromuscular. They may occur at the level of the skeletal muscles, resulting in spasmophilia under all its forms, at the level of the myocardium, resulting in various cardiac disturbances, and also at the level of the smooth (visceral) muscles [1, 3, 6]. The bronchial and pharyngeal muscles are especially affected, leading to bouts of bronchial asthma and to pharyngeal spasm and dysphagia - the "lump in the throat" feeling - respectively, but also the muscles of the urinary bladder, resulting in the neurogenic bladder and enuresis, and the digestive tract muscles.

In the upper segment of the digestive tract, MD may induce esophageal spasm, manifested by dysphagia. Iannello *et al.* [7] reported a case in which severe dysphagia to solid food subsided after 4 months of substitution therapy with Mg only. Both gastro-esophageal and duodeno-gastric reflux disease may occur or aggravate because of MD. The patients complain of heartburn, and in time even esophagitis may occur, aggravating the complaint. In a study of 50 patients with reflux esophagitis and MD we have evidenced that Mg substitution therapy corrected the MD, while the digestive clinical manifestations, namely the endoscopically evidenced gastro-esophageal reflux and esophageal pH respectively, were restored only in 28 patients (46%), which means that though MD is the cause of a significant percentage of reflux esophagitis, there are many cases which are due to other causes [8]. These results correlate with those reported by other authors [9, 10].

In vitro and *in vivo* experimental studies have evidenced the role of Mg in gastric secretion. Through its physiological antagonism it inhibits gastric secretion, an effect that is minimized in MD, the consequence being increased gastric acid secretion [11]. A recent study of Abbasciano shows that in 2 groups of chronic gastritis, one Helicobacter pylori+, the other Hp-, in the first group there were significantly lower values of erythrocytic and gastric Mg levels and no significant difference between seric Mg values.

An interesting phenomenon has been noted by Pasternak *et al.* [12] and Mooren [11], namely the increase of tissular Mg concentration in the patients with gastric cancer as compared to controls, directly related to the evolution of the disease. This has even been proposed as a prognostic factor.

At the level of the small and especially large bowel, MD aggravates the irritable bowel syndrome. The etiopathogenesis of this disease is generally complex, but an important factor is represented by MD. It is manifested by transit disturbances, usually alternation between constipation and diarrhea, distended bowels, meteorism, diffuse abdominal pain having a migrating character. Planche *et al.* [13] studied a group of 41 patients with irritable bowel and MD in which endoscopic and radiological examinations ruled out organic diseases. Many patients also presented other symptoms characteristic of MD. After treatment with Mg, Mn and Co the symptoms disappeared, including the non-digestive ones, serum and erythrocytic Mg levels were restored to normal, electromyography evidenced normal parameters. In our own study of 100 patients with spasmophilia and other 100 with irritable bowel syndrome the results were as follows: In group A (with spasmophilia) 46 patients presented irritable bowel symptoms, and in group B (with irritable bowel) 82 presented MD, predominantly young women. This shows that most cases of irritable bowel syndrome are associated with MD, especially that Mg substitution therapy led not only to the correction of the MD but also the disappearance of irritable bowel complaints [8].

Chronic idiopathic constipation may also be due to MD. There were patients with MD in which even ileus was found [14, 15], which subsided following parenteral Mg administration.

According to Galland [16], MD would also be involved in ulcerative colitis and Crohn's disease, according to [17, 18] also in colorectal cancer, existing an inverse dose-response relationship between Mg intake and risk of colorectal cancer.

MD may also affect the gallbladder muscles. Biliary dyskinesia may occur or become aggravated, manifested by nausea, bitter taste, vomiting, biliary migraine, discomfort in the right hypochondrium to colicky pain. In a group of 81 patients with biliary dyskinesia and MD in which organic diseases of the gallbladder (gallstones, malformation, cholesterolosis, cholecystitis) or parasitosis (by repeated coproparasitological examinations) were ruled out, the symptoms disappeared at the same time with MD after an average period of Mg substitution therapy of 5.8 weeks [19]. The most surprising fact was the 100% efficiency of the treatment, other authors reporting only 61-95% efficiency [20, 21]. This was probably because of the patient selection: biliary dyskinesia + MD not just any biliary dyskinesia.

A study was carried out in 45 patients operated for biliary syndromes [22, 23]. It is known that the biliary dyspeptic syndrome recurs in 20-30% of the cholecystectomized patients after several weeks to even years. After Mg substitution therapy for a mean period of 9 weeks the biliary dyspeptic syndrome disappeared together with MD.

At the level of the liver MD plays a role in the chronic alcoholic liver disease, under all its forms: steatosis, steatofibrosis, alcoholic hepatitis and alcoholic cirrhosis, as well as in its complications.

The correlation between Mg and alcohol is complex [24]. It is known that alcohol consumption determines increased magnesiuria, therefore almost all the patients with chronic alcoholism suffer from MD. The latter may also be due, besides magnesiuria, to insufficient Mg supply or vomiting, often present in alcoholic patients. It may be stated that Mg therapy is recommended in chronic alcoholism. Besides MD, hypokaliemia, hypocalcemia and hypophosphatemia also occur. There have been cases reported with hypoparathyroidism secondary to MD [25, 26]. It may be stated that MD is the most frequent electrolytic disturbance in alcoholic patients, with complex pathogenetic links [27]. In the same way, a period of abstinence of three weeks led to the significant increase of serum and ionic Mg [28].

An interesting double blind study belongs to a Norwegian team [29], in which Mg substitution therapy in a group of alcoholic patients resulted in the significant improvement of the ASAT, ALAT and gamma-GT serum levels as compared to a placebo group, but not the bilirubin level. In our own study of 60 patients with chronic alcoholic hepatitis, 10 patients were in the active phase. In 5 of them ASAT and ALAT were improved exclusively by Magnerot treatment - probably through the effect of the orotic acid, known to be hepato-protective. In all the 60 patients the gamma-GT levels were improved, in 18 of them (30%) to normal [30].

A lot of the manifestations attributed to toxic-deficient polyneuropathy are in fact manifestations of spasmophilia, disturbances of the neuromuscular excitability due to MD respectively, or which enhance further an existing polyneuropathy.

Moreover, in alcoholism the influence of acetaldehyde on a biogenic amine in the brain leads to the formation of pseudoopioid substances with affinity for the endorphin receptors, determining alcohol dependence. MD favours this process and maintains the dependence, thus closing a vicious circle. It has also been demonstrated that MD favours the crises of delirium tremens [31]. Inversely, the detoxifying cures associated with Mg therapy are better tolerated by the patients. Besides MD, delirium tremens is also associated with marked hypokaliemia. The Mg/K ratio (the decrease of Mg respectively) may represent an indicator of the predisposition to delirium tremens at the start of the detoxifying cure [31]. Alcoholic encephalopathy is also favoured by MD.

The evolution from steatosis to liver cirrhosis is also influenced by the combination alcohol - MD, which favours the excessive production of collagen in the liver. A similar phenomenon was noted in alcoholic cardiomyopathy, in which MD contributed to the onset of alcohol-induced myocardial dystrophy and the aggravation of the disease. A proof is the fact the cardiomyopathy is improved by Mg substitution therapy.

Mg administration also proved to be beneficial in the coagulation disturbances in the patients with alcoholic cirrhosis.

A recent, very interesting experimental study [32] shows a significantly higher apoptosis of rat cultured hepatocytes in MD by increasing susceptibility of these cells to oxidative stress (enhanced lipid peroxidation).

A peculiar situation is represented by digestive parasitoses. In an older study [19] it was noted that MD even if treated correctly was not restored as long as the coexisting parasitosis is not cured. The study was prompted by a clinical case of MD resistant to Mg substitution therapy. The phenomenon was studied in a group of 51 patients with lambliasis and MD, treated with Mg, in which MD was not corrected; both biliary symptoms and the extra-biliary manifestations of MD persisted. No satisfactory explanation was found in literature. In other words, when MD coexists with a digestive parasitosis, there is no point in initiating Mg substitution therapy unless the parasitosis is also treated. Moreover, in the correctly treated cases of MD, but resistant to this therapy, the testing for digestive parasitosis is required.

To conclude, in digestive pathology MD does not have as complex and especially not as serious repercussions as for instance in cardiovascular pathology. However, an important part of functional digestive pathology is due to or interferes with MD. The evidence is that these manifestations disappear completely solely by Mg substitution therapy. The important aspect is that in all the manifestations of MD the treatment should be applied correctly: 500 mg Mg^{2+}/day + 180 mg Ca^{2+}/day + vitamin B_6 for an adequate period of time, namely over 6 weeks.

References

1. Durlach J. *Magnesium in clinical practice.* London : Libbey, 1988.

2. Durlach J, Bara M. In : *Le magnésium en biologie et en médecine.* Cachan : Ed Médic Internat, 2000 : 125-6.

3. Porr PJ. Manifestarile clinice ale deficitului de magneziu la adult. In : Miu N, Dragotoiu G, eds. *Magneziul in biologia si patologia umana. Cluj-Napoca : Casa Cartii de Stiinta.* 2000 : 88-97.

4. Reinhardt F, Gabsch HC, Porr PJ, Szántay J. Untersuchungen zu körpereigenen Spareffekten in der Magnesiumutilisation unter Malnutrition. *Magnesium Bull* 1989 ; 11 : 12-7.

5. Johnson S. The multifaced and widespread pathology of magnesium deficiency. *Med Hypotheses* 2001 ; 56 : 163-70.

6. Porr PJ, Marginean C. Digestive manifestations of magnesium deficit. In : Nechifor M, Porr PJ, eds. *Magnesium : involvements in biology and pharmacotherapy. Cluj-Napoca : Casa Cartii de Stiinta.* 2003 : 160-5.

7. Iannello S, Spina M, Leotta P, Prestipino M, Spina S. Hypomagnesemia and smooth muscle contractility : a case of diffuse esophageal spasm in an old female patient. *Miner Electrolyte Metab* 1998 ; 24 : 348-56.

8. Szántay J, Porr PJ, Venczel T, Dumitraşcu D. Is there a correlation between magnesium deficit and biliary and gastrointestinal functional disturbances? *Zschr Gastroenterol* 1991 ; 5 : 114.

9. Classen HG. Stress and magnesium with special regard to the gastrointestinal tract. *Magnesium Bull* 1989 ; 11 : 143-6.

10. Mc. Callum RW. Role of calcium antagonists in digestive disorders. In : Christen MO, Goodfraind T, Mc.Callum RW, eds. *Calcium antagonism in gastrointestinal motility*. Paris : Elsevier, 1989 : 55-70.

11. Mooren FC. Magnesium status and gastric function : experimental and clinical data. *Magnesium Res* 1999 ; 12 : 311-8.

12. Pasternak K, Przyszlak W. Magnesium in stomach cancer. *Magnesium Res* 1999 ; 12 : 139-43.

13. Planche D, Daieff N. Salducci. Controlled clinical trial of a combination of manganese-cobalt oligosol and magnesium oligosol in the clinical, biochemical and electromyographic manifestations of spasmophilia observed in the course of spastic colon syndrome. *Rev Franç Gastro-Entérol* 1987 ; 23 : 341-6.

14. Shils ML. Experimental human magnesium depletion. *Medicine* 1969 ; 48 : 61-85.

15. Hirsh EH, Brandenburg D, Hersh T, Brooks Jr. WS. Chronic intestinal pseudoobstruction. *J Clin Gastroenterol* 1981 ; 3 : 247-54.

16. Galland L. Magnesium and inflammatory bowel disease. *Magnesium* 1988 ; 7 : 78-83.

17. Tanaka T, Shinoda T, Yoshimi N, Niwa K, Iwata H, Mori H. Inhibitory effect of magnesium hydroxide on methylazoxymethanol acetate-induced large bowel carcinogenesis in male F344 rats. *Carcinogenesis* 1989 ; 10 : 613-6.

18. Mori H, Morishita Y, Shinoda T, Tanaka T. Preventive effect of magnesium hydroxide on carcinogen-induced large bowel carcinogenesis in rats. *Basic Life Sci* 1993 ; 61 : 111-8.

19. Porr PJ, Szántay J, Gocan A, Rusu M, Dejica D. Efficacy of the treatment with Tiomag in biliary dyskinesia. *Magnesium Res* 1990 ; 3 : 144.

20. Ritter U. Funktionelle Störungen der Gallenwege. *Dtsch Verdauungs Stoffwechselkr* 1974 ; 34 : 119-26.

21. Shils ME. Magnesium in health and disease. *Annu Rev Nutr* 1988 ; 8 : 429-38.

22. Szántay J, Varga D, Porr PJ. Post-cholecystectomy syndrome and magnesium deficit. In : Mózsik G, *et al.*, eds. *Cell injury and protection in the gastrointestinal tract : From basic sciences to clinical perspectives*. Budapest : Akadémiai Kiadó, 1993 : 391-8.

23. Porr PJ. Psychosomatic pathology in dyselectrolitic disorders. In : Dumitrascu DL, ed. *Psychosomatic medicine. Recent progress & current trends. Cluj-Napoca : Ed Med Univ "Iuliu Hatieganu"*. 2003 : 101-8.

24. Flink EB. Magnesium deficiency in alcoholism. *Alcohol Clin Exp Res* 1986 ; 10 : 590-4.

25. Hermans C, Lefebvre C, Devolgelaer JP, Lambert M. Hypocalcaemia and chronic alcohol intoxication : transient hypoparathyroidism secondary to magnesium deficiency. *Clin Rheumatol* 1996 ; 53 : 71-4.

26. Iannello S, Belfiore F. Hypomagnesemia. A review of pathophysiological, clinical and therapeutical aspects. *Panminerva Med* 2001 ; 43 : 177-209.

27. Elisaf M, Merkouropoulos M, Tsianos EV, Siamopoulos KC. Pathogenetic mechanisms of hypomagnesemia in alcoholic patients. *J Trace Elem Med Biol* 1995 ; 9 : 210-4.

28. Hristova EN, Rehak NN, Cecco S, Ruddel M, Herion D, Eckardt M, Linnoila M, Elin RJ. Serum ionized magnesium in chronic alcoholism : is it really decreased? *Clin Chem* 1997 ; 43 : 394-9.

29. Gullestad L, Dolva LØ, Soyland E, Manger HT, Falch D, Veierod MB, Kjekshus J. Effects of oral magnesium treatment in chronic alcoholics. In : Lasserre B, Durlach J, eds. *Magnesium – a relevant ion*. London : Libbey, 1991 : 405-9.

30. Porr PJ. Orotatul de magneziu in boala hepatica cronica alcoolica. Meeting "Magnerot – progres în terapia cu Mg" Cluj 2000 : 31.

31. Schmickaly R, Nickel B, Kursawe HK, *et al.* Hypomagnesiämie, Hypokaliämie und initialer Anfall beim Delirium tremens. *Magnesium Bull* 1989 ; 11 : 47-52.

32. Martin H, Richert L, Berthelot A. Magnesium deficiency induces apoptosis in primary cultures of rat hepatocytes. *J Nutr* 2003 ; 133 : 2505-11.

Can long term magnesium supplementation stabilize the progression of diabetic retinopathy in magnesium-depleted type 1 diabetic patients?

I. De Leeuw, C. De Block, L. Van Gaal

Antwerp Metabolic Research Unit, Faculty of Medicine, University of Antwerp 1 Universiteitsplein, B2610 Antwerp Belgium

Background

Despite better management and intensive insulin treatment, retinopathy remains one of the major chronic complications of type 1 diabetes [1]. Hypomagnesemia is a risk factor in diabetic retinopathy (DR) [2] and is present in the more severe forms of this complication [3–5]. Since there is a large consensus now that magnesium depletion is present in 20-25% of type 1 diabetic patients (T1DM) [6], magnesium supplementation must be considered in the treatment of diabetes [7]. This looks particularly interesting in regard of the natural evolution of retinopathy since no progression has been observed in insulin-treated patients when plasma Mg concentration was higher than 0.8 mmol/l [8].
To have an objective idea on the effects of Mg supplements on DR some basic principles must be respected:

1. The study must be sufficiently long to observe significant changes in the eye-fundi using a standardized staging protocol (minimum 5 years);
2. The patients must have Mg depletion with repeated low circulating Mg levels and randomized in two matched groups (supplemented or not);
3. Metabolic control must remain stable throughout the study and insulin treatment chronically adapted;
4. Intercurrent diseases or drugs interfering with Mg metabolism or the progression of DR must lead to the drop-out of the participant;
5. For ethical reasons no invalidating or progressive diabetic complications must be present and renal function must be acceptable (GFR > 80 ml/min).

Research design and methods

Patients

From the large population (> 500) of the University Diabetes Outpatient Clinic a cohort of Mg-depleted (last 3 values of erythrocyte Mg < 2.3 mmol/l) T1DM was selected following the principles cited in the previous chapter.
After informed consent, 110 patients (60 men, 50 women) were randomized by gender to receive during a period of 5 years either 300 mg Mg++ as Mg gluconate (Ultramagnesium®) daily per os or no supplement.

After 5 years, 97 patients finished the study (53 men, 44 women) Drop-outs were justified by recurrent diabetic keto-acidosis, gastrointestinal side-effects of oral Mg intake or intercurrent diseases or drugs interfering with the Mg status.

For the study we ended with 2 groups:

- Group A: (Mg supplement) n = 49
  - 27 men: age: 43.5 years (sd 12); duration diabetes: 19.2 years (sd 7.7); HbA1c: 7.86% (sd 0.76),
  - 22 women: age: 38 years (sd 8.5); duration diabetes: 18.5 years (sd 7.9); HbA1c: 7.8% (sd 1.3);
- Group B: (no supplement) n = 48
  - 26 men: age: 37.1 years (sd 12.9); duration diabetes: 19 years (sd 10.6); HbA1c: 8.13% (sd0.98),
  - 22 women: age: 39.9 years (sd 11.5); duration diabetes: 19 years (sd 10.2); HbA1c: 7.9% (sd 0.9).

No significant differences were disclosed between groups at the start.

Methods

1. Every 3 months patients were seen at the clinic by the same treatment team for compliance and adaptation of insulin treatment on the basis of HbA1c (nl values: 4.8-6%) and self monitoring results. Compliance of Mg treatment was measured by comparing the number of powder-bags used to the number prescribed.

2. Every year a complete survey of biochemistry and diabetic complications was done. An ophtalmoscopic eye-fundus was done by the same ophthalmologist, unaware of the Mg treatment of the patient but using the same simplified standardized protocol [9] for staging of the retinopathy (DR):

   R0 = normal R3 = pre-proliferative retinopathy
   R1 = microaneurysms R4 = proliferative retinopathy
   R2 = background retinopathy R5 = end-stage retinopathy (*Figure 1*)

When abnormalities were observed, retinal photographs were taken for later comparison.

3. Serum Mg (MgS) and erythrocyte Mg (Mgrbc) were measured by standard laboratory methods [10].

4. As a marker of endothelial dysfunction, von Willebrand factor antigen (vWF) was measured by enzyme-linked immunosorbent assay [11].

5. All statistical tests were done with the SPSS package 9.0 for Windows. Kolmogarov-Smirnov testing was used for normality of distribution, Student t-test or ANOVA (analysis of variance) were used for comparison of means. The Chi-square or Fisher's Exact test were used for frequency distributions. Multiple logistic regression, stepwise forward, was used to assess the strength of associations and to test interrelations between the co-variates. All tests were performed two-sided and a P-value < 0.05 was considered significant.

Results

The levels of metabolic control (HbA1c) and circulating Mg (MgS and Mgrbc) at the start and the end of the 5 year study in both groups are summarized in *Table I*. Mean HbA1c remained unchanged in Groups A and B. In Group A MgS raised significantly ($p < 0.01$) but remained low in group B. Mgrbc increased to a mean of 2.37 mmol/l (sd 0.16) considered as a near normal value ($p < 0.001$) but did not change significantly in group B.

Insulin needs (U/kg body weight) did not change significantly in both groups (data not shown).

In group A there is an improvement in 6% of the patients, a *statu quo* in 80% and a worsening in 14%. In contrast, in group B there was no improvement, a *statu quo* in 63% and a worsening in 37% of the patients (Fisher Exact test: $p < 0.05$).

Logistic regression at the end of the study shows that only duration of diabetes ($\beta = 0.29$, $p < 0.0001$, OR = 1.34, CI = 1.18-1.52) and MgS ($\beta = -3.86$, $p < 0.05$, OR = 0.02, CI = 0.0007-0.581) were determinant parameters in the evolution of DR. In both groups the staging of blood pressure or nephropathy remained unchanged at the end of the study as compared to the start.

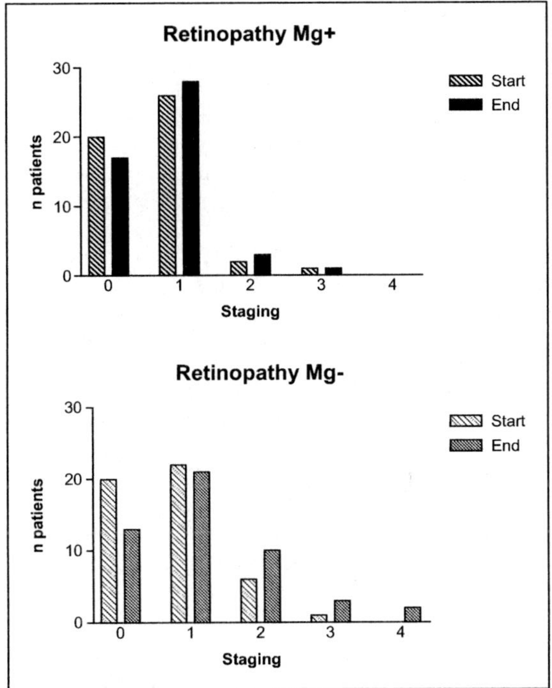

Figure 1. Staging of retinopathy in group A (Mg+) and group B (Mg-) at the start and the end of the study.

In both groups, vWF was increased at the start: 191% (sd 57) in group A and 188% (sd 62) in group B. Only in group A a significant decrease to 162% (sd 51) ($p < 0.01$) was observed related to the normalisation of the Mg status.

Discussion

Retinopathy is a common microvascular complication of T1DM. Insufficient metabolic control, duration of diabetes, uncontrolled hypertension and/or associated nephropathy are well-known risk factors for progression of retinopathy [12]. The mechanism for retinal blood flow changes remain however uncertain and multifactorial. Capillary basement membrane thickening, sorbitol pathway activation, hyperaggregation of erythrocytes and platelets, endothelial dysfunction and release of vascular growth factors are generally considered to be direct consequences of chronic hyperglycemia [11].
However Mg depletion can enhance most of these mechanisms:

Table I. Metabolic control and Mg status at the start and end of the study (paired t-test).

Mean (sd)	HbA1c start %	HbA1c end %	MgS start mmol/l	MgS end mmol/l	Mgbc start mmol/l	Mgbc end mmol/l
A	7.83 (0.09)	7.80 (0.80)	0.70 (0.09)	0.74 (0.09) $p < 0.01$	2.02 (0.14)	2.37 (0.16) $p < 0.001$
B	8.00 (0.95)	7.75 (0.75)	0.73 (0.08)	0.73 (0.09)	2.08 (0.16)	2.15 (0.17)

- Hypertension and retinal vasospasm [13];
- Reduced vascular reactivity [14];
- Increased basal membrane thickening [15];
- Thrombophilia of diabetes [16];
- Inhibition of Na/K ATPase activity [17].

Therefore the restoration of a normal Mg status in T1DM by supplementation could theoretically help to slow the progression of microangiopathic complications as was already demonstrated for peripheral polyneuropathy [18].
In this study metabolic control remained stable and there was no change in the blood pressure or nephropathy-staging in the patients over a period of 5 years.
In the supplemented group however, normalisation of intracellular Mg resulted in stabilisation of the retinopathy in contrast to a significant worsening in the matched group without supplements where the circulating Mg levels remained low.
In this regard, the possible role of Mg in restoring endothelial function must be emphasized. Indeed, the levels of vWF, a marker of endothelial dysfunction and platelet adhesion, were elevated at the start in both groups but decreased only significantly in the supplemented group. It is well-known that qualitative and quantitative abnormalities of vWF have been reported in diabetic patients [19] and that high plasma vWF is associated with a breakdown of the blood-retinal barrier even in early minimal DR [20]. The recent observation that high vWF in T1DM is associated with a decrease of retinal blood flow which can lead to local hypoxia only emphasizes the importance of this marker [11].

Conclusion

The exact mechanism by which Mg influences the pathogenesis and progression of chronic diabetic complications merits further studies but meanwhile it looks reasonable to supplement Mg-depleted T1DM together with optimalisation of insulin treatment, normalisation of blood pressure and dyslipidemia, an individualized diet and a perfect teaching not forgetting to include parameters of Mg status in the biochemistry follow-up.

References

1. DCCT. The effect of intensive treatment of diabetes on the development and progression of long-term complications in insulin-dependent diabetes mellitus. *N Engl J Med* 1993 ; 329 : 977-86.
2. McNair P, Christiansen C, Madsbad S, *et al.* Hypomagnesemia, a risk factor in diabetic retinopathy. *Diabetes* 1978 ; 27 : 1075-7.
3. Ceriello A, Giugliano D, Dello Russo P, Passariello N. Hypomagnesemia in relation to diabetes retinopathy. *Diabetes Care* 1982 ; 5 : 558-9.
4. Fujii S, Takemura T, Wada M, Akai T, Okuda K. Magnesium levels in plasma, erythrocyte and urine in patients with diabetes mellitus. *Horm Metab Res* 1982 ; 14 : 161-2.
5. Vanroelen W, Van Gaal L, Van Rooy P, De Leeuw I. Serum and erythrocyte magnesium levels in type 1 and type 2 diabetics. *Acta Diabetol Lat* 1985 ; 22 : 185-90.
6. Tosiello L. Hypomagnesemia and diabetes mellitus. *Arch Intern Med* 1996 ; 156 : 1143-8.
7. American Diabetes Association. Magnesium supplementation in the treatment of diabetes. *Diabetes Care* 1992 ; 15 : 1065-7.
8. De Valk H, Hardes P, Van Rijn H, Erkelens D. Plasma magnesium concentration and progression of retinopathy. *Diabetes Care* 1999 ; 22 : 864-5.
9. Porta M, Bandello F. Diabetic retinopathy : a clinical update. *Diabetologia* 2002 ; 45 : 1617-34.
10. Engelen W, Bouten A, De Leeuw I, De Block C. Are low magnesium levels in type 1 diabetes associated with EMG signs of polyneuropathy? *Magnesium Res* 2000 ; 13 : 197-203.
11. Feng D, Bursell S, Clermont A, *et al.* Von Willebrand factor and retinal circulation in early-stage retinopathy of type 1 diabetes. *Diabetes Care* 2000 ; 23 : 1694-8.
12. Eurodiab Prospective Complications Study Group. Risk factors for progression to proliferative diabetic retinopathy. *Diabetologia* 2001 ; 44 : 2203-9.

13. Cohen L, Laor A, Kitzes R. Reversible retinal vasospasm in magnesium-treated hypertension despite no significant change in blood pressure. *Magnesium* 1984 ; 3 : 159-63.

14. Ewald U, Tuvemo T. Reduced vascular reactivity in diabetic children and its relation to diabetic control. *Acta Paediatr Scand* 1985 ; 74 : 77-85.

15. Dyckner T, Hägg E, Lithner F, Nyhlin H, Wester P. Muscle magnesium and capillary basement membrane thickness in diabetes mellitus. *Diabete Metab* 1988 ; 14 : 619-22.

16. Nadler J, Malayan S, Luong H, Shaw S, Natarayan R, Rude R. Intracellular free magnesium deficiency plays a key role in increased platelet reactivity in type 2 diabetes mellitus. *Diabetes Care* 1992 ; 15 : 835-41.

17. Grafton G, Bunce C, Sheppard M, Baxter M. Effect of Mg^{++} on Na^{+}-dependent inositol transport. Role for Mg^{++} in etiology of diabetic complications. *Diabetes* 1992 ; 41 : 35-9.

18. De Leeuw I, Engelen W, De Block C, Van Gaal L. Long-term magnesium supplementation influences favourably the natural evolution of neuropathy in Mg-depleted type 1 diabetic patients. *Magnesium Res* 2004 ; 17 : 109-14.

19. Pasi K, Enayat M, Horrocks P, Wright A, Hill F. Qualitative and quantitative abnormalities of von Willebrand antigen in patients with diabetes mellitus. *Thromb Res* 1990 ; 59 : 581-91.

20. Porta M, Townsend C, Clover G, *et al.* Evidence for functional endothelial cell damage in early diabetic retinopathy. *Diabetologia* 1981 ; 20 : 597-601.

Advances in Magnesium Research: New Data 2006: 135-8

Magnesium influence on lipid-lowering effect of fenofibrate in non-insulinodependent diabetes mellitus patients

M. Nechifor[1], S. Bistriceanu[2], M. Scutaru[3], D. Chelărescu[1], C. Nechifor[4]

1. Dept. of Pharmacology, University of Medicine and Pharmacy "Gr. T. Popa" Iasi, Romania; 2. General Ambulatory Service Botosani, Romania; 3. Human Anatomy Dept; 4. Student, University of Medicine and Pharmacy "Gr. T. Popa" Iasi, 700115, Romania

Non insulin dependent diabetes mellitus (NIDDM) is a widespread disease, frequently related with dyslipidemias. NIDDM therapy contained especially orally antidiabetic dongs associated with antidylipidemic drugs when it is necessary.
Magnesium has multiple influences on lipidic and glucidic metabolism [1-3]. The magnesium modulator role on plasma lipid level and their antiatherogen effect were sustained by many authors [4]. Low magnesium plasmatic level is associated frequently with NIDDM.
In this paper we followed the influence of magnesium intake on the fenofibrate administration in NIDDM patients with dyslipidemia.

Patients and method

We worked on 2 groups of patients of both gender, aging between 20 to 62 years, diagnosed with diabetes mellitus type II (NIDDM) during 2003-04 and with dyslipidemia. Repartition of patients was randomized. Dyslipidemia was diagnosed in the presence (before therapy) of total serum cholesterol level > 200 mg/dl, serum LDL cholesterol level > 160 mg/dl. The initial values of plasma triglyceride were variable. In each group were 8 patients with initial serum triglyceride level > 400 mg/dl. In both groups NIDDM treatment was performed with glibenclamide (Maninil® – Berlin Chemie) 5 mg/kg/day *per os* in 2 equal daily dose. For dyslipidemia treatment was administrated fenofibrate (Lypanthyl®), 300 mg/kg/day *per os*. First group (28 patients) received glibenclamide and fenofibrate. Second group (26 patients) received glibenclamide and fenofibrate and magnesium orotate (Magnerot® – Worwag Pharma) 500 mg/day *per os* (2.7 mEq or 32.6 mg magnesium/day). Including criteria were: NIDDM, fasted plasmatic glucose level > 1.3 g/l, dyslipidemia.

There were not included in trial patients with diabetic nephropathy, chronic heart failure, renal chronic failure, liver cirrhosis, malabsorbtion syndrome, treatment with diuretics, with magnesium- (or other cations) containing drugs, chronic ethanol intake.

There were determined in all patients plasmatic magnesium level, fasted plasmatic glucose level, total cholesterol, LDL, HDL, triglycerides plasma levels before starting lipid lowering therapy and after 45 days of treatment. Data obtained were statistically interpreted with t test. The trial was approved by an ethical committee and respected the rules for clinical trials.

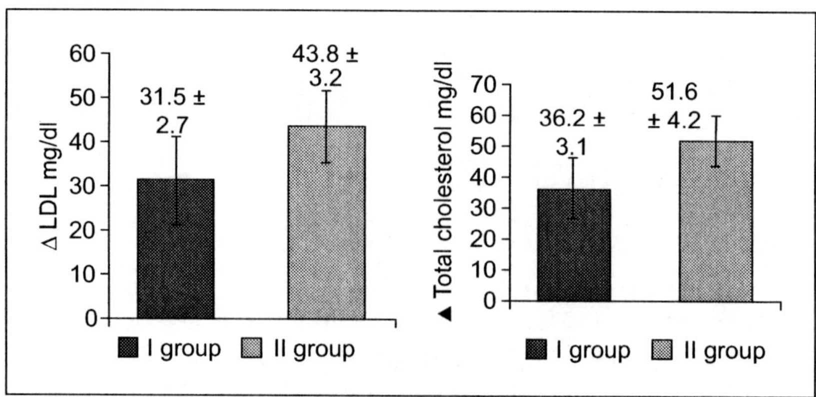

Figure 1. The influence of magnesium orotate (500 mg/day) on fenofibrate action (300 mg/day) on LDL and total cholesterol in patients with NIDDM. Group I: Lypanthyl® 300 mg/day. Group II: Maninil® + Lipanthyl® 300 mg/day + Mg Orotat 500 mg/day. Δ LDL = difference between initial and after 45 days LDL plasma level. ▲ Total cholesterol = difference between initial and after 45 days total cholesterol plasma level.

Results

Administration of magnesium orotate 500 mg/day increased significantly plasmatic level of total Mg (*Figure 1*).

Association between magnesium and fenofibrate significantly increased fenofibrate action on decreasing LDL and total cholesterol plasma levels (*Figure 2*).

If is considered hypotriglyceridemiant fenofibrate action on all patients, magnesium administration doesn't change this effect. If are considered only patients with triglyceride level > 400 mg/dl, magnesium administration increases significantly this effect (*Figure 3*).

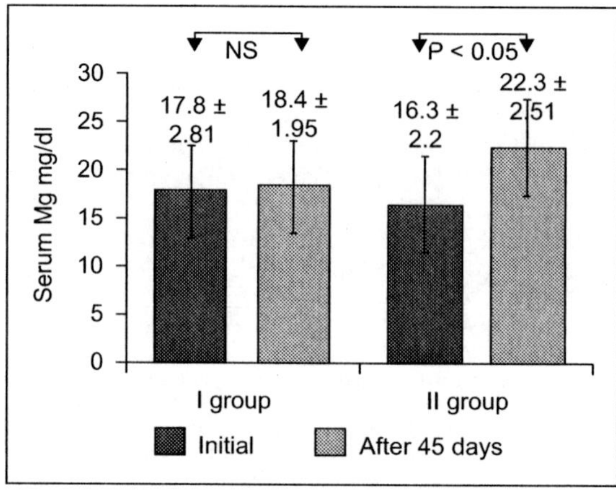

Figure 2. Variations in plasmatic level of total magnesium in patients treated with fenofibrates.

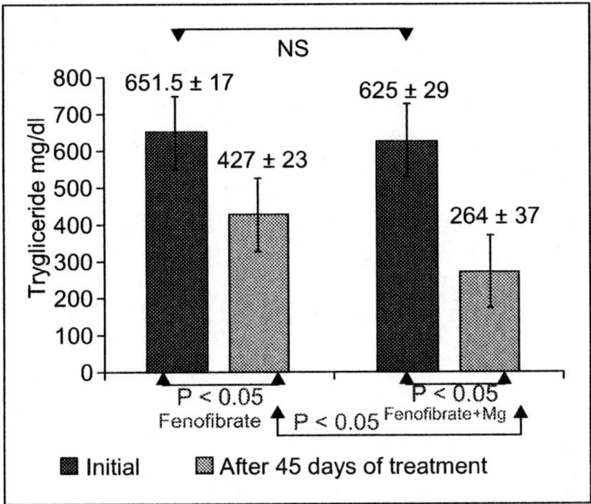

Figure 3. Mg influence on fenofibrate lowering effect in patients with high triglyceride plasmatic level. Each group had 8 patients with initial triglyceride level > 400 mg/dl.

Discussions

Fibrates (gemfibrozil, fenofibrate, bezafibrate, ciprofibrate) represent one of the main group of lipid lowering drugs. Their main effect is decreasing plasmatic lipoproteins, increasing HDL, and decreasing plasmatic concentrations of triglycerides.

Their mechanism of action is not fully understood. It is considered that their effect on plasmatic lipids could be done on the level of peroxisome proliferators-activated receptors (PPARs) [5]. Fibrates increase lipoprotein lipase (LPL) synthesis. On this way the clearance of triglycerides and lipoproteins are increased by fibrates therapy.

There are many data showing that in diabetes mellitus type II (NIDDM) exists a decrease of plasmatic magnesium level [6]. In diabetic patients with NIDDM with microalbuminemia, the level of serum magnesium is lower than in diabetic patients with normal albuminemia [7].

In all diabetic patients, ionized plasmatic magnesium concentrations is negatively correlated with the level of serum triglycerides and concentrations of glycozilated hemoglobin (HbA1C). Data regarding correlation of serum magnesium level with the plasma glucose level in patients with NIDDM are controversy.

There are few data regarding relations between lipid lowing medication administrated to patients with NIDDM and serum and intracellular magnesium. Haenni *et al.* (2001 [8]) show that antidyslipidemic treatment with gemfibrozil, 600 mg b.i.d., 4 months and simvastatin 10 mg/day 4 months in patients with NIDDM decreases serum LDL level and in the same time magnesium level (with 0.02 mM/l). Changes in glucose tolerance or in sensitivity to insulin were not correlated with changes in concentrations of serum magnesium level [8].

Our data show that magnesium administration (Magnerot®) increases significantly statistic fenofibrate (Lypanthyl®)-induced effect on decreasing total cholesterol and LDL plasma concentrations. In the case that is considered fenofibrate -induced hypotriglyceridemiant action on all patients, magnesium administration doesn't change this effect (after 45 days of treatment). If are considered only patients with triglyceride level > 400 mg/dl, magnesium administration significantly increases this effect.

There are relatively few data which show that magnesium intake (on few months) in oral administration in mice significantly decreases plasmatic level of cholesterol and triglycerides [9]. In experimental atherogenic diet in mice and rabbits, magnesium decreases cholesterol level and its deposit in vascular wall and decreases lipid peroxidation [4, 10]. In areas were tap water contains less magnesium, coronary atheromatose incidence and myocardial infarct is increased [11]. Magnesium deficit favors the increase of serum cholesterol and increases lipid peroxidation and is associated with dyslipidemia [12-14].

Busserolles *et al.* (2003 [15]) have shown that sucrose feeding in magnesium deficit rats was associated with a higher plasma triglycerides levels and a higher susceptibility to lipid peroxidation compared with magnesium deficient rats feed with normal diet.

We considered that magnesium association with lipid lowering fibrates therapy may enhance moderately but significantly the effect of this therapy.

References

1. Kummerow FA, Mahfouz M, Zhou Q. Cholesterol metabolism in human umbilical arterial endothelial cells cultured in low magnesium media. *Magnes Res* 1997 ; 10 : 355-60.
2. Rayssiguier Y. Magnesium and lipids interrelationships in the pathogenesis of vascular diseases. *Magn Bull* 1981 ; 9 : 165-71.
3. Durlach J. Les relations entre magnésium et glucides. *Diabete* 1971 ; 19 : 99-113.
4. Altura BT, Brust M, Bloom S, Barbour RL, Stempak JG, Altura BM. Magnesium dietary intake modulates blood lipid levels and atherogenesis. *Proc Natl Acad Sci USA* 1990 ; 87 : 1840-4.
5. Kersten S, Desvergne B, Wahli W. Roles of PPARs in health and disease. *Nature* 2000 ; 405 : 421-4.
6. Walti MK, Zimmermann MB, Spinas GA, Hurrell RF. Low plasma magnesium in type 2 diabetes. *Swiss Med Wkly* 2003 ; 133 : 289-92.
7. Corsonello A, Ientile R, Buemi M, *et al.* Serum ionized magnesium levels in type 2 diabetic patients with microalbuminuria or clinical proteinuria. *Am J Nephrol* 2000 ; 20 : 187-92.
8. Haenni A, Ohrvall M, Lithell H. Serum magnesium status during lipid-lowering drug treatment in non-insulin-dependent diabetic patients. *Metabolism* 2001 ; 50 : 1147-51.
9. Ravn HB, Korsholm TL, Falk E. Oral magnesium supplementation induces favorable antiatherogenic changes in ApoE-deficient mice. *Arterioscler Thromb Vasc Biol* 2001 ; 21 : 858-62.
10. Yamaguchi Y, Kitagawa S, Kunitomo M, Fujiwara M. Preventive effects of magnesium on raised serum lipid peroxide levels and aortic cholesterol deposition in mice fed an atherogenic diet. *Magnes Res* 1994 ; 7 : 31-7.
11. Crawford T, Crawford MD. Prevalence and pathological changes of ischaemic heart-disease in a hard-water and in a soft-water area. *Lancet* 1967 ; 1 : 229-32.
12. Mahfouz MM, Zhou Q, Kummerow FA. Cholesterol oxides in plasma and lipoproteins of magnesium-deficient rabbits and effects of their lipoproteins on endothelial barrier function. *Magnes Res* 1994 ; 7 : 207-22.
13. Itoh K, Kawasaka T, Nakamura M. The effects of high oral magnesium supplementation on blood pressure, serum lipids and related variables in apparently healthy Japanese subjects. *Br J Nutr* 1997 ; 78 : 737-50.
14. Durlach J, Bara M. *Le magnésium dans la biologie et en médicine.* Paris : Éditions médicales Internationales, 2000 ; (p17).
15. Busserolles J, Gueux E, Rock E, Mazur A, Rayssiguier Y. High fructose feeding of magnesium deficient rats is associated with increased plasma triglyceride concentration and increased oxidative stress. *Magnes Res* 2003 ; 16 : 7-12.

Effects of risk factors for chronic degenerative diseases on magnesium metabolism in human

M. Nishimuta, N. Kodama, E. Morikuni, N. Matsuzaki

Laboratory of Mineral Nutrition, Division of Human Nutrition, The Incorporated Administrative Agency of Health and Nutrition, 162-8636, Tokyo, Japan

Many scientists, who are interesting in the magnesium (Mg) metabolism, seek the scientific evidences concerning the relationship between Mg deficit and chronic degenerative diseases, such as hypertension, diabetes mellitus, hyperlipemia and so in human. However, plasma Mg content is not a good indicator for Mg status because it is kept within narrow ranges. Nishimuta *et al.* [1] demonstrated that some risk factors for chronic degenerative diseases, such as overeating, physical and mental stresses and so on, induced Mg uresis accompanied with isomolar calcium (Ca) uresis in human. Nishimuta *et al.* [2] also demonstrated that an anaerobic physical exercise increased in urine excretions of Mg and Ca and that mild exercise decrease in them. And serum Mg increased after the anaerobic exercise and kept high for more than 12 hours [3]. To understand these phenomena, characteristic of Mg and Ca as elements should be considered *(Table I)*.

The concept [intracellular, extracellular and bone minerals]

Minerals can be divided into two categories of intracellular and extracellular minerals based on physiological sites of accumulation.

Essential minerals rich in intracellular space compared with in extracellular one, such as potassium (K), Mg, phosphorus (P), iron (Fe) and zinc (Zn), are proposed to define as "intracellular minerals (ICMs)". On the contrary, those rich in extracellular space, such as sodium (Na) chloride (Cl) and Ca, are also proposed to define as "extracellular minerals (ECMs)" [4]. Essential minerals, which compose the bone, and whose physiological pool is including the bone, are proposed to define as "bone minerals (BMs)". BMs consist of both intracellular (Mg, P and Zn) and extracellular (Na and Ca) minerals [5]. In these definitions, Mg is belong to intracellular and bone minerals, and Ca is belong to extracellular and bone minerals.

Homeostasis of serum Ca and Mg after isomolar uresis of Mg and Ca

Plasma contents in Mg and Ca are controlled within narrow ranges and have no circadian variations. Urine contents in Mg and Ca are usually isomolar or high in Mg compared with that of Ca. Urinary excretion of both Mg and Ca have obvious circadian rhythm, high in daytime and low at night [3]. Mechanism to keep plasma Mg and Ca levels constant to compensate their urine loss may be important in considering both the metabolism of Mg and Ca, and etiology of chronic degenerative diseases. Under the condition of no food supply, Ca must be released from the bone, sole Ca pool in the body. However, it is difficult to clear which organ, the bone or the cells, release Mg into the blood stream to keep plasma Mg constant, because these two organs are both the pool of Mg in the body. If all Mg was released from the bone where is major pool of Mg, by the osteophagocytosis, more than ten times of Ca must be released at the same time from the bone, because the contents of Ca in the bone is far higher than that of Mg. So, Mg in the cell is concluded to be the major source of the compensation in case of enough Mg

Table I. Speciation of essential elements (20).

Major elements	(4) : H, C, N, O
Minerals	All elements except major elements

Speciation of essential minerals (16)

	Intake/day	Intracellular minerals	Extracellular minerals	Others
Major minerals (7)	more than 100 mg	K, P, Mg	Na, Cl, Ca	P
Trace elements (9)	less than 100 mg			
Trace element I (4)	more than 1 mg	Fe, Zn		Cu, Mn
Trace element II (5)	less than 1 mg			Co, Cr, I, Mo, Se
Bone minerals (5)		P, Mg, Zn	Na, Ca	

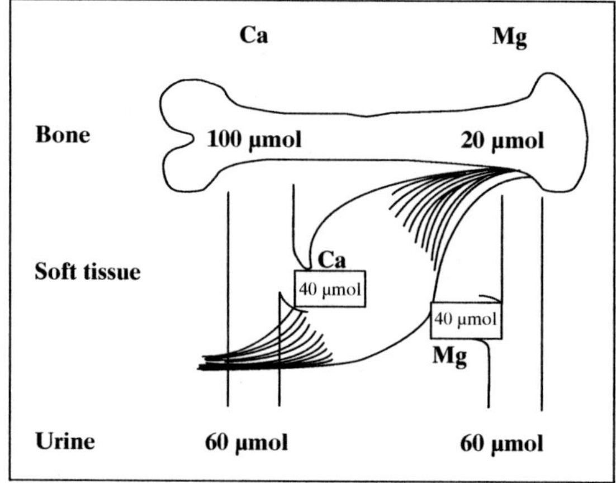

Figure 1. Trans tissue transport of calcium (Ca) and magnesium (Mg) after isomolar uresis of Ca and Mg. Actually, more Ca than that excreted are absorbed from the bone, enters into the soft tissue.

reservation. However, Mg in the bone may release into blood stream when Mg in the cell is short. At that case, excess Ca released from the bone entered into blood stream, then entered into the muscle, causing heart muscle infarction, skeletal muscle contracture and tonic contraction of blood vessel. Excess Ca also forms stone outside of the bone as well as eminent Ca uresis without accompanied Mg uresis. In this case molar ratio of Ca/Mg exceeds more than one.

So, molar ratio of Ca/Mg is a possible indicator for the Mg nutrition. If this value is more than one, Mg in the bone is supposed to be resorped with large amount of Ca.

When sodium (Na) in the body is short, Na in the bone is released with Ca and Mg, because the bone is only physiological pool of Na [6] *(Figure 1)*.

References

1. Nishimuta M, Kodama N, Ono K. Magnesium uresis by risk factors for chronic degenerative diseases. In: Itokawa Y, Durlach J, eds. *Magnesium in Health and Disease.* London: John Libbey & Co Ltd, 1989: 279-84.

2. Nishimuta M, Kodama N, Takeyama H, Toyooka F. Magnesium metabolism and physical exercise in humans. In: Theophanides T, Anastassopoulou J, eds. *Magnesium; Current Status and New Developments.* Netherlands: Kluwer Academic Pub, 1997: 109-13.

3. Meludu SC, Nishimuta M, Yoshitake Y, *et al.* Magnesium homeostasis before and after high intensity (anaerobic) exercise. In: Rayssiguier Y, Mazur A, Durlach J, eds. *Advance in Magnesium Research: Nutrition and health.* London: John Libbey & Co Ltd, 2001: 443-6.

4. Nishimuta M. The concept (intra and extra cellular minerals. In: Collery P, Poirier LA, Manfait M, Etienne JC, eds. *Metal Ions in Biology and Medicine.* Paris: John Libbry Eurotext, 1990: 69-74.

5. Nishimuta M. The concept of intracellular-, extracellular- and bone-minerals. *Biofactors* 2000; 12: 35-8.

6. Kodama N, Nishimuta M, Suzuki K. Negative balance of calcium and magnesium under relatively low sodium intake in human. *J Nutr Sci Vitaminol (Tokyo)* 2003; 49: 201-9.

The behaviour of circulatory magnesium in experimental goiter

D. Simu[1], G. Drăgotoiu[2], L. Gozariu[2], O. Rotaru[3]

1. Clinical Center of Diabetes, Nutrition and Metabolic Diseases, Cluj-Napoca, Romania; 2. Clinical of Endocrinology, UMF, Cluj-Napoca, Romania
3. Faculty of Veterinary Medicine, Cluj-Napoca, Romania

Background

In the animal world magnesium is absolutely necessary for the synthesis and the use of macroergic compounds, the synthesis of hydrogen and electrons' transporters as well as the activation of more than 300 enzymes. It is involved in the three main types of metabolism of carbohydrates, lipids and proteins [1, 2].
Until now, many factors have been identified to play an important role in the etiopathogeny of endemic goiter, through the interference of the iodine absorption, its capture, its incorporation in thyroid hormones or its use at the tissue level, together or apart from iodine deficiency [3].
To identify new factors implied in the etiopathogeny of endemic goiter, we considered important to observe the behaviour of circulatory magnesium in experimental goiter. This aspect has not been studied until now, as only the association of the goiter with hypomagnesaemia at children had been notice before.

Study design

The experimental goiter has been obtained through administration of Methiltiouracil, 10 mg/dose, for 30 days. As an antithyroid of synthesis, it inhibits the formation of the thyroid hormones; the effect becomes obvious clinically after a period of 2-3 weeks of delay. Magnesium supplementation was made by using Trimagant tablets. This brings about 30 mg ions of magnesium/tablet. The administrated daily dose was 50 mg. To assure the right intake of iodine necessary for the thyroid function altered by Methiltiouracil, the tablet of potassium iodide of 1 mg/dose was used once a week. Out of the respective tablets we obtained a powder which was administrated daily, differently on groups, under the form a fresh suspension in milk, per os, physiologically, to avoid the stress of the use of an endo-gastric catheter.

Animal groups

60 white Wistar rats, 52 days old, weighting 125-150 g, fed with standard food, were used after an accommodation period of 7 days. The green mass and the drinking water were allowed with no restriction. After the accommodation period the animals were weighted with laboratory scales. The rats were then parted into 6 groups of 10 animals each, isolated in separated cages having the same conditions of food, light and temperature. For 30 days they were treated as following:

– Group 1 - M – control group without treatment (N = 10);
– Group 2 – Mt – which was administrated Methiltiouracil 10 mg/rat/day (N = 10);
– Group 3 – Mt+Mg which was administrated Methiltiouracil 10 mg/rat/day + Trimagant 50 mg/rat/day (N = 10);
– Group 4 - Mt+Mg+I which was administrated Methiltiouracil 10 mg/rat/day + Trimagant 50 mg/rat/day + potassium iodide 1 mg/rat/week (N = 10);

– Group 5 – Mt+I which was administrated Methiltiouracil 10 mg/rat/day + potassium iodide 1 mg/rat/week (N = 10);
– Group 6 – Mg - which was administrated Trimagant 50 mg/rat/day (N = 10).

After a month of treatment, the animals were weighted with the same laboratory scales. Being under anesthesia with ether, they were sacrificed by beheading. Blood samples for laboratory analyses were taken as well as the thyroid gland of each animal in order to weight and get histological tests.

Calculation of thyroid index

$$TI = \frac{\text{Thyroid weight}(\text{mg}) \times 1,000}{\text{Body weight}(\text{g})}$$

Hemoglobin dosage (Drabkin method)

Normal value = 13.3 g/d.

Hematocrit determination (standard method)

Normal value = 46%.

Plasma calcium and magnesium dosage (spectrophotometry of atomic absorption) (4)

Normal values: Plasma calcium = 10.2 ± 1.4 mg/dl.
Plasma magnesium = 2.85 ± 1.2 mg/dl.

Erythrocyte Magnesium dosage (spectrophotometry of atomic absorption)

We have not found normal values for the laboratory animals.

The preparation of the histological tests

Was made according to a standard procedure and the examination was done under a microscope with a power of magnifying 400 times.
The statistic processing of the data was made by help of computerized program EXCEL, by using the t-Student test, as well as indicators of correlation and regression.

Results

Animal weight (*Figure 1*)

By analyzing these data we observe that the most important increasing in weight was present in the animals in group 1 (M) and 6 (Mg). The final weight of the animals in group 1 and 6 is significantly higher ($p < 0.05$) than their initial weight. The final weight of the animals in group 2 (Mt), treated only with Methiltiouracil is significantly lower than that of the control group.

The thyroid weight of the animals in the 6 studied groups (*Figure 2*)

We notice that the average value of the thyroid weight of the animals in group 2 (Mt) is significantly higher ($p < 0.001$) than that of the animals in the control group (M), showing the increase in volume of the thyroid gland for the rats treated with Methiltiouracil. We obtained a significantly increase of the thyroid gland in all groups in which Methiltiouracil was administrated, alone or in association.

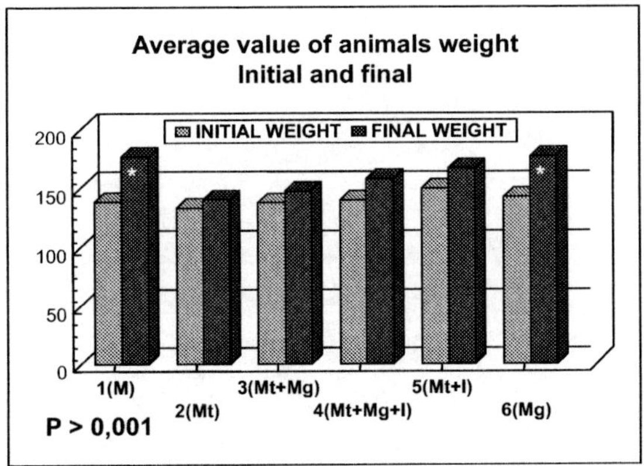

Figure 1. Average value of animal weight.

Thyroid index (*Figure 3*)

By comparing the thyroid index of group 2 with the thyroid index of the group 3 and 4 we notice that the last two are significantly lower ($p < 0.01$) than in group 2, but not with the same level of significance as compared with the control group. The average value of the thyroid index of group 6 supplemented only with magnesium is fairly equal with that of the control group.

Hemoglobin and hematocrit (*Figures 4,5*)

By comparing the average values of Hb in all the groups which were administrated anthyroidian alone or in association, we observed no significant differences. The average value of the Hct in groups 2 (Mt) and 3 (Mt + Mg) is significantly lower that than of the control group ($p < 0.01$).

Blood plasma calcium and magnesium (*Figure 6*)

The average value of blood plasma Ca and Mg in all the treated groups does not significantly differ from that of the control group.

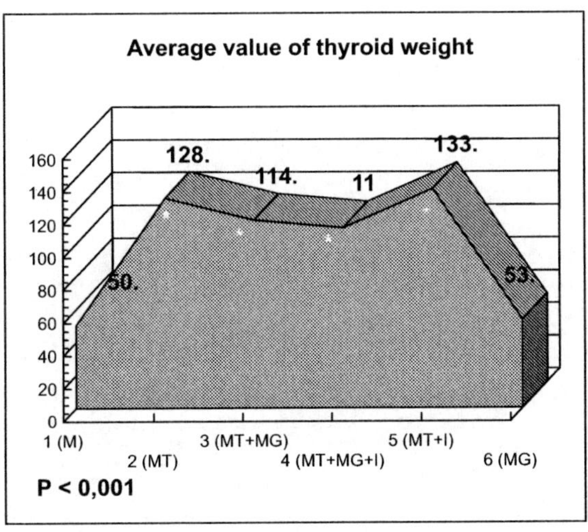

Figure 2. Average value of thyroid weight.

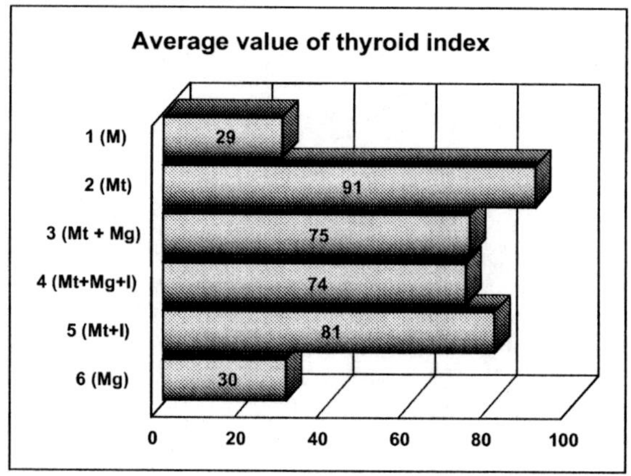

Figure 3. Average value of thyroid index.

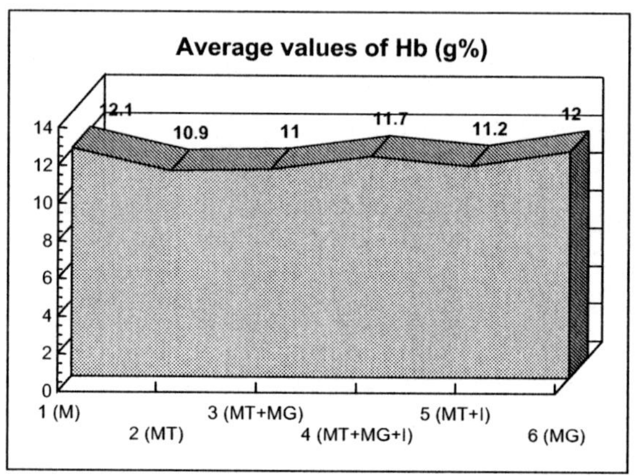

Figure 4. Average values of Hb.

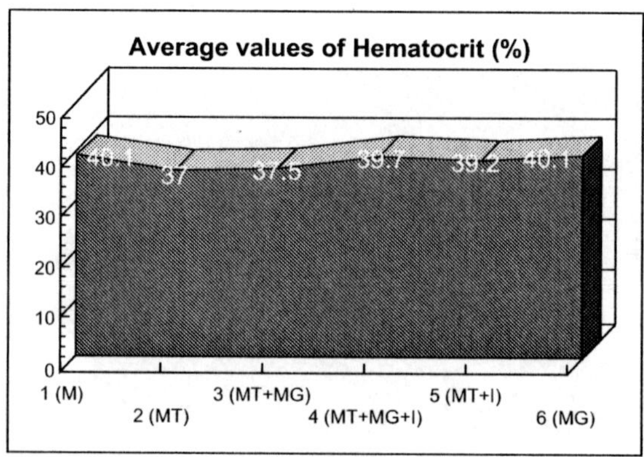

Figure 5. Average values of Hct.

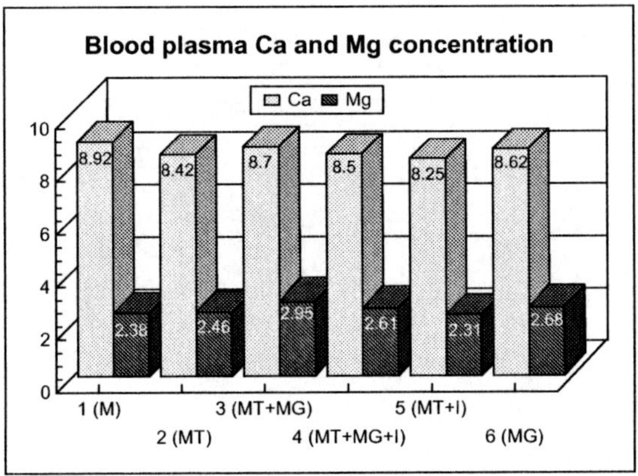

Figure 6. Blood plasma Ca and Mg concentration.

Erythrocyte magnesium (*Figure 7*)

By analyzing the obtained data we observe the fact that the average value of the erythrocyte magnesium in group 2 (Mt) and 3 (Mt + Mg) increases significantly in comparison with the control group ($p < 0.01$). In all the other group the average value of the erythrocyte magnesium does not differ significantly from the control group.

There are positive correlation between the average weigh of the thyroid and the average value of the erythrocyte magnesium in the case of groups 2 (Mt) and 3 (Mt + Mg) (*Figure 8*).

Histological aspects

The histological examination was meant to compare the modifications appearing at the level of thyroid follicles with the control group. They referred to:

- The height of the follicle epithelium;
- The characteristics of the colloid in the follicle lumen;
- The relationship between stroma thyroid follicles and the thyroid vascularization;

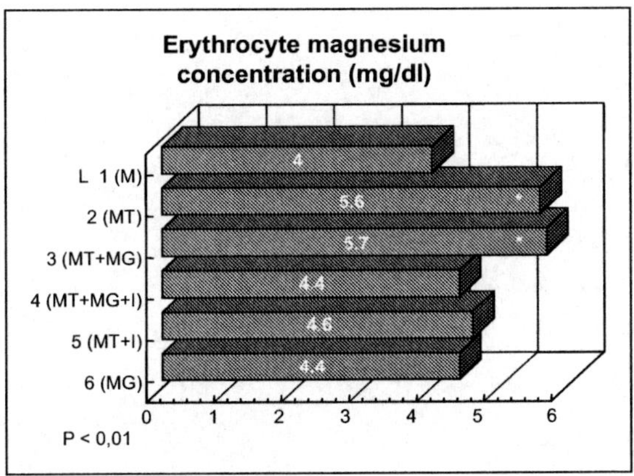

Figure 7. Erythrocyte magnesium concentration (mg/dl).

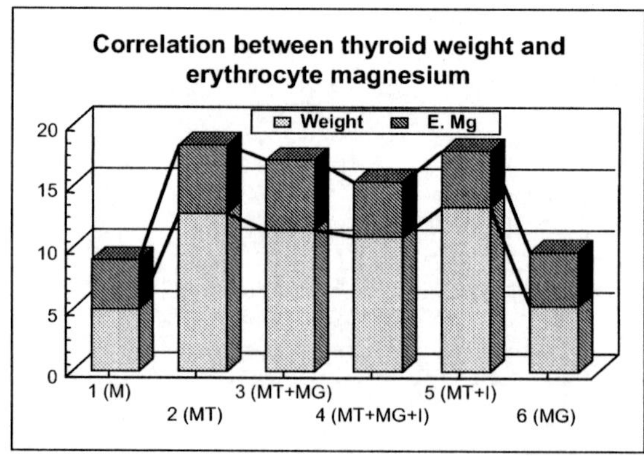

Figure 8. Correlation between thyroid weight and erythrocyte magnesium.

Group 1 – control

The follicle epithelium character and the configuration of the colloid storage in the lumen plead for the morphologic aspects of a thyroid parenchyma within physiological limits (*Figure 9*).

Group 2 – Mt (*Figure 10*)

For this group the histological aspects render under the action of the antithyroid substances and they may be synthesized as following:

- The quantitative decrease of the thyroid follicle;
- The increase of the height of the thyroid follicle cells;
- Diffused follicular epithelial hyperplasia with metaplasia;
- Increased mitotic activity;
- Active perifollicle vascularization.

Group 3 (Mt + Mg) (*Figure 11*)

We notice that at the level of some follicular forms the proliferation of epithelium lead to the formation of intrafollicular papiliform excrescences specific to parenchymatous struma forms.

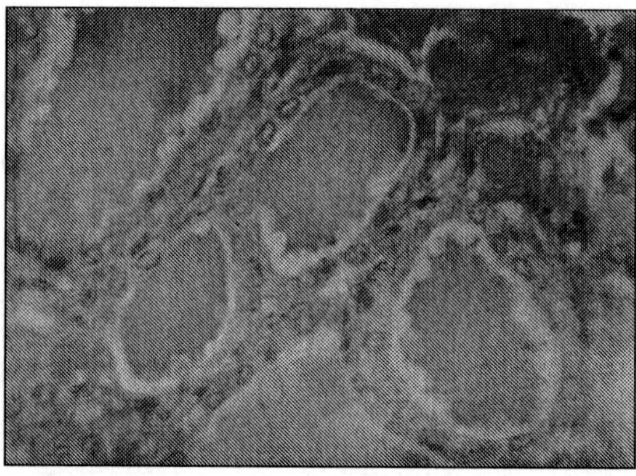

Figure 9. Thyroid section control group.

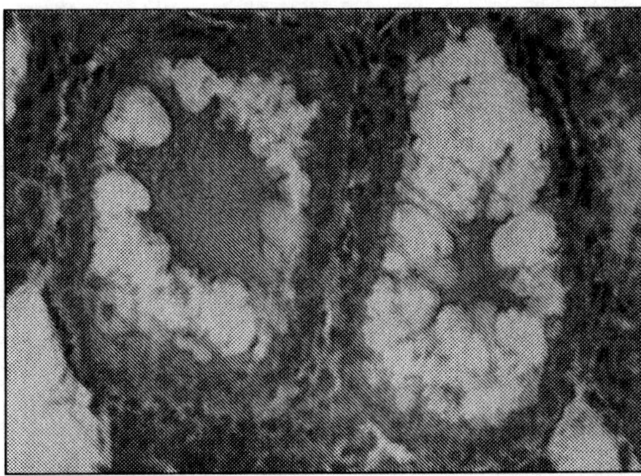

Figure 10. Thyroid section group 2, treated by Methiltiouracil. Parenchymatous goiter aspect.

Group 4 (Mt + Mg + I) (*Figure 12*)

The thyroid sections suggest a status of hypofunction materialized through with follicles with high aspects; the highly reduced lumen appears empty, without colloid or with a reduced colloid mass, with a compact structure, without tendency to mobilization.

Group 5 (Mt + I) (*Figure 13*)

Histological modifications are those met at people treated at the same time with antithyroid substances and with compounds based on iodine. These aspects are:

- The follicles have a large lumen;
- The follicular epithelium has flat cells, pressed on the basement membrane;
- Hypochromic nuclei are stuck to the basal limitant.

Group 6 (Mg) (*Figure 14*)

The histopathological examination shows large follicles, with an almost cylindric high epithelium, but colloid less and there where it exists, is on its way to resorption.

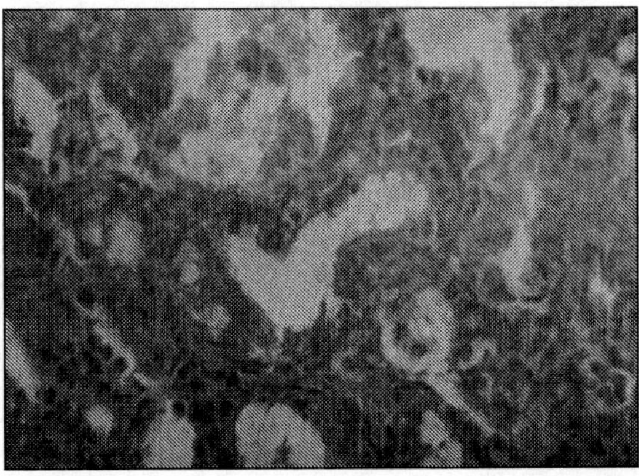

Figure 11. Thyroid section group 3, treated by Methiltiouracil and Magnesium. Parenchymatous struma aspect.

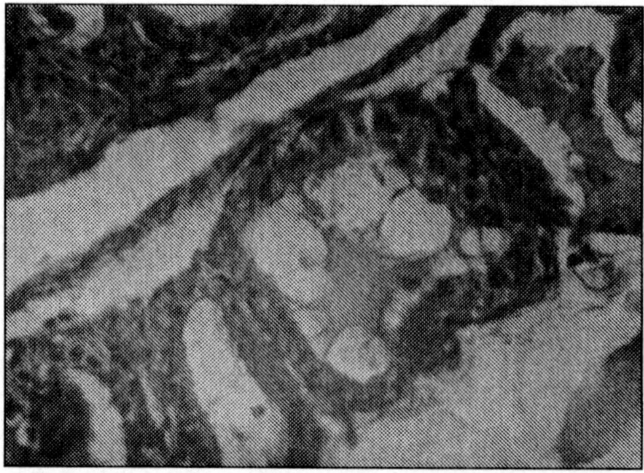

Figure 12. Thyroid section group 4, treated by Methiltiouracil, Magnesium and Potassium Iodide Thyroid hypofunction aspect.

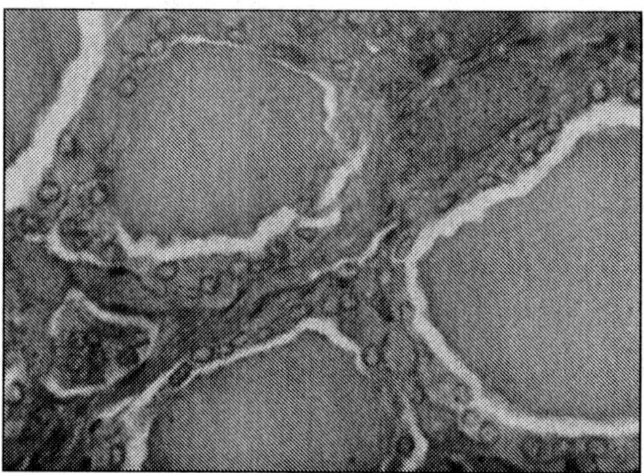

Figure 13. Thyroid section group 5, treated by Methiltiouracil and Potassium Iodide. Transitory thyroid hypofunction aspect.

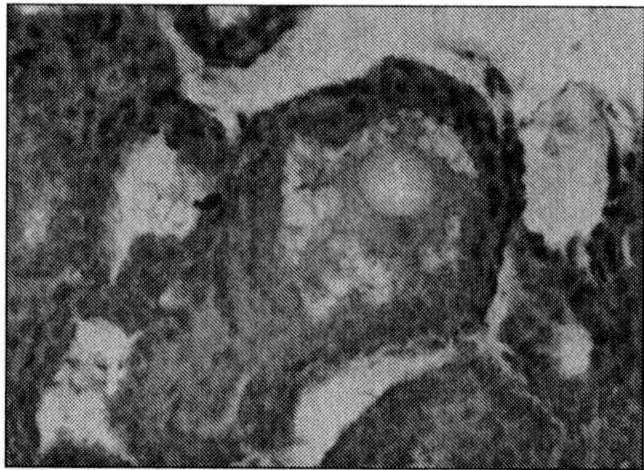

Figure 14. Thyroid section group 6, treated by Magnesium Thyroid hypofunction with tendency to parenchymatous struma aspect.

Discussions

The values of the thyroid weight and the histological examination of the gland sections confirm the rightfulness of the experimental model used. The modification of the erythrocyte magnesium in the groups treated with Methiltiouracil shows the fact that, as at the biochemical level, disturbances are produced in the experimental goiter. Higher increasing in weight in the control group and in one supplemented with magnesium suggests discretely the role of thyroid hormones in growth and development; in the other groups, Methiltiouracil inhibits there formation [1, 5].
As a result of the administration of antithyroid of synthesis, the average weight of the thyroid increased significantly ($p < 0.001$) at the animals in group 2-5 in comparison whit group1 and 6, but differently. Iodine supplementation at the same time with an antithyroid of synthesis does not influence the hypothyroidian effect produced by Methiltiouracil. This can be counteracted only by the administration of thyroid hormones. It seems that the association between iodine and magnesium offers the best protection against the action of Methiltiouracil [4, 6].
Usually, in hypothyroidia normochroma anemia appears, as an expression of the accommodation of the body to the low oxygen consumption at peripherical level [7, 8]. During the experiment we obtained significantly lower average values of Hb and Hct in group 2 and 3 in comparison with the control group without speaking, nevertheless, about anemia. These results may confirm a hypothyroidia.
Blood plasma magnesium concentration does not significantly differ for any group in comparison with the control group, as it is well known that the regulation of its plasmatic metabolism does not depend on a hormone system, the kidney being the one which maintains stable magnesemia, despite the high variations in food intake [8, 9].
The different behaviour in the groups of erythrocyte magnesium with significantly increases even in the groups with the biggest thyroid hypertrophies, suggests its certain involvement in the thyroid function, as two things are well known:

- Magnesium is an important cofactor in energetic metabolism and in the protein synthesis [10];
- Urinary excretion is diminished in hypothyroidism [11, 12];

The erythrocyte magnesium concentration seems to be genetically regulated [13] and is not correlated with the storage of magnesium in other tissues. The absence of an exchange between erythrocyte magnesium and the serum one is the subject of many debates [14]; our data plead for the existence of this phenomenon.
The fact that magnesium cannot replace the iodine in synthesis and the mechanism of thyroid secretion results from the histopathological configuration of the experimental variant in which the association Methiltiouracil and magnesium was used.
The histopathological results show without any doubt a status of hypofunction, even with a tendency to struma. The same thyroid hypofunction also appears in the group treated exclusively with magnesium, confirming the data in the specialized literature, which consider magnesium as an additional factor in the endemic goiter. The limit between the real needs of the body in magnesium and the food or therapeutic excess of this macroelement in the conditions of iodine deficiency cannot be established exactly yet.

Conclusions

1. The size of the experimental goiter induced through antithyroidian of synthesis differs in group, according to the used therapeutic associations.
2. Only iodine supplementation at the same time with an antithyroidian of synthesis does not hinder the appearance of thyroid hypertrophia.
3. The association Methiltiouracil and magnesium maintains and even stresses the status of this function, leading to structural modifications which justify the diagnosis of parenchymatous goiter.
4. The experimental variant through which we tested the association Methiltiouracil, magnesium and iodide does not modify the status of thyroid hypofunction, but it amplified it even more, demonstrating that magnesium seems to cancel the effect of the iodine. Between these two ions there is a state of incompatibility at thyroid level.

5. The thyroid hypofunction with a tendency to goiter also appears in the group treated exclusively with magnesium, confirming the data in the specialized literature, which consider magnesium ion in excess as an additional factor in the endemic goiter.

6. The increase of erythrocyte magnesium without the modification of the serum circulatory magnesium plead for its involvement in the cell reaction mechanism of the experimental goiter.

7. The real role of the magnesium ion at thyroid level may be known only through biochemical studies at subcellular thyroid level.

References

1. Arthur JR. The role of Selenium in thyroid hormone metabolism. *Can J Physiol Pharmacol* 1991 ; 69 : 1648-52.
2. Beckers C, Delange F. In : *Iodine deficiency. Endemic goiter and endemic cretinism.* New York : Wiley and Sons, 1980 : 199.
3. Anke M, Groppel B, Schwartz S, Krause U, Arnhold W, Janreis G. In : *Influence of Zinc deficiency on Iodine status of pigs. International Trace Elements Symp.* 1989 : 6.
4. Dragatoiu G. Dinamica magneziului intra si extracelular in conditii de stres si tratamente medicamentoase. Teza de Doctorat, UBB, Cluj-Napoca, 1995 : 22-28, 333-48, 117-24.
5. Adams DD, Kennedy TH, Stewart JC, Utiger RE, Vidor FI. Hyperthyroidism in Tasmania following iodine supplementation : mesurements of thyroid-stimulating autoantibodies and thyreotropin. *J Clin Endocrinol Metab* 1975 ; 41 : 221-2.
6. Aikawa JK. In : *The biochemical and cellular functions of Magnesium. Ist Intern Symp on Magnesium Deficiency in Human Pathology.* 1971 : 39-54.
7. Bretter E, Sinca A. In : *Iodine and co-pathogenic elements in endemic goiter. Intern Symp on Iodine and other trace elements. Yena.* 1986 : 14-6.
8. Dorofteiu M. Tiroida in fiziologie. In : *Coordonarea organismului uman.* Cluj-Napoca : Ed. Dacia, 1992 : 223-33.
9. Alabbassy A, Delbrige J, Eckstein R. Microfollicular thyroid adenoma and congenital goitrus hypothyroidism. *Arch Dis Child* 1992 ; 67 : 1294-5.
10. Abbasciano V, Mazzotta D, Vecchiatti G, Tassinari D, Niesen I, Sartori S. Changes in serum, erythrocyte and urinary magnesium, after a single dose of cisplatin combination chemotherapy. *Magnesium Res* 1991 ; 2 : 123-5.
11. Amir SN, Ingebar SH. *New concepts in thyroid disease. New York.* 1993.
12. Barlet JP. Role of the thyroid gland in Mg-induced hypocalcemia in bovine. *Horm Metab Res* 1971 ; 3 : 63-4.
13. Durlach J. *Magnesium.* London : J. Libbey, 1993.
14. Gozariu L. Metabolismul calciului si magneziului in gusa endemica. *Clujul Medical* 1989 : 1.

VII. Magnesium and cell functions

Molecular biology of the CorA-Mrs2-Alr1 family of magnesium transport proteins

R.J. Schweyen

Max F. Perutz Laboratories, Department of Genetics, University of Vienna, Austria

Abstract. The CorA-Mrs2-Alr1 superfamily of proteins form Mg^{2+} selective channels for the transit of Mg^{2+} through membranes. CorA, the first Mg^{2+} transport protein to be described, is ubiquitous in bacteria and constitutes their major Mg^{2+} uptake system. Alr1 is a distant homologue in the plasma membrane of lower eukaryotes. Mrs2 proteins, which are distantly related to CorA and to Alr1, form the major mitochondrial Mg^{2+} channel from yeast to plants and mammals. Plants also express Mrs2-type proteins in the plasma membrane. The yeast mitochondrial Mrs2 protein forms a homo-oligomeric channel with high Mg^{2+} conductivity. This channel protein may sense intramitochondrial Mg^{2+} concentrations and exert control on the influx of Mg^{2+}.

Membranes constitute tight barriers for the passage of cations, preventing their free movement along the electrochemical gradients. With the help of proteins inserted into membranes cations overcome this barrier. Additionally, the proteins exert control on ion transport, balancing ion uptake and ion extrusion such that intracellular or intra-organellar concentrations remain within ranges required for physiological processes.

Uptake of cations into cells or from the cytoplasm into mitochondria essentially occurs via channels (pores) and is driven by the inside negative membrane potentials. The channel forming proteins themselves, or associated proteins, limit ion fluxes by opening or closing the channels. Extrusion of cations from the cytoplasm to the exterior of cells or from mitochondria to the cytoplasm occurs against the electrochemical gradient and consumes energy. Proteins mediating such extrusion processes either are consuming ATP or act as exchangers, making use of concentration gradients of one ion to drive extrusion of another one.

Physiology of Mg^{2+} transport across the plasma membrane and the mitochondrial membrane has been studied in detail over the past three decades (Romani *et al.*, 1993; Jung and Brierley, 1994; for review) [6, 12]. However, it remained obscure: i) which and how many genes and proteins are involved in these processes, ii) which transport mechanisms are being used and how Mg^{2+} homeostasis is controlled and iii) to which extend their activity is essential for cellular functions.

Various genetic approaches have led meanwhile to the identification and characterization of two classes of cellular Mg^{2+} channel proteins, the large CorA-Mrs2-Alr1 family and the TRPM6 / TRPM7 family. Members of the former appear to be ubiquitous from bacteria to plants and mammalia (see below). Genes encoding members of the latter family have been found in the animal world only. In humans they appear to constitute a major uptake system of Mg^{2+} from the exterior into the intracellular milieu of cells (Schmitz *et al.*, 2004; Konrad *et al.*, 2005, for recent reviews) [11, 13].

The CorA-Mrs2-Alr1 protein family

Pioneering work in bacteria led to the first identification of genes involved in Mg^{2+} transport (reviewed in Silver, 1978; Smith and Maguire, 1998) [15, 16]. The constitutively expressed CorA protein was recognized as constituting the major Mg^{2+} influx system in Bacteria and Archaea. Genetic screens led to the identification of two distant homologues of the CorA protein in the yeast *Saccharomyces cerevisiae*, Alr1 (MacDiarmid and Gardner, 1998) [10] and Mrs2 (Bui *et al.*, 1999) [1]. Alr1 is located in the plasma membrane where it forms the major pore for Mg^{2+} influx into the cytoplasm (Graschopf *et al.*, 2001) [3]. While Alr1 related proteins are restricted to lower eukaryotes, Mrs2 type proteins are encoded by nearly all eukaryotic genomes whose sequences are known today. Mrs2 proteins are found in mitochondria of yeast, of plants as well as of mammalia [1,5,18]. In plants Mrs2-related proteins also are found in the plasma membrane and probably in other membranes as well [8,14].
Sequence variation within the CorA-Mrs2-Alr1 superfamily of proteins is very high. The only common features are the presence of two predicted transmembrane (TM) domains near the C-terminus, separated by a short sequence stretch which is preceded by a GMN motif at the end of first TM domain (*Figure 1*). The long N- and short C-terminal sequences of the CorA, Mrs2 and Alr1 subfamilies are variable in size and have no sequence motifs in common. The conclusion that members of these subfamilies are functional orthologs rests essentially on the findings i) that CorA in *Salmonella typhimurium*, Mrs2 and Alr1 in yeast are mediating influx of Mg^{2+} into cells or mitochondria (Smith

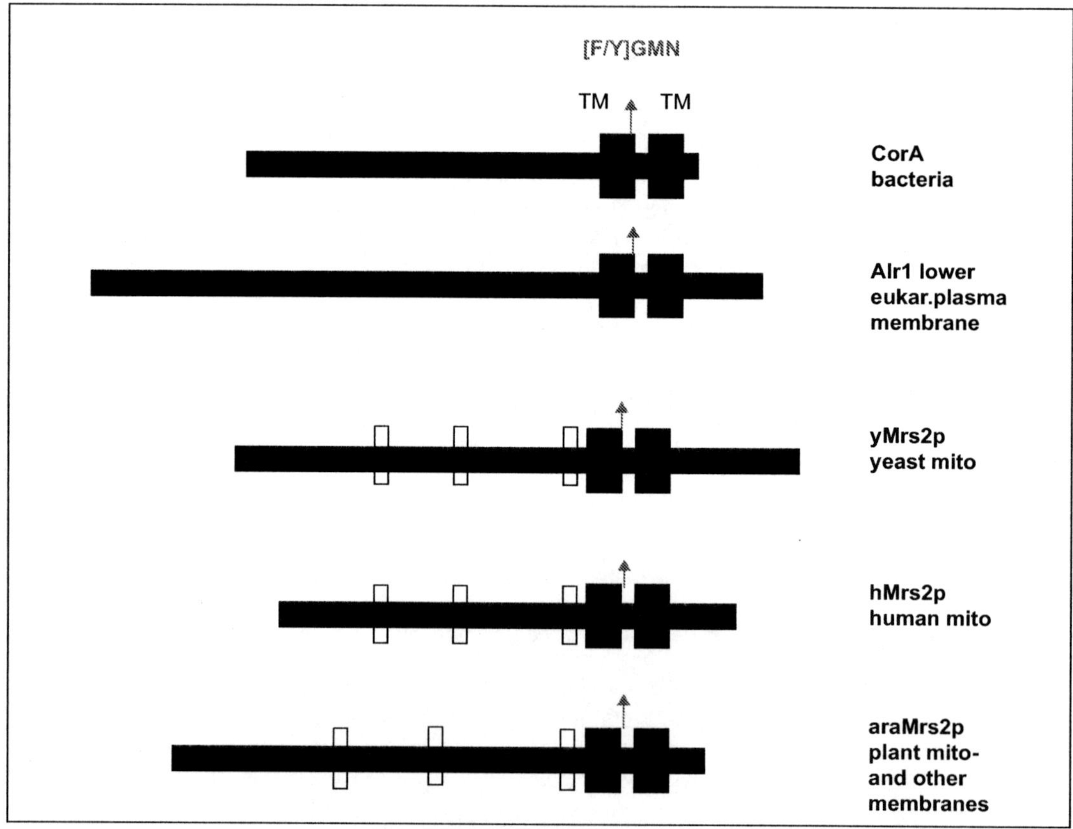

Figure 1. Conserved elements in the CorA-Mrs2-Alr1 superfamily of proteins. In common to all these proteins are two adjacent transmembrane domains (black boxes) near the C-terminus. The first of these domains terminates with a GMN motif, the only primary sequence motif conserved between all of them. Within the Mrs2 subfamily there are several regions of high sequence conservation (grey boxes).

and Maguire, 1998; Liu *et al.*, 2002; Kolisek *et al.*, 2003) [7, 9, 16] and ii) CorA can partially substitute for either Mrs2 or for Alr1 when expressed in yeast (Bui *et al.*, 1999; Graschopf *et al.*, 2001) [1, 3] (*Figure 2A*).

Bacterial CorA as well as yeast Mrs2 and Alr1 have been shown to form homo-oligomeric (tetra- or pentameric) complexes in their cognate membranes (Kolisek *et al.*, 2003; Warren *et al.*, 2004) [7, 17]. Circumstantial evidence indicates that they may form channels for the influx of Mg^{2+} from extracellular milieus into the cytoplasm (CorA, Alr1) or from the cytoplasm into mitochondria (Mrs2). This ion flux is driven by the inside negative membrane potential and in physiological terms therefore constitutes a uniport (Kolisek *et al.*, 2003; Smith *et al.*, 1998; Szegedy *et al.*, 1999; Liu *et al.*, 2002; Graschopf *et al.*, unpubl.; Froschauer *et al.*, 2004) [2, 7, 9, 16].

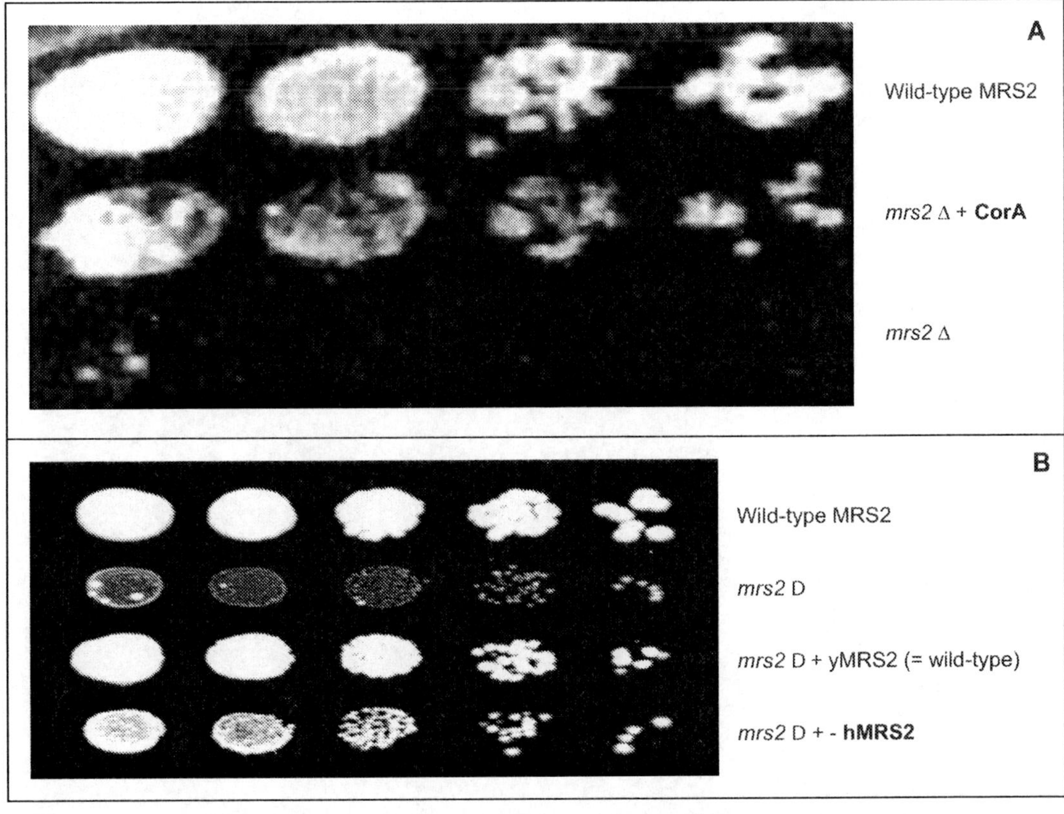

Figure 2. Growth defect of yeast MRS2 knock-out cells (mrs2Δ), and restoration of growth by expression of the bacterial CorA protein or the human Mrs2 protein. Serial dilutions of yeast cells were applied to agar plates with non-fermentable substrate (YPG). Wild-type cells grew well (from left to right: confluent growth to single colony growth). Mutant mrs2Δ cells fail to grow, except for few papilla which represent spontaneous suppressor mutants. Growth was fully restored when the yeast Mrs2 protein (yMRS2) was expressed from a plasmid in the mutant cells. Growth was partly restored when the bacterial CorA protein (of *Salmonella typhimurium*) was expressed in mrs2Δ yeast (provided with a mitochondrial targeting sequence) or when the human Mrs2 protein (with its own mitochondrial targeting sequence) was expressed from a plasmid in mrs2Δ yeast.

The Mrs2 subfamily of proteins

Proteins of this subfamily are characterized by a considerable degree of sequence conservation, particularly in their central part, N-terminal to the TM domains. The short C-terminus, in contrast, is highly variable in length and sequence. Consistent with this sequence conservation is our finding that the single human MRS2 gene as well as some of the fifteen MRS2 genes of *Arabidopsis thaliana* functionally substitute for Mrs2p when expressed in a yeast knock-out mutant deleted for MRS2 (mrs2Δ; Zsurka *et al.*, 2001; Schock *et al.*, 2000) [14, 18] (*Figure 2B*).

Yeast mrs2Δ mutant cells are mitochondrially defective, but viable when provided with fermentable carbon sources (Bui *et al.*, 1999) [1]. Human cells with strongly reduced expression of Mrs2 (siRNA mediated knock-down), in contrast, are inviable (Piskacek *et al.*, unpubl.). The bacterial CorA protein or the human Mrs2 protein restore growth of the yeast mrs2Δ mutant, when expressed in yeast cells (*Figures 2A and B*).

Making use of the fluorescent dye mag-fura 2 entrapped in isolated mitochondria we have shown that wild-type yeast as well as human mitochondria exhibit a rapid increase in intramitochondrial [Mg^{2+}] when the external [Mg^{2+}] is raised (*Figure 3*). Mutant yeast and human cells (MRS2 knock-out or knock-down, respectively) lack the rapid uptake of Mg^{2+} (Kolisek *et al.*, 2003; Piskacek *et al.*, unpubl.) [7]. In the absence of the Mrs2 protein in mrs2Δ yeast mutant cells there is a residual, slow uptake of Mg^{2+} which over extended periods of time brings intramitochondrial [Mg^{2+}] to levels which are only twofold below wild-type levels and apparently are sufficient to support basic functions of mitochondria which are not related to oxidative energy conservation. The nature of this transport system remains presently unknown. It also remains obscure why knock-down of Mrs2 in human cells is lethal while knock-out in yeast cells is not. Either cells can survive without energy conservation by mitochondria.

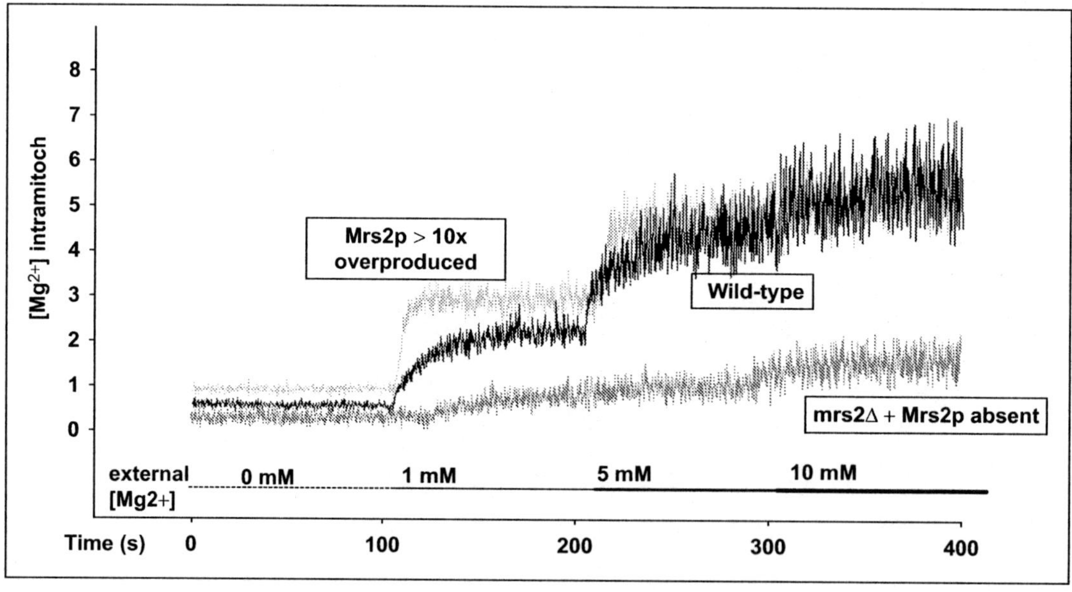

Figure 3. Influx of Mg^{2+} into isolated yeast mitochondria. Isolated yeast mitochondria were loaded with the Mg^{2+} sensitive fluorescent dye mag-fura2 and intramitochondrial [Mg^{2+}] was determined by ratiometric measurements (Kolisek *et al.*, 2003) [7] at resting conditions (nominally Mg^{2+} free buffer) or upon addition of Mg^{2+} to the buffer to final concentrations of 1 mM, 5 mM and 10 mM. Mitochondria were isolated from cells expressing a single, chromosomal MRS2 copy (wild-type) or expressing Mrs2 from a high copy number vector (Mrs2p > 10x) or not expressing any Mrs2p at all (mrs2Δ). The rapid increase of intramitochondrial [Mg^{2+}] upon addition of external Mg^{2+} is dependent on Mrs2 expression and reflects Mg^{2+} influx. Steady state [Mg^{2+}] is similar in mitochondria with low or high expression of the Mrs2 protein. This points to a pronounced control of Mg^{2+} influx.

Studies with isolated yeast mitochondria revealed that intramitochondrial [Mg^{2+}] increased within seconds up to ten-fold above basic levels when external [Mg^{2+}] were raised (*Figure 3*). This led us to postulate a high capacity influx via a channel involving the Mrs2 protein (Kolisek *et al.*, 2003) [7].

The Mrs2 ion channel

This channel appears to consist of a tetra- or pentamer of the Mrs2 protein (Kolisek *et al.*, 2003) [7]. The Mrs2 protein with its two adjacent transmembrane domains is inserted in the mitochondrial membrane such that both N- and C-termini are protruding towards the matrix while the short sequence connecting the transmembrane domains only is oriented towards the intermembrane space (*Figure 4A*). This short connecting sequence on the outer side of the membrane is likely to form the mouth of the channel. Interestingly, it contains a surplus of negatively charged amino acids, particularly two conserved glutamic acid residues at position six and seven after the GMN motif (*cf. Figure 4A*), which may serve to attract cations via electrostatic interaction to the mouth of the channel.

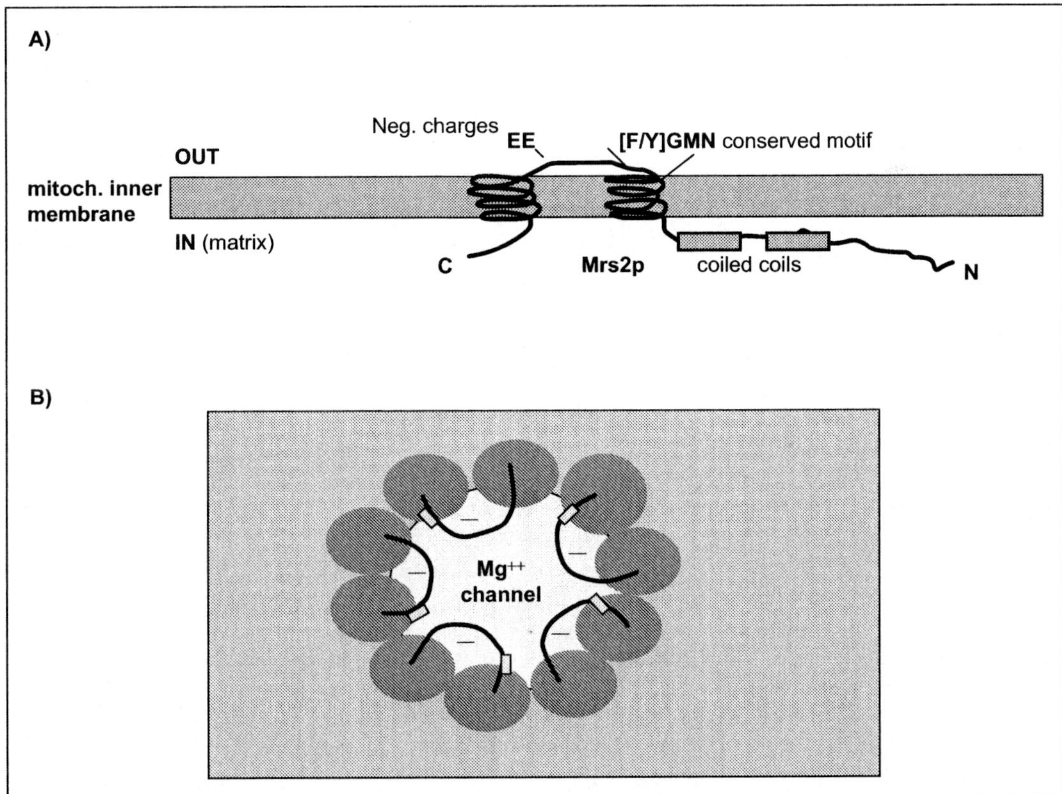

Figure 4. Topology of the Mrs2 protein.
A) Most eukaryotes have members of the Mrs2 subfamily in their mitochondria. The yeast and the human Mrs2 proteins have both the N- and the C-terminus exposed to the mitochondrial matrix (inside) and only the short sequence connecting the two transmembrane domains oriented towards the intermembrane space (outside). The connecting sequence contains two glutamic acid residues at positions 6 and 7 after the GMN motif. The long N-terminal sequence contains two predicted coiled coil domain regions which may participate in oligomerization of Mrs2.
B) View from outside on the 'mouth' of the putatively pentameric Mrs2 channel. The GMN motif (box) and the negative charges of two glutamic acid residues (-) are characteristic features of the putative Mg^{2+} entry site.

Overexpression of the MRS2 gene in yeast leads to a several fold increase in Mrs2 protein in the mitochondrial membrane and a concomitant increase of Mg^{2+} influx into isolated mitochondria (*Figure 3*). This strongly supports the notion that the uptake system is made up of a homo-oligomeric complex of Mrs2 proteins only. Interestingly, steady state intramitochondrial [Mg^{2+}] levels reached shortly after a raise of external [Mg^{2+}] are only slightly increased when Mrs2 is overexpressed (*Figure 3*). And independent of the level of Mrs2 expression, the increase in intramitochondrial [Mg^{2+}] reaches a maximum at about 5 mM, far below electrochemical equilibrium, which would thousand fold higher, given the high, inside negative membrane potential of mitochondria. This saturation of intramitochondrial [Mg^{2+}] at low mM levels is interpreted to reflect an efficient system of Mg^{2+} homeostasis control. In part this may be mediated through a H+/Mg^{2+} exchanger, but its continuous activity (with constant external [Mg^{2+}] of 10 mM) would result in the partial dissipation of the proton gradient, which we have not observe. We rather assume that the Mrs2 protein itself exerts a control of Mg^{2+} influx into mitochondria, *e.g.* by closing the channel when intramitochondrial [Mg^{2+}] increases. This is supported by our findings that mutations in the middle part of the Mrs2 protein affect steady state intramitochondrial [Mg^{2+}] (Gregan *et al.*, 2001; Kolisek *et al.*, 2003; Weghuber *et al.*, in prep.) [4, 7].

What next?

So far, evidence for CorA, Mrs2 or Alr1 forming cation conducting channels is mostly circumstantial. Electrophysiology has been involved to characterize Alr1 mediated Mg^{2+} uptake of yeast protoplasts (cells devoid of cell walls) (Liu *et al.*, 2002) [9]. Results were fully consistent with Mg^{2+} influx through channels, but electrophysiology on yeast protoplasts is challenging and does not allow a full biophysical characterization of channels. We recently set out now to perform single channel patch clamping on mitochondrial inner membrane material (collaboration with C. Romanin, Linz, Austria). Preliminary data provide clear evidence for a high conductance Mg^{2+} channel, composed of Mrs2 protein units only (Weghuber *et al.*, in prep.).

While the basic role of Mrs2 as forming a membrane channel now seems to be clear, the role of certain domains of the protein (*cf. Figure 3*) for the oligomerization of this protein and, most importantly, for ion conductance and flux control of the channel remain widely obscure. To address these questions, we will make use of the excellent genetic methods in yeast to introduce mutations in various domains and test their effects on the formation and stability of Mrs2 oligomers and on Mg^{2+} influx into the organelles.

Why would mitochondria need to have a high conductance Mg^{2+} channel? Cations are not metabolized. One might think, therefore, that uptake would occur at the same rate as growth of cells and organelles, *i.e.* it would double once during several hours. A low conductance channel would perfectly match this requirement. Yet mitochondria have a high conductance channel which in few seconds can double the intramitochondrial [Mg^{2+}] (*cf. Figure 3*). We may speculate that there are physiological situations where mitochondria have an immediate requirement for large amounts of Mg^{2+}. Mitochondrial Mg^{2+} does not seem to play a role in signalling as prominent as that of Ca2+, but we may not exclude it *a priori*. We rather speculate that the requirement for high influx rates is connected with the most prominent role of mitochondria, ATP synthesis. ATP levels may vary rapidly and to a considerable extend. Mg^{2+} is the binding partner of choice for ATP (with higher affinity to ATP than to ADP). A burst in ATP synthesis thus may require rapid influx of Mg^{2+}.

Finally, why is the function of Mrs2 essential for mitochondrial function in yeast and for life of human cells? Steady state mitochondrial [Mg^{2+}] fall only to about 50% of normal levels when Mrs2 protein is lacking, because other, slow uptake systems are still active. There may be components in mitochondria whose function is severely affected by a two-fold reduction in [Mg^{2+}]. Alternatively, not the reduced steady state [Mg^{2+}] causes problems to the mitochondria, but the absence of the rapid, high capacity, Mrs2-mediated Mg^{2+} influx.

References

1. Bui DM, Gregan J, Jarosch E, Ragnini A, Schweyen RJ. The bacterial magnesium transporter CorA can functionally substitute for its putative homologue Mrs2p in the yeast inner mitochondrial membrane. *J Biol Chem* 1999; 274: 20438-43.
2. Froschauer EM, Kolisek M, Dieterich F, Schweigel M, Schweyen RJ. Fluorescence measurements of free $[Mg^{2+}]$ by use of mag-fura 2 in *S. typhimurium*. *FEMS Microbiol Lett* 2004; 237: 49-55.
3. Graschopf A, Stadler J, Hoellerer M, *et al.* The yeast plasma membrane protein Alr1p controls Mg^{2+} homeostasis and is subject to Mg^{2+} dependent control of its synthesis and degradation. *J Biol Chem* 2001; 276: 16216-22.
4. Gregan J, Kolisek M, Schweyen RJ. Mitochondrial magnesium homeostasis is critical for group II intron splicing in vivo. *Genes Dev* 2001; 15: 2229-37.
5. Gregan J, Bui DM, Pillich R, Fink M, Zsurka G, Schweyen RJ. The mitochondrial inner membrane protein Lpe10p, a homologue of Mrs2p, is essential for magnesium homeostasis and group II intron splicing in yeast. *Mol Gen Genet* 2001; 264: 773-81.
6. Jung DW, Brierley GP. Magnesium transport by mitochondria. *J Bioenerg Biomembr* 1994; 26: 527-35.
7. Kolisek M, Zsurka G, Samaj J, Weghuber J, Schweyen RJ, Schweigel M. Mrs2p is an essential component of the major electrophoretic Mg^{2+} influx system in mitochondria. *EMBO J* 2003; 17: 1235-44.
8. Li L, Tutone AF, Drummond RS, Gardner RC, Luan S. A novel family of magnesium transport genes in Arabidopsis. *Plant Cell* 2001; 13: 2761-75.
9. Liu GJ, Martin DK, Gardner RC, Ryan PR. Large Mg(2+)-dependent currents are associated with the increased expression of ALR1 in Saccharomyces cerevisiae. *FEMS Microbiol Lett* 2002; 213: 231-7.
10. MacDiarmid CW, Gardner RC. Overexpression of the *Saccharomyces cerevisiae* magnesium transport system confers resistance to aluminium ion. *J Biol Chem* 1998; 273: 1727-32.
11. Konrad M, Schlingmann KP, Gudermann T. Insights into the molecular nature of magnesium homeostasis. *Am J Physiol Regul Integr Comp Physiol* 2005; 288: R782-R795.
12. Romani A, Marfella C, Scarpa A. Cell magnesium transport and homeostasis: role of intracellular compartments. *Miner Electrolyte Metab* 1993; 19: 282-9.
13. Schmitz C, Perraud AL, Fleig A, Scharenberg AM. Dual-function ion channel/protein kinases: novel components of vertebrate magnesium regulatory mechanisms. *Pediatr Res* 2004; 55: 734-7.
14. Schock I, Gregan J, Steinhauser S, Schweyen RJ, Brennicke A, Knoop V. A member of a novel Arabidopsis thaliana gene family of candidate Mg^{2+} ion transporters complements a yeast mitochondrial group II intron splicing mutant. *Plant J* 2000; 24: 489-501.
15. Silver S. Transport of cations and anions. In: Rosen BP, ed. *Bacterial transport*. New York: Marcel Dekker, Inc, 1978: 221-324.
16. Smith RL, Maguire ME. Microbial magnesium transport: unusual transporters searching for identity. *Mol Microbiol* 1998; 28: 217-26.
17. Warren MA, Kucharski LM, Veenstra A, Shi L, Grulich PF, Maguire ME. The CorA Mg^{2+} transporter is a homotetramer. *J Bacteriol* 2004; 186: 4605-12.
18. Zsurka G, Gregan J, Schweyen RJ. The human mitochondrial Mrs2 protein functionally substitutes for its yeast homologue; a candidate magnesium transporter. *Genomics* 2001; 72: 158-68.

Magnesium and calcium-modulated mitochondrial functions

N.E.L. Saris[1], V.V. Teplova[2]

1. Department of Applied Chemistry and Microbiology, Viikki Biocenter 1, PO Box 56, 00014 University of Helsinki, Helsinki, Finland
2. Institute of Theoretical and Experimental Biophysics, Russian Academy of Sciences, Pushchino, Moscow Region, Russia

Abstract. The chemical bond formation properties of Mg and Ca are compared with emphasis on protein binding according to RJP Williams. This will determine the effect of Mg on Ca-modulated functions. Ca is a signal for activating the cell with activation of enzymes needed for increased ATP synthesis. Mitochondrial Ca homeostasis is regulated by the activities of the uptake mechanism, the calcium uniporter, and efflux mechanisms, the most important being the Ca^{2+}/nNa^{+} and the Ca^{2+}/nH^{+} antiporters. Mg also influences concentration of Ca in cell compartments.When Ca is taken up over a certain threshold, mitochondria get uncoupled by opening of a large pore in the permeability transition (MPT). This occurs in cellular Ca overload as in ischemia-reperfusion. Then Ca activated phospholipase A_2 produces fatty acids that also, especially palmitic acid, promote MPT. Cytochrome c and apoptosis-inducing factor are liberated and are signals for cell death, apoptosis or necrosis when cellular [ATP] is lowered by decreased mitochondrial synthesis after MPT. Mg increases the Ca^{2+} threshold and thus keeps the pore closed, and also inhibits Ca-activation of PLA_2, and thus protects against damage in ischemia-reperfusion, Apoptosis is of central interest and has made mitochondrial calcium handling and its regulation of renewed interest.

The earth alkali cations magnesium (Mg) and calcium (Ca) form similar electrostatic bonds, while solubilities with various anions like phosphates, sulphates and organic anions like oxalate may differ considerably. Chelate formation properties also are different as described in detail by RJP Willliams [1-3]. They are due to different ionic radii, greater flexibility for Ca in coordination bond number (6 or 8), bond lengths and direction as well as the chelate form. In addition to its restrictions in bond formation, the difficulty of Mg to penetrate hydrophobic domains is also due to the difficulty of removing the last water molecule. Thus, Mg may be bound with less affinity to Ca-binding sites, if at all, but may be bound well to sites of the right dimension. Mg is bound to neutral nitrogen groups such as $-NH_2$, imidazole, while Ca is bound preferentially by multidentate anions and strong acid anions. Of interest is the high affinity to ATP, the stability constants (log) being slightly higher for Mg, 4.2, than for Ca, 4.0 [1].

ATP thus is an efficient Mg buffer and its transformation to ADP would increase the free Mg concentration in the cytosol considerably, and thereby affect its interactions with Ca binding. The electrostatic binding of Mg to membrane surface negative bonds will lower the negative surface potential (screening) and thereby decrease the concentration Ca near the surface, which affects the membrane transport [4].

The greater flexibility of Ca aids its ability to reach suitable binding sites deeper in protein pockets and membranes, and to penetrate membranes more easily. Its binding changes protein conformation, often exposing hydrophobic domains which aids binding of the protein to membranes and other proteins. Ca in contrast to Mg may join two negatively charged sites, which is important in membrane fusion and exocytosis, but also for protein binding and interactions.

These general properties are of interest in the effects of Mg on activities of Ca in mitochondria.

Effects of Mg on Ca binding of proteins

Of special interest is the interaction of Mg with Ca binding of proteins that are important in Ca-modulated functions, Ca transport and signalling. This is influenced by the relative binding constants to the binding sites, and the on and off rate constants [5]. Ca is bound preferably to carboxylates, the number of these strongly affect the affinities. Thus, thus carp albumin with 4 carboxylates has a $K > 10^6$, thermolysin with 3 carboxylates $> 10^4$, and concanavalin with 2 $> 10^2$ [2]. The effect of an antagonistic cation like Mg on Ca binding to protein binding sites is given by the equation [2]:

$$\text{Activity} = \alpha \frac{[\text{Ca}] \cdot \text{K}_{\text{Ca}} \cdot \text{p}_{\text{Cax}}}{[\text{Mg}] \cdot \text{K}_{\text{Mg}} \cdot \text{p}_{\text{Mgx}}}, \text{ where}$$

[Ca] and [Mg] are the free concentrations in the aqueous medium, the K values give the binding constants to the binding site involved in activity, while the partition coefficients of the complexes to a separate medium, as the membrane phase, is given by the p values.

In regard to enzyme activities it should be kept in mind that the common substrate for ATPases and kinases is MgATP rather than the free ATP.

Mg modulation of Ca contents extra- and intracellularly

Parathyroid hormone, vitamin D and calcitonin not only affect Ca homeostasis but also Mg, see review [6]. There is also a feedback with both Mg and Ca affecting the secretion of parathyroid hormone and calcitonin [7], that regulate also the free Ca in the extracellular fluid. Recently interest has focused on the role of the Ca-sensitive receptor [CaR] in plasma membranes in the entry of Ca and Mg in cells and in the secretion of parathyroid hormone [8, 9]. The CaR binds primarily Ca, but also other bivalent cations like Mg. From the point of view of mitochondrial functions the free Ca in the cytosol is of interest. In the activation of the cell, both entry of Ca via channels in the plasma membrane and release from intracellular stores are important [10]. Mg not only is involved in these transport mechanisms, but may also be involved directly in Ca efflux, as reported for the heart, where withdrawal of Mg caused increase in cytosolic Ca [11].

Ca signalling in mitochondria

Interest in Ca handling by mitochondria was increased by the finding that matrix Ca stimulates mitochondrial metabolism, see review [12]. This is by stimulating the citric cycle dehydrogenases citric acid dehydrogenase and 2-oxoglutarate dehydrogenase and also pyruvate dehydrogenase by activating a phosphatase removing the phosphate of the inhibited enzyme. This will make possible a stimulation of mitochondrial respiration and increased synthesis of ATP needed by the activated cell. That may be promoted also by stimulating electron flow in the respiratory chain itself [13]. We have also found a direct effect on ATP synthase mediated by a Ca and Mg-dependent phosphorylation of subunit c [14], reported in earlier Magnesium symposia [15, 16].

Another area, which more recently has stimulated interest in mitochondrial Ca handling, is the role of mitochondria and Ca in the signalling involved in **apoptosis**, see review [17]. Mitochondria are central in the signalling, with proapoptotic substances, cytochrome c and AIF [apoptosis-inducing factor] being released on mitochondrial Ca overload or/and oxidative stress [18]. These signals may also be released without Ca overload and MPT through other means or channels (*Figure 1*).

There are members of the Bcl protein family that form such channels, see reviews [19, 20]. Bcl-2 also raises the Ca threshold for opening of the MPT pore [21]. Tumor cells frequently have higher expression of bcl-2, preventing apoptosis, which promotes oncogenesis [22]. Indeed, we have found higher expression of bcl-2 in some hepatomas, while in HeLa cells apoptosis was prevented by high mitochondrial content of Mg [23]. Stanniocalcin, with high amount of receptors in mitochondria [24], also is a protein that prevents Ca overload-induced damage in ischemia-reperfusion and is overexpressed in neuroblastoma cells [25]. We also found it to prevent MPT and apoptosis in these cells when this was induced by potassium ionophores [26].

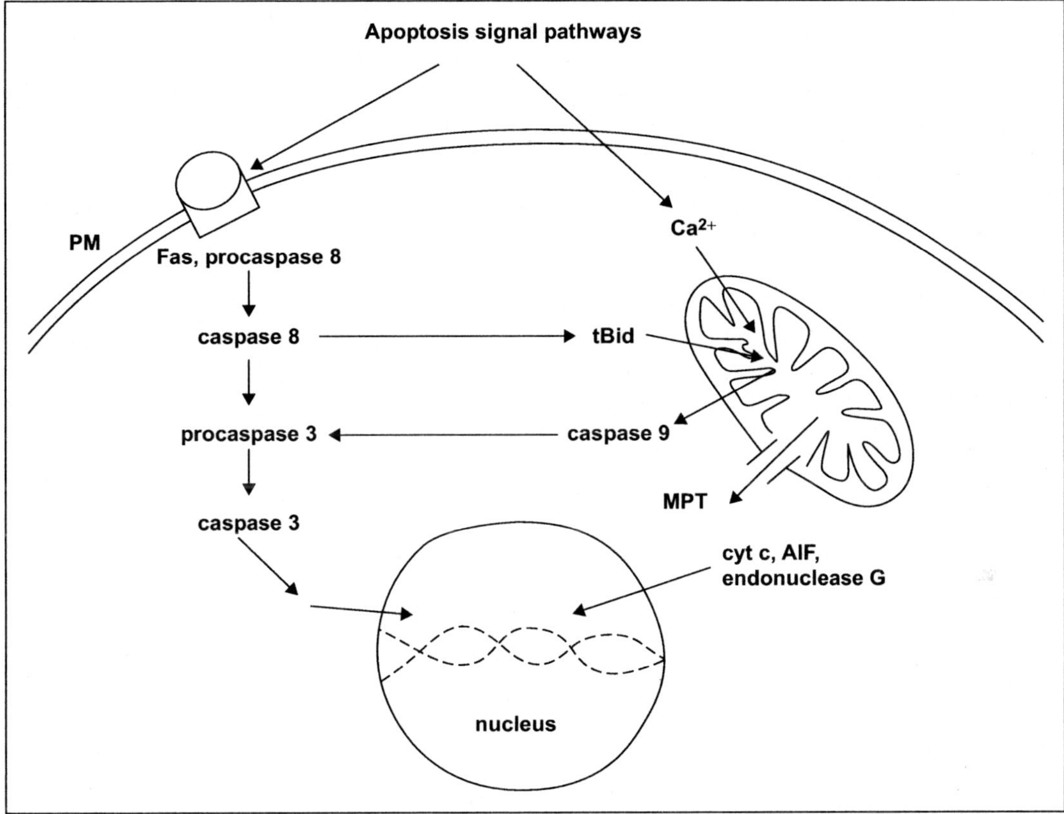

Figure 1. Ca signalling in apoptosis. Apoptosis signals may be through plasma membrane (PM) receptors leading to activation of various caspases and endonuclease G via or without participitation of mitochondria, which may release proaptotic signals cyt. c and apoptosis-inducing factor (AIF) from intermembrane space after Ca-mediated MPT or by other pathways.

Mitochondrial Ca handling

Mitochondrial Ca uptake was discovered more than 40 years ago, see review [27]. At the end of my stay in Dr Britton Chance's laboratory in Philadelphia, University of Pennsylvania, in 1958-59 I found that Ca, when added to energized mitochondria, caused formation of H^+ and Ca gradients, which were equilibrated when Ca was taken up over a certain threshold. That threshold was lowered by inorganic phosphate [Pi], and increased by addition of Mg or ADP. Unfortunately this was not published in generally read journals [28, 29]. We now call the Ca uptake system the **calcium uniporter**, and the Ca overload-induced gradient equilibration the **mitochondrial permeability transition** [MPT] [27]. About the same time two groups independently discovered the Ca uptake system, studying the accumulation of Ca under conditions of maximal uptake, *i.e.* in the presence of Mg, ATP and Pi, when Ca and Mg forms precipitates in the matrix with Pi [30, 31]. We now know that the uniporter is a membrane potential-driven uptake of Ca^{2+} alone, and that the MPT is due to opening of a large pore in the inner membrane [27].

Efflux of Ca from mitochondria was found to be by antiport (exchange) against 2 Na^+ in heart mitochondria [32, 33]. This means that there is a futile cycling of Ca in the presence of Na^+, with uptake on the uniporter consuming energy in the form of the membrane potential [$\Delta\Psi$], and efflux on the antiporter [27, 33, 34]. The $Ca^{2+}/2Na^+$ antiporter is present in many excitable cells [35], but in liver and some other non-excitable tissues, Ca efflux is mainly by a Na-independent mechanism, believed to be a

$Ca/2H^+$ antiporter [36]. Both these antiporters are generally assumed to be electroneutral, but there is evidence that their activity may be influenced by the membrane potential, which excludes an electroneutral exchange. Thus, the H^+ antiporter was not observed in deenergized mitochondria [37], nor was Ca efflux changed by changing medium or matrix pH, indicating varying stoichiometries and absence of a passive antiporter [38]. The $Ca/2Na^+$ antiporter generally operates electroneutrally, indicating the stoichiometry above [39], but may be driven by the $\Delta\Psi$ under some conditions [40].
Ca efflux may also be by short-term opening of the MPT pore as a part of the Ca signalling [41]. Such opening would allow ions like Ca to diffuse through the pore, resulting in release of accumulated Ca. A long-term opening would result in MPT.

Ca overload-induced cellular damage

Cellular Ca overload may occur under pathological conditions, such as ischemia-reperfusion. The cellular damage (*Figure 2*) is caused mainly by MPT, but also by activation of hydrolytic enzymes, like phospholipase A_2 and Ca-activated proteases and nucleases. Formation of fatty acids, especially palmitic acid, by phospholipase A_2 may cause Ca efflux by cyclosporin A-insenstive channels [42, 43]. The protection by Mg by inhibiting phospholipase A2 and MPT has been treated in an article in the volume of the Romanian Magnesium Symposium in 2003 [44]. Lysosomal production of reactive oxygen species may contribute by sensitizing mitochondria to MPT [45].

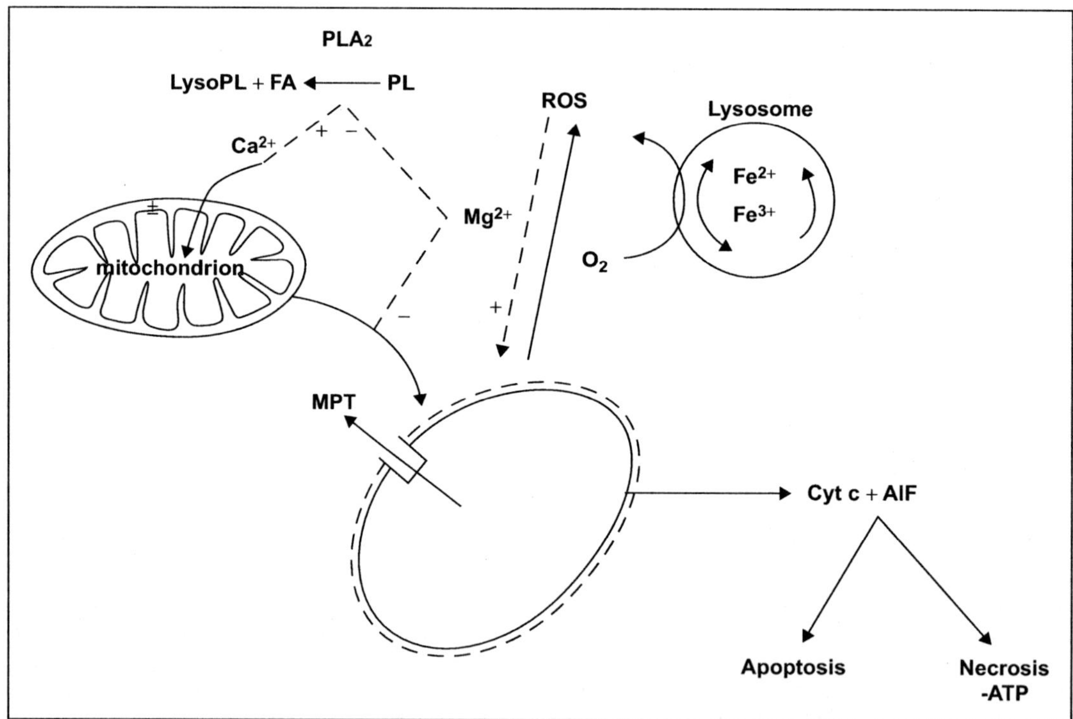

Figure 2. Cellular damage mechanisms in Ca overload. From Saris N-EL in Nechifor M, Porr PJ, editors. *Magnesium Involvement in Biology and Pharmocotherapy.* Cluj Napoca: Casa Cartii de Stiinta, 112-119. Ca overload may activate phospholipase a2 (PLA2) producing fatty acids (FA) from phospholipids (PL). Lysosomes may produce reactive oxygen species (ROS) when degrading iron-containing proteins like cytochromes. Ca uptake may cause MPT, mitochondrial swelling and release of cyt c and AIF. MPT is promoted by reactive oxygen species (ROS) which may be produced by lysosomes in autophagocytosis of mitochondria which promotes MPT. If the ATP level is reduced, cell death is by necrosis.

Ca overload of the cell may result in mitochondrial accumulation of Ca over the threshold inducing MPT pore opening. This may be physiological in apoptosis, see above, but may result in pathological cell death by necrosis when the cellular ATP has been reduced by the uncoupling of some of the mitochondria, which then hydrolyze ATP instead of synthesizing it [46, 47].
MPT pore opening is affected by many parameters in addition to Bcl-2 family members, see [36, 42], with Pi, pro-oxidants, increased pH and decrease in $\Delta\Psi$ lowering the Ca threshold, while Mg, antioxidants, lowered pH and higher $\Delta\Psi$ increase it, and cyclosporin A inhibits it. Opening of the mitochondrial ATP-sensitive potassium channel has a protective effect against damage in ischemia-reperfusion and occurs in preconditioning – the increased tolerance against one attack of ischemia-reperfusion [48]. The lowered $\Delta\Psi$ and competition with K^+ for uptake decreases the Ca uptake and thereby postpones MPT pore opening [49]. Openers of this channel thus prevent the damage and are used therapeutically, we have studied one new opener, levosimendan, in mitochondria [50].

References

1. Williams RJP. The biochemistry of sodium, potassium, magnesium, and calcium. *Quarterly Rev Chem Soc* 1970; 24: 331-65.
2. Williams RJP. Calcium ions: Their ligands and their functions. *Biochem Soc Symp* 1974; 39: 133-8.
3. Williams RJP. Calcium chemistry and its relation to biological function. *Symp Soc Exp Biol* 1976; 30: 1-17.
4. Bara M, Guiet-Bara A, Durlach J. Analysis of magnesium membraneous effects: binding and screening. *Magnes Res* 1988; 1: 29-33.
5. Williams RJP. Calcium-binding proteins in normal and transformed cells. *Cell Calcium* 1994; 16: 339-46.
6. Saris N-E, Mervaala E, Karppanen H, Khawaja JA, Lewenstam A. Magnesium - an update on physiological, pathological and analytical aspects. *Clin Chim Acta* 2000; 294: 1-26.
7. De Rouffinac C, Quamme G. Renal magnesium handling and its hormonal control. *Physiol Rev* 1994; 74: 305-22.
8. Penniston JT, Filoteo AG, McDonough CS, Carafoli E. Purification, reconstitution and regulation of plasma membrane Ca^{2+}-pumps. *Methods Enzymol* 1988; 157: 340-51.
9. Mayan H, Hourvitz A, Schiff E, Farfel Z. Symptomatic hypocalcemia in hypomagnesemias-induced hypoparathyroidism, during magnesium tocolytic therapy - possible involvent of calcium-sensing receptor. *Nephrol Dial Transpl* 1999; 14: 1764-6.
10. Carafoli E. Intracellular calcium homeostasis. *Annu Rev Biochem* 1987; 56: 395-433.
11. Vierling W, Stampfl A. Magnesium-dependent calcium efflux in mammalian heart muscle. *Cell Calcium* 1994; 15: 175-82.
12. McCormack JG, Halestrap AP, Denton RM. Role of calcium ions in regulation of mammalian intramitochondrial metabolism. *Physiol Rev* 1990; 70: 391-425.
13. Murphy AN, Kelleher JK, Fiskum G. Submicromolar Ca^{2+} regulates phosphorylating respiration by normal rat liver and AS-30D hepatoma mitochondria by different mechanisms. *J Biol Chem* 1990; 265: 10527-34.
14. Azarashvily TS, Tyynelä J, Baumann M, Evtodienko Y, Saris N-E. Ca^{2+}-Modulated phosphorylation of a low-molecular-mass polypeptide in rat liver mitochondria: Evidence that it is identical with subunit c *of FoF1-ATPase. Biochem Biophys Res Commun* 2000; 270: 741-4.
15. Saris N-E, Krestinina OV, Azarashvili TS, Odinokova IV, Tyynelä J, Evtodienko Y. Regulation of ATP synthase by Ca^{2+} and Mg^{2+}-dependent phosphorylation of subunit c. In: Rayssiguer Y, Mazur A, Durlach J, eds. *Advances in Magnesium Research: Nutrition and Health.* London: John Libbey & Co, 2001: 101-6.
16. Saris NEL, Azarashvili TS, Odinokova IV, Krestinina OV, Gadjieva SS, Evtodienko YV. Regulation of protein phosphorylation in brain and liver mitochondria by magnesium and calcium with emphasis on subunit c of ATP synthase. In: Escanero JF, Alda JO, Guerra M, Durlach J, eds. *Advances in Magnesium Research. Physiology, Pathology and Pharmacology.* Zaragoza: ES: Prensas Universitarias de Zaragoza, 2003: 49-54.
17. Zamzami N, Hirsch T, Dallaporta B, Petit PX, Kroemer G. Mitochondrial implication in accidental and programmed cell death: apoptosis and necrosis. *J Bioenerg Biomembr* 1998; 29: 185-93.
18. Lemasters JJ, Qian T, He L, *et al.* Role of mitochondrial inner membrane permeabilization in necrotic cell death, apoptosis, and autophagy. *Antioxid Redox Sign* 2002; 4: 769-81.
19. Mattson MP, Kroemer G. Mitochondria in cell death: novel targets for neuroprotection and cardioprotection. *Trends Mol Med* 2004; 9: 196-205.
20. Sharpe JC, Arnoult D, Youle RJ. Control of mitochondrial permeability by Bcl-2 family members. *Biochim Biophys Acta* 2004; 1644: 107-13.
21. Murphy RC, Schneider E, Kinnally KW. Overexpression of Bcl-2 suppresses the calcium activation of a mitochondrial megachannel. *FEBS Lett* 2001; 497: 73-6.
22. Williams GT. Programmed cell death: Apoptosis and oncogenesis. *Cell* 1991; 65: 1097-8.

23. Evtodienko Y, Teplova VV, Azarashvily TA, *et al.* The Ca^{2+} threshold for the mitochondrial permeability transition and the content of proteins related to Bcl-2 in rat liver and Zajdela hepatoma mitochondria. *Mol Cell Biochem* 1999; 194: 251-6.
24. McCudden CR, James KA, Hasilo C, Wagner GF. Characterization of mammalian stanniocalcin receptors. Mitochondrial targeting of ligand and receptor for regulation of cellular metabolism. *J Biol Chem* 2002; 277: 45249-58.
25. Zhang K-Z, Lindsberg PJ, Tatlisumak T, Kaste M, Olsen HS, Andersson LC. Stanniocalcin: A molecular guard of neurons during cerebral ischemia. *Proc Natl Acad Sci USA* 2000; 97: 3637-42.
26. Teplova VV. Jääskeläinen E, Salkinoja-Salonen M, Saris N-EL, Li F-Y, Andersson LC. Differentiated Paju cells have increased resistance to toxic effects of potassium ionophores. *Acta Biochim Polon* 2004; 51: 539-44.
27. Carafoli E. Historical review: Mitochondria and calcium: ups and downs of an unusual relationship. *Trends Biochem Sci* 2003; 28: 175-81.
28. Saris NE. Om oxidativ fosforylering. (On oxidative phosphorylation). *Finska Kemistsamfundets Meddelanden* 1959; 68: 65-72.
29. Saris N-E. The calcium pump in mitochondria. *Soc Sci Fenn; Comment Phys -Math* 1963; 28(11): 1-77.
30. DeLuca HF, Engstrom GW. Calcium uptake by rat kidney mitochondria. *Proc Natl Acad Sci USA* 1961; 47: 1744-50.
31. Vasington FD, Murphy JV. Ca^{++} Uptake by rat kidney mitochondria and its dependence on respiration and phosphorylation. *J Biol Chem* 1962; 237: 2670-7.
32. Crompton M, Capano M, Carafoli E. The sodium-induced efflux of calcium from heart mitochondria. A possible mechanism for the regulation of mitochondrial calcium. *Eur J Biochem* 1976; 69: 453-62.
33. Carafoli E. The calcium cycle of mitochondria. *FEBS Lett* 1979; 104: 1-5.
34. Gunter KK, Gunter TE. Transport of calcium by mitochondria. *J Bioenerg Biomembr* 1994; 26: 471-85.
35. Gunter TE, Gunter KK, Sheu S-S, Gavin CE. Mitochondrial calcium transport: physiological and pathological relevance. *Am J Physiol* 1994; 267: C313-C339.
36. Bernardi P. Mitochondrial transport of cations: channels, exchangers, and permeability transition. *Physiol Rev* 1999; 79: 1127-55.
37. Saris N-E. Non-respiring rat liver mitochondria do not have a Ca^{2+}/H^{+} antiporter. *Acta Chem Scand* 1987; B41: 79-82.
38. Gunter TE, Chace JH, Puskin JS, Gunter KK. Mechanism of sodium independent calcium efflux from rat liver mitochondria. *Biochemistry* 1983; 22: 6341-51.
39. Affolter H, Carafoli E. The Ca^{2+}-Na^{+} antiporter of heart mitochondria operates electroneutrally. *Biochem Biophys Res Commun* 1980; 95: 193-6.
40. Jung DW, Baysal K, Brierley GP. The sodium-calcium antiport of heart mitochondria is not electroneutral. *J Biol Chem* 1995; 270: 672-8.
41. Ichas F, Mazat JP. From calcium signaling to cell death: two conformations for the mitochondrial permeability transition pore. Switching from low- to high-conductance state. *Biochim Biophys Acta* 1998; 1366: 33-50.
42. Mironova GD, Gateau-Roesch O, Levrat C, *et al.* Palmitic and stearic acid bind Ca^{2+} with high activity and form nonspecific channels in black-lipid membranes. Possible relation to Ca^{2+}-activated mitochondrial pores. *J Bioenerg Biomembr* 2001; 33: 319-31.
43. Sultan A, Sokolove PM. Palmitic acid opens a novel cyclosporin A-insensitive pore in the inner mitochondrial membrane. *Arch Biochem Biophys* 2001; 386: 37-51.
44. Saris N-E. The protective functions of magnesium on calcium overload-induced cellular damage Magnesium: involvements in biology and pharmacotherapy. In: Nechifor M, Porr PJ, eds. *Magnesium Involvement in Biology and Pharmocotherapy.* Cluj Napoca: Casa Cartii de Stiinta, 2003: 112-9.
45. Brunk UT, Terman A. The mitochondrial-lysosomal axis theory of aging. Accumulation of damaged mitochondria as a result of imperfect autophagocytosis. *Eur J Biochem* 2002; 269: 1996-2002.
46. Leist M, Nicotera P. The shape of cell death. *Biochem Biophys Res Commun* 1997; 236: 1-9.
47. Gottlieb RA. Mitochondria: execution central. *FEBS Lett* 2000; 482: 6-12.
48. Chen Q, Camara AKS, An J, Riess ML, Novalija E, Stowe DF. Cardiac preconditioning with 4-h, 17°C ischemia reduces $[Ca^{2+}]_i$ load and damage in part via K_{ATP} channel opening. *Am J Physiol* 2002; 282: H1961.
49. Hausenloy DJ, Maddock HL, Baxter GF, Yellon DM. Inhibiting mitochondrial permeability transition pore opening: a paradigm for myocardial preconditioning? *Cardiovasc Res* 2002; 55: 534-43.
50. Kopustinskiene D, Pollesello P, Saris N-EL. Potassium-specific effects of levosimendan on heart mitochondria. *Biochem Pharmacol* 2004; 328: 807-12.

Magnesium and the muscle cell

R. Hunger
Lürlibadstrasse 80, CH-7000 Chur, Switzerland

The function of the muscle cell is regulated by the Ca concentration in the cytosol (Cai) and by the membrane potential (MP). Intracellularly the Ca is located in the cytosol and the sarcoplasmic reticulum (SR) (Ca stores). The Ca ATPases (150-250 Ca/sec) in those two membranes uses ATP for the distribution of the Ca *(Figure 1)*. In the membrane of the SR SERCA pumps the Ca from the cytosol into the SR, in the cell membrane PMCA pumps the Ca from the cytosol into the serum. The Na/Ca exchange (NCX) (up to 5000 Ca/sec) exchanges Ca from the cytosol against Na from the serum without using energy. The Na is needed by the Na/K ATPase (150-250/sec) to maintain the MP in order to guide the irritability and the functions of the muscle cells. Cai induces in the skeletal muscle and heart muscle a conformation of the troponin. The now accessible myosin ATPase initiates with Mg ATP the stroke. The resulting complex of actomyosin has to be dissolved before the next stroke. In smooth muscle cells Cai enhances the activity of the myosin ATPase. The Ca of the Ca stores can reach the cytosol by diffusion and by the Ip3 receptor Ca channels (initiated by angiotensin II and alpha/beta receptors of the cell membrane) and ryanodin receptor channels (initiated by AP of the cell membrane) and modulate therefore the activity of the muscle cell. In addition the Ca of the ribosomes influences the translation of proteins in myofibrilles and influences thus the build up of muscle cells.

With the three basic functions of Mg: formation of MgATP, competition of Ca and with PO4 activation the Na/K exchange in the Na/K ATPase, Mg influences the regulation of muscle cells with the three membrane proteins Ca ATPase, Na/Ca exchange and Na/K ATPase. A deficiency of Mg leads first of all to a lack of Ca competition and the Ca ATPase pump quickly Ca from the cytosol to the SR or to the serum. Thus the Na/Ca exchange does not gets enough Na towards the Na/K ATPase. The MP is sinking and the irritability of the cell enhances the possibility of cramps of the muscle cell. As during Mg deficiency at the Na/K ATPase the relation pump/exchange is shifted towards the pump, cramps may be inhibited at low Mg deficiency. An excess of Mg competes the Ca from the ATPases. Ca is exported by the very potent Na/Ca exchange and the exchanged Na increases the MP although the Mg++ concentration together with PO4 the relation pump/exchange is shifted towards the exchange. The Mg ATP concentration in the cytosol is mainly determined by the amount of ATP. At Mg deficiency it is trapped in the cell as MgATP. At Mg excess it increases as Mg++ the Ca competition.

The interaction between Ca, Mg and ATP with the three membrane mechanisms can be analysed by the measurement of the retraction of the blood clot [1–4] *(Figure 2)*. 0.1 ml freshly isolated whole blood is diluted with 4.9 ml saline and the clotting is induced by adding thrombin. After 5 hours at 37°C the length of the blood clot is determined. The more ATP is formed in the platelets the shorter the clot. By changing the concentration of the ions their influence on the formation and consumption of ATP in the platelets can be determined. The platelets need ATP to maintain their structure and only the excess can be used for the contraction of the blood clot. If 1.5 mmol CaCl is added to saline the retraction is deteriorated by 9 mm. By the addition of 1-12 mmol Mg the retraction is improved by 9 mm. Mg has shifted the output of Ca from the ATPases to the exchange mechanisms and the whole energy the cell used for the output of Ca can be used for the retraction. In the muscle cell Mg competes Ca from troponin and although there is more ATP in the cell the muscle cell is paralysed at increasing Mg concentrations leading to heart arrest. The nerve cell can use the additional Na to stabilize the MP leading to a decreased irritability of the nerve cell as far as CNS depression and coma.

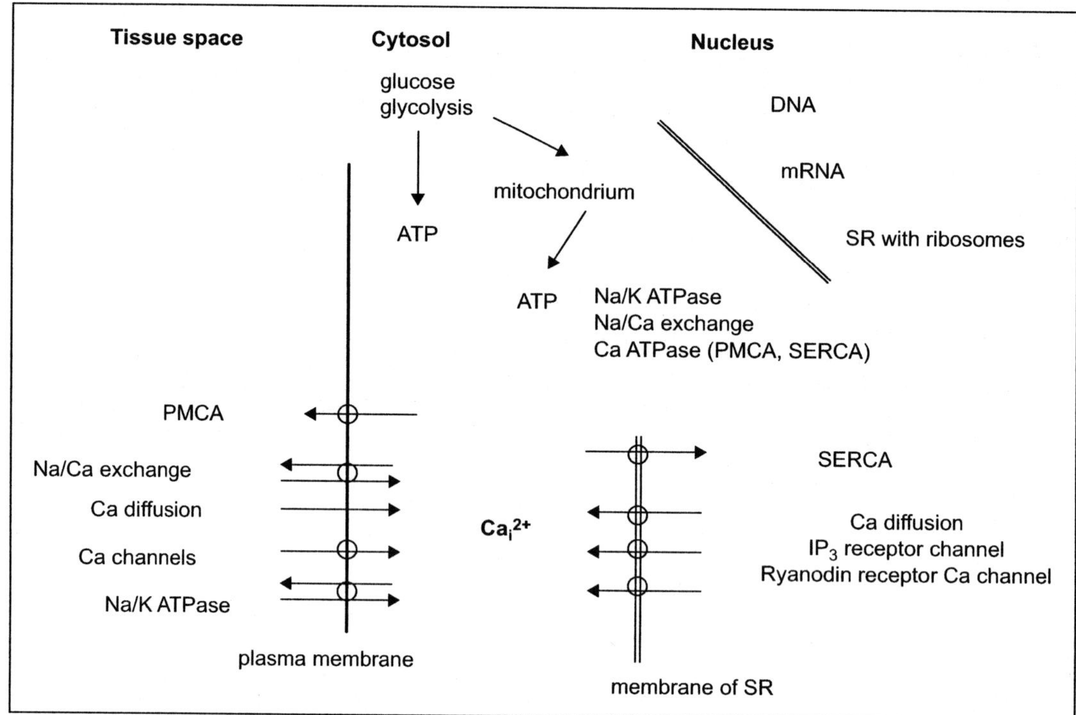

Figure 1. The cytosol is the battle field of life.

Skeletal muscle cell: at rest without AP there is too little Ca and Na entering the cell by diffusion. All Ca is pumped out by the ATPases with ATP but without the exchange mechanisms. As a part of the Ca is relocated to the serum, the Ca stores are only badly filled. This leads to a decreased power of the muscle and to a decreased translation of proteins in the ribosoms of the myofibrils (danger of muscle atrophy). The lack of Na inhibits the formation of a stable MP with irritability and the danger of muscle cramps (*e.g.* Nightly muscle cramps). The addition of Mg competes Ca from the Ca ATPases, the Na/Ca exchange is active and puts Na to the Na pump and stabilizes the MP. In the elderly with mitochondrial dysfunction there is a special high danger of muscle cramps, but also the effect of Mg is limited because of a lack of ATP. In addition the cell loses even more Ca into the serum through the Na/Ca exchange. The most efficient manner to help during muscle cramps is to have slow movements (with few AP) and a soft behaviour (to leave the Ip3 receptor channels closed).
At high load of the muscle cell with lots of AP there is an overload of the cytosol with Na and Ca. On one hand there is a relieve of the Ca ATPases by the Na/Ca exchange *(Figure 1)*. On the other hand at a decompensation of the Na/K ATPase there is an increase first of the Nai and later to an increase of the Cai. The Na/Ca exchange is inhibited by the increasing Nai. A high level of Na in the serum accelerates the decompensation of the Na/K ATPase. The increase of the Cai may be delayed With Mg at good ATP production in the mitochondria. But at high Mg concentration the cell loses Ca and together with the competition of Ca the power of the muscle decreases. After workout Mg may delay the recovery of the muscle until all ca stores are replenished. Mg reduces the probability of cramps (by increasing the MP) but also the power of the muscle cell (by competing the Ca at troponin).
Smooth muscle cell: it is less organized than the skeletal muscle cell. It is less overloaded with AP. But also at rest the balance between pumps and exchange mechanisms are granted. Under stress an increase of the Cai can be initiated by angiotensin II and the alpha/beta receptors via proteinkinase C pathway by the Ip3 receptor Ca channels. However, only the decompensation leads to an increase of the Cai. This may easily happen at high serum levels of Na. The addition of Mg competes Ca from the

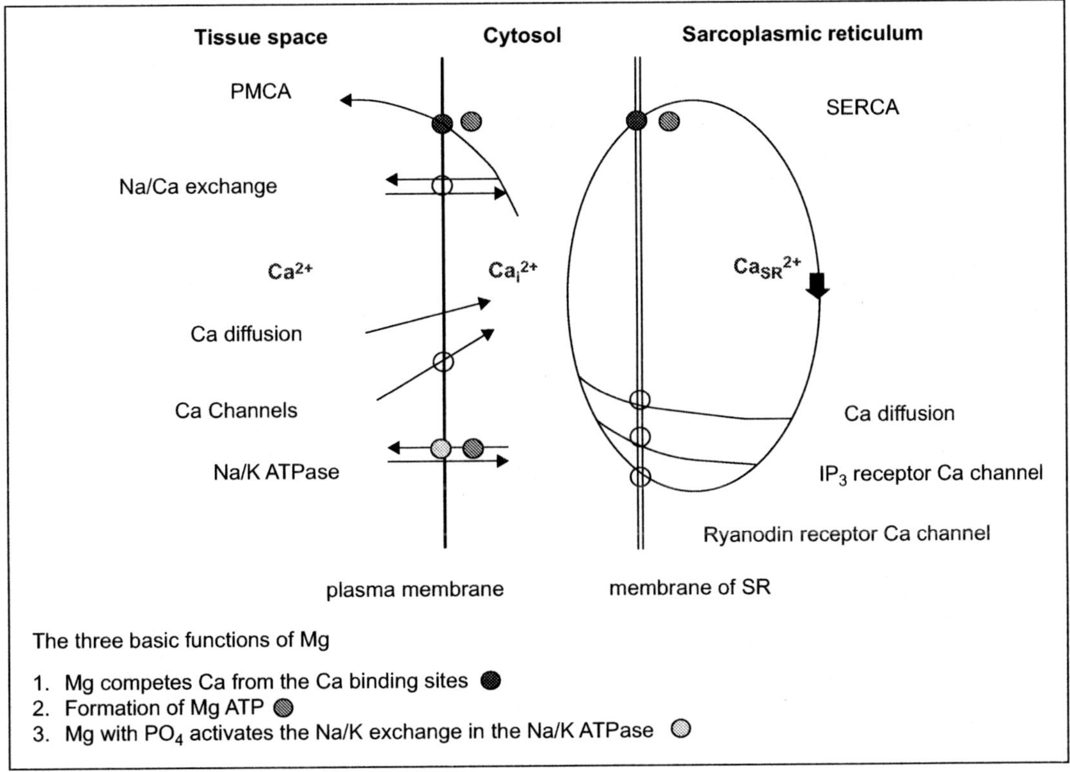

Figure 2. Effects of Magnesium.

ATPases and leads to a decrease of the ATP consumption by activating the exchange mechanisms. Simultaneously the power of the smooth muscle is decreasing and the blood pressure decreases. This may lead to decompensation of the circulation and to heart arrest.

Essential hypertension: Na and Ca are elevated in the cytosol. Patients suffering from hypertension have a reduced ATP production in the mitochondria. The ATPases are, therefore, less active and more Na and Ca is exported by the exchange mechanisms.

In the retraction test the decompensation of the Na/K ATPase can be demonstrated [4]. The addition of K has no effect in young health persons. However, in elderly people the addition of K improves the retraction symmetrically by 2 mm in case of low or high concentrations of NaCl. However in patients with hypertension the addition of K has an asymmetrical effect, at low concentration of NaCl the improvement of the retraction is 2.5 mm, at high NaCl concentration the improvement is only 1 mm. The difference in the length of the blood clot at low and at high NaCl concentration is only marginally (with in the variations of the method) in healthy people, in contrast in patients with hypertension there is a difference in the length of 1-4 mm. This means that in patients with hypertension the Na/K ATPase can not be activated in the presence of K - the Na/ATPase is decompensated. During a loss of ATP production in the mitochondria of elderly people a hypertension will evolve as soon the Na/K ATPase decompensates and as soon as the Na/K ATPase can not be activated by K. But also with Mg the decompensated protein can not be activated. As a consequence of the lack of Mg there is too much Na at the Na/K ATPase and it can with 150-250 changes of conformation per second neither as a pump with ATP nor as exchange mechanism increase this rate. The limitation at the stabilization of the Cai concentration of the cytosol is the rate of the changes of the conformations of the Na/K ATPase protein. A growing rigidity of the proteins during aging together with mitochondrial dysfunctions lead to an increase of the Cai and to a hypertension.

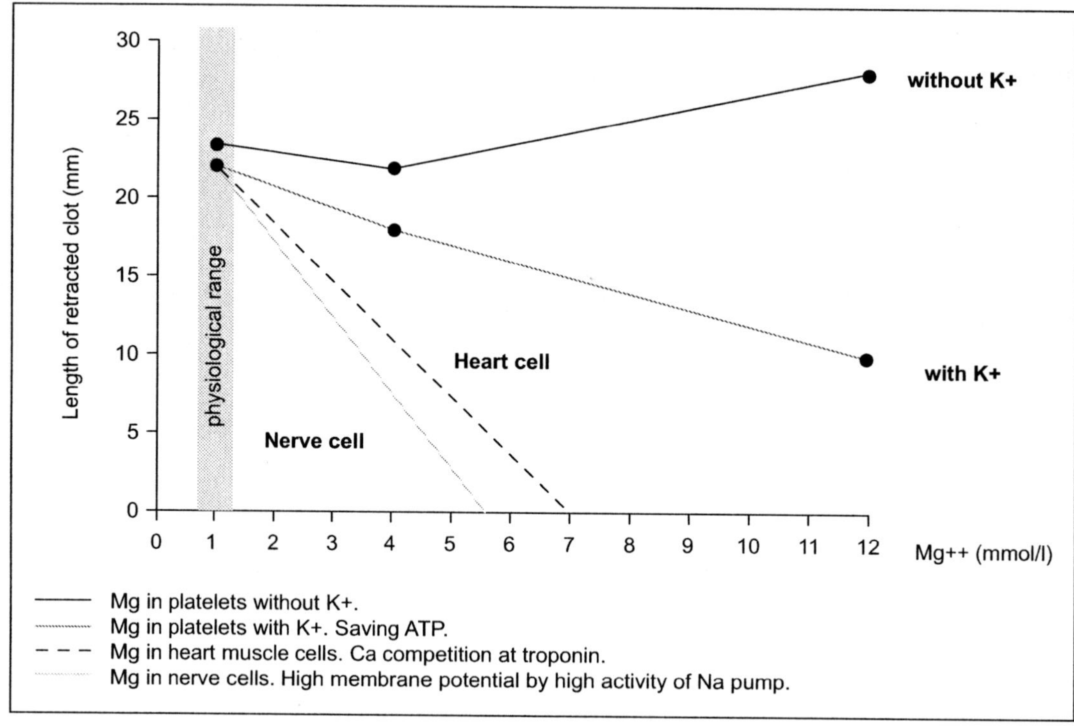

Figure 3. Ca competition in nerve and muscle cells. Effects of Mg.

In the heart muscle there are similar processes. At low ATP production because of mitochondrial dysfunction, O2 deficiency or insulin resistance etc. for the output of Ca there is too a shift from the pumps to the exchange mechanism with a corresponding loss of Ca in the Ca stores [1,5–9]. As ions as the Na/K ATPase has a good performance the Cai is constant. At decompensation of the Na/K ATPase there is an increase of the Cai, although the total Ca is reduced. The decompensation of the Na pump leads to a reduction of the MP and therefore to an increase of the heart rate at rest. An increase of the heart rate at rest is a clinical predictor for a heart insufficiency in higher age. In healthy people with high ATP production Mg induces for the Ca output a shift from the pumps to the exchange mechanisms, but the Na pump is able to export the excess of Na, leading to an increase of the MP and the Cai may stay low. When the Mg concentration rises over the ATP concentration and the Mg++ competes all ca from the Ca pumps bradycardia, heart failure and cardiac arrest are the results. Although Mg reduces the ATP consumption of the heart muscle cell. However in cells of a heart attack the decreased ATP consumption may increase the time of survival at O2 deficiency *(Figure 3)*.

In summary we can conclude that for muscle cells optimal Mg concentrations are between 0.7 and 1.1 mmol/l Mg++. Below 0.7 mmol/l Mg there is a danger of MP reduction with increased irritability. Over 1.1 mmol/l Mg there is a danger of weakness or even palsy because of competition of Ca at the Ca ATPases.

References

1. Petersen KF, Befroy D, Dufour S, *et al.* Mitochondrial Dysfunction in the elderly: possible role in insulin resistance. *Science* 2003; 300: 1140-2.

2. Michikawa Y, Mazzicchelli F, Bresolin N, Scarlato G. Attardi. *Science* 1999; 286: 774.

3. Hunger R. Die zellulären und humoralen Wirkungen von Magnesium und deren Einfluss auf die Leistung der Thrombozyten, gemessen an der Retraktion des Blutgerinnsels. *Magnesium Bulletin* 1994; 3: 87-8.

4. Hunger R. The intracellular Calcium antagonism of Magnesium. Proceeding Book First International Symposium on Trace Elements in Human, Athen 1998: 435-441.

5. Hunger R. The influence of ionized Magnesium and ionized Calcium on the performance of Thrombocytes. In: Smetana IR (ed). *Advances in Magnesium Research.* Vienna: John Libbey, Fifth European Magnesium Congress 1995: 586-90.

6. Hunger R. In: *Magnesium und die Doppelfunktion der Na/K ATPase.* Jena: Mengen und Spurenelemente, 2000: 497-501.

7. Proceeding Book 3rd International Symposium on Trace Elements in Human. Athen, 2000: 827-42.

8. Sodium-Calcium Exchange. Annals of the New York Academy of Sciences. Volume 775 1996.

9. Na/K ATPase and Related Transport ATPases. Annals of New York Academy of Sciences. Volume 834, 1997.

Magnesium in the nutrition of man

M. Anke[1], M. Glei[1], J. Vormann[2], R. Müller[3], C. Hoppe[3], U. Schäfer[1]

1. Institute of Nutrition and Environment, Friedrich Schiller University of Jena, D-07743 Jena; Dornburger Straße 24
2. Institute for Prevention and Nutrition, Ismanning
3. Society of Ecology and Environmental Chemistry Ltd, Erfurt, Germany

Abstract. The magnesium consumption of adults with self-selected mixed and ovo-lacto-vegetarian diets in Central Europe (Germany) and Mexico was determined by means of duplicate portion technique (21 test populations). German and Mexican women consumed 200 and 300 mg/day, respectively; men, 250 and 320 mg/day. Ovo-lacto-vegetarians of both genders take in 375 and 475 mg Mg daily, respectively. Supplementation of the mixed diet of women in a double-blind placebo controlled study with 100 mg Mg/day did not change their magnesium balance. The Mg-supplemented women increased their faecal excretion by 95 mg/day. The normative magnesium requirement of women was satisfied by an intake of > 200 mg/day. The normative magnesium requirement of persons with mixed diet and without genetic disorders of magnesium homeostasis amount to 200 mg/day for women and 250 mg/day for men, which is < 3.0 mg/kg body weight or 650 mg Mg/kg consumed dry matter. Persons with genetic disorders of magnesium homeostasis have a much higher requirement than humans without such disorders. Four different disease entities with primary hypomagnesaemia have been described so far, three of which are also associated with a disturbance in calcium homeostasis. Women and men with mixed diets and without these genetic disorders are recommended to take in 300 and 350 mg/day, respectively. Compared to the results obtained by the duplicate method, the calculation of the magnesium intake (basket method) overestimates the Mg intake by 30 to 50%.

The planet Earth appeared about 4.6 milliard years ago. One milliard years later, life arose from nonliving matter. Organisms evolved from living cells in water and on land, which were able to capture solar energy and use it for the synthesis of organic molecules. The miracle of photosynthesis appeared to support life on Earth. Chlorophyll captures the energy of the sun and converts it to chemical energy in the form of adenosine triphosphate (ATP). The energy of ATP is used to reduce carbon dioxide and water to carbohydrate and oxygen (3 milliard years ago). All enzyme reactions that are known to be catalysed by ATP show a requirement for magnesium compounds. Magnesium (Mg) participated in the earliest biochemical steps necessary for the evolution of life. Almost nine decades have passed since Willstätter and Stoll demonstrated the central position of the magnesium atom in the chlorophyll molecular. Magnesium is thus an essential element and various deficiently symptoms in plants, animals and humans are well known. On the other hand, magnesium is practically nontoxic (Aikawa 1991, Shils 1997, Vormann 2004) [1, 22, 23]. The transfer of magnesium from several rocks to soils, plants, animals and man is of great importance for health and wellness of man (Anke *et al.*, 1976; Arnhold *et al.*, 1998) [6, 10].

Materials and Methods

Table I informs about materials analysed and their number. The analysis of the samples took place after dry ashing at 450°C. The white ash was solved in 25% HCl and the magnesium content analysed by flame absorption spectroscopy (AAS, Carl Zeiss Jena) and inductively coupled plasma atomic emission spectroscopy (ICP AES) (Spectroanalytical Instruments Kleve, Spectroflame – D). Both analytical

Table I. Number of the participants of the duplicate studies and analysed samples.

Diet	Country	Time	Women	Men	Duplicates	Urine	Faeces	Milk	Blood	Blood-serum	Hair
Md	G[1]	1988	28	28	392	-	-	-	-	-	-
	G	1992	42	42	588	-	-	-	-	-	-
	G	1996	31	31	434	434	434	-	-	-	-
	M	1996	14	14	196	-	-	-	-	-	-
	G[2]	1997	14	-	98	98	98	-	28	28	28
	G[3]	1997	14	-	98	98	98	22	42	42	-
V	G	1996	10	10	140	140	140	-	-	-	-
Number	T(n)		153	125	1946	770	770	22	70	70	28
Total number			278					3676			

Md = Mixed diet; V = Ovo-lacto-vegetarians; G or G[1] = Germany; M = Mexico; G[2] = Placebo controlled double blind study with young women; G[3] = Study with pregnant and beast-feeding women.

methods delivered similar results with a deviation of ± 5% (± 3.3%) (Glei, 1995) [11]. The precision of the analytical methods was checked with the reference material "ARC/CL, total diet reference material (HDP)" (Kumpulainen and Tahvonen, 1990) [19].

For the duplicate studies of adults with mixed and ovo-lacto-vegetarians diets in Germany and Mexico, 21 test populations were available. They consisted of at least 7 women and 7 men and collected all consumed foodstuffs and beverages as visual estimates on 7 successive days (Anke *et al.*, 1997) [7]. The balance studies were carried out with 7 test populations (women and men) and mixed diet.

The individual results were analysed with the help of Fox Pro database system (version 2.6 for Windows, Microsoft Corporation). The statistical calculation was carried out with SPSS/PC (version 6.01 for Windows, SPSS Inc), using the t-test, variance analysis, and smallest limit difference (SLD).

Results

Magnesium intake

Magnesium intake per day depending on gender, time, form of diet and habitation

The magnesium consumption of adults with self-selected mixed and ovo-lacto-vegetarians diet in Central Europe and Mexico was determined with 21 test populations in duplicate studies.

On an average, men consume 24% more magnesium than women. The significantly higher magnesium intake of men results from their 24% higher dry matter consumption (Anke *et al.*, 1997) [7]. Both sexes eat a dry matter with the same magnesium concentration (*Table IV*). German and Mexican women consumed 200 and 300 mg Mg/day, respectively, men 250 and 320 mg Mg/day. The reunification of Germany (1990) and the transition from local food production and market to world open trade and supermarket did not practically influence the magnesium intake (+ 6 to 7% more magnesium). In Mexico, peoples with mixed diet eat on an average 20 to 50% more magnesium ($p < 0.001$) than people in Central Europe (Gonzalez *et al.*, 1999) [16]. Ovo-lacto-vegetarians consumed 1996 approximately 80% more magnesium than omnivores (*Table II*). The reason for this high magnesium intake is the high consumption of magnesium-rich vegetables, which contain lower energy concentration and increase the dry matter intake of ovo-lacto-vegetarians by about one third in comparison to peoples with mixed diet. The influence of time (1988 to 1996) is weak but that of the dwelling-place important and significant (*Table III*).

The reason for this variation is the magnesium concentration of the local drinking water. Its magnesium content varies between 7 and 57 mg/l (Anke *et al.*, 1998a) [3]. Especially the drinking water coming from Keuper weathering soils (dolomit) (Bad Langensalza) is rich in magnesium, that of the Pleistocene sands and boulder clay delivers only 10 to 15 mg Mg/l (Wusterhausen, Vetschau). The magnesium supply of women and men in the regions of the diluvial sands through local drinking water consumption is poor.

Table II. Magnesium intake by adults in Germany and Mexico depending on gender, time and form of diet (mg/day) (n 1946).

Form of diets	Country/year		Women		Men		Fp[5]	
			s[3]	x[4]	x	s	P[6]	%
People with mixed diet (Md)	G[1]	1988	56	193	248	69		128
	G	1992	79	213	260	104	< 0.001	122
	G	1996	72	205	266	92		130
	M[2]	1996	95	301	318	122	> 0.05	106
Vegetarians (V)		1996	101	376	474	199	< 0.001	126
Fp (Md)	G	1988-96		< 0.001				
p	G: M	1996		< 0.001				
	G: V	1996		< 0.001				
% (Md)	G	1988-96	106		107		-	
	G: M	1996	147		120			
	G: V	1996	183		178			

1. G or G[1] = Germany; 2. M = Mexico; 3. standard deviation; 4. arithmetic mean; 5. significant level in one factorial or multiple variance analyse; 6. significance level of the t-test to Student.

Table III. Magnesium intake of adult peoples with mixed diets depending on the region of Germany.

Dwelling place/year	Women		Men		Fp	%
	s	x	x	s		
Bad Langensalza, 1992	89	259	352	127		136
Jena, 1996	49	256	318	70		124
Chemnitz, 1992	67	208	286	84		138
Jena, 1988	54	217	272	55		125
Bad Langensalza, 1988	55	204	271	66	< 0.001	133
Bad Liebenstein, 1992	81	203	261	84		129
Freiberg, 1992	56	219	238	82		109
Greifswald, 1992	69	237	208	86		88
Rositz, 1996	62	184	249	71		135
Ronneburg, 1996	68	184	249	93		135
Wusterhausen, 1988	58	167	238	63		143
Steudnitz, 1996	75	175	226	106		129
Vetschau, 1988	43	183	213	75		116
Wusterhausen, 1992	73	154	214	85		139
Fp		< 0.001				
%[1]	59		61			

1. Bad Langensalza, 1992 = 100%; Wusterhausen, 1992 = x%.

The magnesium concentration of the consumed dry matter depending on gender, time, form of diet and habitation

Women and men from Germany and Mexico consume a dry matter with the same magnesium concentration (*Table IV*). The sex difference disappears. Both sexes consume the same foodstuffs. Neither men nor women prefer magnesium-rich or magnesium-poor food groups.

Table IV. Magnesium concentration of the dry matter consumed by adults in Germany and Mexico depending on gender, time and form of diet (mg/kg DM).

Form of diets	**Country/year**		**Women**		**Men**		**Fp**	
			s	**x**	**x**	**s**	**p**	**%**
People with mixed diet (Md)	G	1988	151	655	654	138	> 0.05	100
	G	1992	187	690	681	187		99
	G	1996	188	713	743	163		104
	M	1996	182	1008	930	196	> 0.05	92
Vegetarians (V)		1996	257	987	1016	333	> 0.05	103
Fp (Md)	G	1988-96			< 0.001			
p	G: M	1996			< 0.001			
	G: V	1996			< 0.001			
%	G	1988-96		109	114		-	
	G: M	1996		141	125			
	G: V	1996		138	137			

Table V. Magnesium concentration of the consumed food dry matter depending on the region of Germany.

Region/year	**Women**		**Men**		**Fp**	**%**
	s	**x**	**x**	**s**		
Bad Langensalza, 1992	192	821	798	179		97
Steudnitz, 1996	222	770	773	183		100
Jena, 1996	130	733	769	164		105
Bad Liebenstein, 1992	163	733	752	161		103
Bad Langensalza, 1988	165	734	729	136		99
Jena, 1988	103	743	704	111	> 0,05	95
Ronneburg, 1996	171	652	730	141		112
Rositz, 1996	218	688	687	148		100
Greifswald, 1992	124	705	660	163		94
Freiberg, 1992	153	686	636	199		93
Chemnitz, 1992	135	629	680	145		108
Vetschau, 1988	102	576	609	143		106
Wusterhausen, 1988	126	565	575	96		102
Wusterhausen, 1992	231	568	559	178		98
Fp		< 0.001			-	
%		69	70			

The reunification increases the magnesium concentration of the consumed dry matter by approximately 10%. The Mexicans consume a dry matter with one third more magnesium than the Germans. The ovo-lacto-vegetarians eat a dry matter with a third more magnesium than the omnivores. All differences are significant.

The dwelling place influences the magnesium intake significantly (*Table V*). The Keuper and Muschelkalk regions of the Triasic time deliver a magnesium-rich, the Pleistocene sands a magnesium-poor food dry matter for man. The differences are significant and vary the magnesium concentration of the consumed dry matter about one third. The magnesium concentration of the consumed dry matter is a good indicator of the magnesium supply of humans.

Magnesium intake per kg body weight depending on gender, time, form of diet and habitation
The magnesium intake per kg body weight is significantly varied by the gender, but very weak (*Table VI*). Men consume on an average of the 5 test groups only 1.4% more magnesium/kg body weight than women. In nutrition, this difference is without practical importance. Also the reunification of Germany did not influence the magnesium consumption per kg body weight. Mexicans consume per kg body weight a quarter to third more magnesium than Germans.
Ovo-lacto-vegetarians eat the double amount of magnesium than omnivores. The difference of the magnesium intake between omnivores and vegetarians is significant. The weight of the vegetarians is lower than those of the peoples with mixed diet. The magnesium intake per kg body weight is also significantly varied by the habitation (*Table VII*).

Table VI. Magnesium intake of adults per kg body weight in Germany and Mexico depending on gender, time and form of diet (mg/kg body weight).

Form of diets	**Country/year**		**Women**		**Men**		**Fp**	
			s	**x**	**x**	**s**	**p**	**%**
People with	G	1988	1.0	3.0	3.4	0.99		113
mixed diet (Md)	G	1992	1.3	3.4	3.3	1.3	< 0.01	97
	G	1996	1.2	3.1	3.4	1.2		110
	M	1996	1.7	5.1	4.3	1.9	< 0.001	84
Vegetarians (V)		1996	2.0	6.6	6.8	2.8	> 0.05	103
Fp (Md)	G	1988-96		< 0.05				
P	G: M	1996		< 0.001				
	G: V	1996		< 0.001				
% (Md)	G	1988-96	103		100			-
	G: M	1996	165		126			
	G: V	1996	213		200			

Table VII. Magnesium intake of adults/kg body weight depending on the region of Germany (mg/kg body weight).

Region/year	**Women**		**Men**		**Fp**	**%**
	s	**x**	**x**	**s**		
Bad Langensalza, 1992	1.2	4.0	4.4	1.4		110
Jena, 1996	0.87	4.0	4.2	0.85		105
Freiberg, 1992	1.2	4.0	3.4	1.2		85
Jena, 1988	0.96	3.4	3.5	0.73		103
Bad Langensalza, 1988	0.97	3.3	3.5	0.94	> 0.05	106
Chemnitz, 1992	0.91	3.1	3.7	1.2		119
Greifswald, 1992	1.2	3.7	2.9	1.2		78
Vetschau, 1988	0.75	3.3	3.0	1.0		91
Bad Liebenstein, 1992	1.3	3.3	3.1	1.0		94
Rositz, 1996	0.81	2.6	3.2	1.1		123
Ronneburg, 1996	1.0	2.7	3.0	0.86		111
Wusterhausen, 1988	0.75	2.0	3.6	1.1		180
Steudnitz, 1996	1.3	2.8	2.7	1.2		96
Wusterhausen, 1992	1.2	2.3	2.5	1.0		109
Fp		< 0.001				
%	58		57		-	

The Keuper and Muschelkalk regions of the Triasic time deliver most magnesium/kg body weight. Those of the Pleistocene sands and Holocene flood plains deliver only 2.0 to 2.8 mg/kg body weight to women and 2.5 to 3.6 mg/kg body weight to men.
In summary, it is to point out, that the habitation and with it the geological origin of the site vary the magnesium intake significantly (Glei *et al.*, 1996, 1997, 1998) [12, 14, 15].

Magnesium intake of breast-feeding women
The magnesium concentration of the consumed dry matter of breast-feeding mothers (28 to 35 day of lactation) differs only insignificantly (*Table VIII*). Their daily magnesium intake was only 18% higher than those of the non-breast-feeding women. Their higher magnesium intake resulted from their 17% higher dry matter intake (Anke *et al.*, 1997) [7].
Their magnesium intakes /kg body weight was one third higher than those of the non-breast-feeding women. The differences are significant.

Influence of body weight on the magnesium intake
In both sexes the magnesium concentration of the consumed dry matter was not varied by body weight and gender. The influence of the body weight on the daily magnesium intake is also not distinct (*Table IX*) and only significant for men. Contrary to the daily magnesium intake, the body weight influences the magnesium intake/kg body weight (*Table X*). It decreases with increasing body weight by about 10% in men and 40% in women. In summary, the influence of the body weight on the magnesium intake of adult omnivores is small and without practical importance.

Influence of age on the magnesium intake
The daily intake of magnesium and the magnesium concentration of the consumed dry matter were not significantly influenced by peoples with mixed diet. Only the magnesium consumption/kg body weight was significantly lowered by about 11% in women and 6% in men (*table XI*). The influence of the age on magnesium intake is very limited and without practical importance.

Influence of the provinces in the Federal Republic of Germany on the magnesium intake
Contrary to body weight and age, the provinces of the Federal Republic of Germany and therefore the geological origin of their soils influence the magnesium concentration of the consumed dry matter (*Table XII*), the magnesium intake /day and the magnesium consumption/kg body weight by approxi-

Table VIII. Magnesium concentration of consumed dry matter, magnesium intake per day and magnesium intake/kg body weight of non-breast-feeding and breast-feeding women (n 98).

Parameter	Non-breast-feeding		Breast-feeding		p	%
	s	x	x	s		
mg/kg DM[1]	188	713	672	181	> 0.05	94
mg/day	72	205	242	85	< 0.01	118
mg/kg body weight	1.2	3.1	4.0	1.5	< 0.001	129

1. dry matter.

Table IX. Magnesium intake of adults with mixed diet in Germany depending on body weight (mg/day).

Body weight (kg)	(n; n)	Women		Men		Fp	%
		s	x	x	s		
< 60	(252; 35)	64	209	188	61		90
61-70	(238; 105)	74	217	250	84		115
71-80	(133; 308)	74	177	261	81		147
> 80	(84; 259)	71	208	269	106	< 0.001	129
Fp			< 0.001			-	
%			100	143			

Table X. Magnesium intake of adults with mixed diet in Germany depending on body weight (mg/kg body weight).

Body weight kg	(n; n)	Women		Men		Fp	%
		s	x	x	s		
< 60	(252; 35)	1.1	3.8	3.4	1.1		89
61-70	(238; 105)	1.1	3.3	3.8	1.3		115
71-80	(133; 308)	1.0	2.4	3.5	1.1		146
> 80	(84; 259)	0.86	2.4	3.1	1.2	< 0.001	129
Fp			< 0.001				
%		63		91		-	

Table XI. Magnesium intake per kg body weight depending on age (mg/kg).

Age in years	(n; n)	Women		Men		Fp	%
		s	x	x	s		
20-29	(182; 161)	1.1	3.5	3.4	1.3	< 0.001	97
30-39	(161; 133)	1.4	3.2	3.4	1.3		106
40-49	(161; 147)	1.3	3.0	3.6	1.1		120
50-69	(203; 266)	1.1	3.1	3.2	1.1		103
Fp			< 0.01			-	
%		89		94			

Table XII. Magnesium concentration of the consumed food dry matter depending on the provinces of Germany (mg/kg DM).

Region	(n; n)	Women		Men		Fp	%
		s	x	x	s		
Thuringia	(413; 413)	177	734	744	157		101
Saxonia	(98; 98)	146	658	658	175		100
Brandenburg[1]	(196; 196)	164	604	601	152	< 0.05	100
Fp			< 0.001				
%		82		81		-	

1. and Mecklenburg-Pommerania.

mately 20% in the both sexes. The Pleistocene sandy soils and the boulder clay in the northern counties of Germany, which were formed by the ice time, deliver low magnesium amounts into the food chain of human, whereas the Triasic formations in Thuringia produce a significantly Mg richer food and drinking water. The dwelling place and its geological origin influence the magnesium intake of man markedly (Anke *et al.*, 2000; Glei *et al.*, 1994) [4, 13].

Influence of the season (summer, winter) on the magnesium intake

In contrast to zinc and vanadium for instance (Anke *et al.*, 1998, 2004) [5, 9], the magnesium intake is not markedly varied by the hot and cold season. During winter, the dry matter intake of women with mixed diet is 10% and of men 5% higher than during summer (Anke *et al.*, 1997) [7]. The magnesium concentration of the consumed dry matter was during winter 4 and 5% lower than in summer and the daily magnesium intake on the average of both sexes 3% higher. Related to body weight, the magnesium consumption in wintertime was 6 and 3% higher in men and women, respectively. The season influences the magnesium intake only very limited.

Influence of the measuring method on the magnesium intake of man
The duplicate portion technique, which analyses the magnesium content in the daily duplicates on 7 successive days and the market basket method which calculates the magnesium consumption involve different results. In 1988 the market basket method overestimated magnesium intake on the average of both genders by about 36% (*Table XIII*).
This overrating of the magnesium intake led to increased recommendation figures for magnesium. The basket method should not be used for measuring the intake of macro, trace and ultratrace elements, but it can be used for the calculation of the origin of the consumed foodstuffs (Anke, 2004) [2].
In case of magnesium, 49% of the amount consumed was found to originate from vegetable food, 23% from animal food and 28% from beverages (*Table XIV*). The importance of the magnesium content in beverages is intelligible (Anke, 2004) [2].

Magnesium balance

Magnesium excretion, apparent absorption rate and balance
The magnesium balance, was estimated in four test populations with mixed diets, one population of ovo-lacto-vegetarians and two teams of breast-feeding mothers. Most of the ingested magnesium was faecally excreted (*Table XV*). Women and men with mixed diet excreted 65 and 66% with faeces and both sexes of ovo-lacto-vegetarians eliminated 73 and 72%, respectively, through faeces. The renal excretions rate amounted in both genders of omnivores to 35 and 34% and in ovo-lacto-vegetarians to only 27 and 28%. The apparent absorption rate add up to 33 and 34% in peoples with mixed diets and to 25 and 30%, respectively, in vegetarians. The magnesium balance of the test persons with both diet forms was with – 3% in women and + 2% in men equalized.

Table XIII. Magnesium intake of adults with mixed diets determined by the duplicate portion technique and calculated by the basket method.

Parameter	Women		Men		%
	Duplicate	Basket	Duplicate	Basket	
Mg mg/day	211	278	259	360	136

Table XIV. Intake of magnesium through animal and vegetable foodstuffs and beverages by people with mixed diet in percent.

Parameter	Animal foodstuffs	Vegetable foodstuffs	Beverages
Percent	23	49	28

Table XV. Magnesium intake, magnesium excretion, magnesium apparent absorption rate and magnesium balance of German peoples with mixed and ovo-lacto-vegetarians diets.

Parameter (n; n)		Women		Men	
		Mixed diet	Vegetarian diet	Mixed diet	Vegetarian diet
Intake, mg/day		205	376	266	474
	Urine, mg/day	75	104	90	130
	Faeces, mg/day	137	283	175	333
Excretion	Urine, %	35	27	34	28
	Faeces, %	65	73	66	72
Apparent absorption, %		33	25	34	30
Balance	mg/day	- 7	- 11	+ 1	+ 11
	%	- 3	- 3	+ 0.4	+ 2

The ovo-lacto-vegetarians of both sexes consumed approximately 80% more magnesium than the peoples with mixed diet (*Table XV*). They excreted approximately only 40% more magnesium than the omnivores. Their faecal excretion rate amounts to approximately double in comparison to the peoples with mixed diet. The bioavailability of the magnesium in the food of the ovo-lacto-vegetarians is lower than that of the peoples with mixed diet.

In a placebo-controlled double-blind study with young women (20 to 25 years), the effect of a supply of 100 mg Mg/day on excretion rate, apparent absorption rate and magnesium balance was examined (*Table XVI*). The magnesium intake of the ladies was 248 and 363 mg/day during 21 days.

The women with the 100 mg Mg supply excreted 95 mg of the supply through faeces and 20 mg through urine. The normative magnesium requirement of the women was satisfied and their apparent magnesium absorption rate obtained only 18%. Their magnesium balance was equalized. The supplementation of young women during 21 days did not significantly influence their magnesium content of blood serum and head hair (*Table XVII*). The magnesium values in the two tissues were normal (Anke and Risch, 1979; Vormann and Anke, 2002) [8, 24].

The daily supply of 100 mg magnesium/day from the 8th month of pregnancy to 35th day of lactation did not significantly influence the magnesium concentration in blood and blood serum of the supplemented women (*Table XVIII*). All blood parameters were in the normal range.

The intake of 242 or 421 mg Mg/day over three months did not influence the magnesium concentration of the breast milk. Its magnesium content was with 33 mg/L relatively high in both groups (Iyengar, 1982; Renner, 1989; Harzer and Haschke, 1989) [17, 18, 21]. The breast-feeding mothers secreted

Table XVI. Magnesium intake, magnesium excretion, magnesium apparent absorption rate and magnesium balance of a placebo-controlled double-blind study with magnesium supplementation.

Parameter		Placebo			Preparation			p	%
		s		x	x		s		
Intake, mg/day		79		248	363		76	< 0.001	146
Excretion	Urine, mg/day	36		77	97		35	< 0.05	126
	Faeces, mg/day	153		203	298		198	< 0.05	147
	Urine, %		28			25			
	Faeces, %		72			75			
Apparent absorption, %			18			18			
Balance	mg/day		- 32			- 32			-
	%		- 13			- 9			

Table XVII. Magnesium content in blood, blood serum and hair of a placebo-controlled double-blind study with young women.

Blood mmol/l	x	x	p	%
Before study	1.2	1.2	> 0.05	100
After study	1.3	1.2	> 0.05	92
P	> 0.05	> 0.05		
%	108	100		-
Blood serum (mmol/l)				
Before study	0.69	0.74	> 0.05	107
After study	0.71	0.69	> 0.05	97
P	> 0.05	> 0.05		
%	103	93		-
Head hair (mg/kg DM)				
After study	270	318	> 0.05	118

Table XVIII. Magnesium content of blood and blood serum at the beginning of the study (8th month of pregnancy), at birth and during nursing (35th day of lactation) (mmol/L).

Parameter (n)	8th month of pregnancy	At birth	35th day of lactation	Fp	%[1]
Blood (42)	1.1	1.2	1.2	> 0.05	109
Blood serum (42)	0.67	0.65	0.71	> 0.05	106

1. 8th month = 100%; 35th day of lactation = x%.

Table XIX. Magnesium intake, magnesium excretion, magnesium apparent absorption rate and magnesium balance of breast-feeding mothers without and with magnesium supplementation.

Parameter (n; n)		Without		With		p	%
		s	x	x	s		
Intake, mg/day (49; 49)		85	242	421	67	< 0.001	174
Excretion	Milk, mg/day (22; 22)	8	32	28	6	> 0.05	88
	Urine, mg/day (49; 49)	29	61	104	38	< 0.001	170
	Faeces, mg/day (49; 49)	123	117	285	217	< 0.001	244
	Milk, %	15		7			
	Urine, %	29		25			
	Faeces, %	56		68			
Apparent absorption, %		52		32			
Balance	mg/day	+ 32		+ 4			
	%	+ 13		+ 1			

approximately 32 and 28 mg magnesium into the milk, these are 7 and 13% of the consumed magnesium (*Table XIX*). Both groups of the breast-feeding mothers excreted 25% of the ingested magnesium renally. The mothers with a supply of 100 mg magnesium eliminated this amount faecally.

The apparent magnesium absorption rate of the mothers with a magnesium intake of 242 mg/day amounted to 52%, those of the mothers with a magnesium intake of 421 mg/day to 32%. The magnesium balance of breast-feeding mothers was equalised. The intake of 242 mg Mg/day covered the normative requirement of breast-feeding women and allowed magnesium incorporation.

Magnesium requirement of man

Women with mixed diets and a magnesium intake of approximately 200 mg/day on the average of 7 days have equalised magnesium balance, the same is true for men with an intake of approximately 250 mg Mg/day. The normative magnesium requirement of adult women, without genetic disorders of magnesium homeostasis amounts to < 200 mg Mg/day (180 mg/day), those of men < 250 mg/day (225 mg/day), which is < 3.0 mg/kg body weight or a magnesium concentration of 650 mg/kg consumed dry matter. A dry matter with this magnesium concentration delivers men (intake of 380 g dry matter) 250 mg Mg/day and women (300 g dry matter intake) 195 mg Mg/day.

Fiigure 1 shows the distribution of the magnesium intake of women and men and demonstrates, that approximately one third of both genders took in < 200 (women) or < 250 mg Mg/day (men). The same is true for the consumed magnesium concentration (*Figure 2*). One third of women and men take in a dry matter with a magnesium concentration < 650 mg Mg/kg dry matter.

Persons with genetic disorders of magnesium homeostasis have a much higher requirement than humans without such disorders. Four different disease entities with primary hypomagnesaemia have been described so far, three of which are also associated with disturbances in calcium homeostasis which is also found in cows. Primary hypomagnesaemia can be caused either by an intestinal or a renal defect. Both intestinal and renal defects can be (partially) corrected by oral magnesium (and calcium)

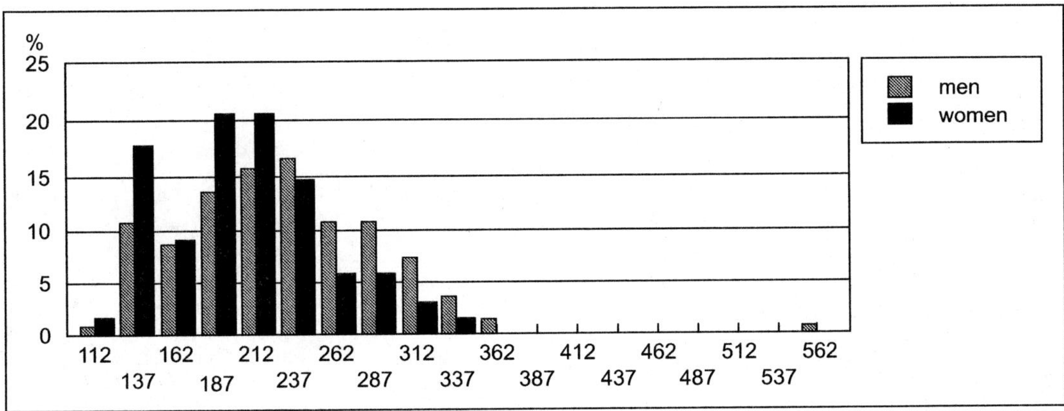

Figure 1. Distribution of the magnesium intake (average of 7 days).

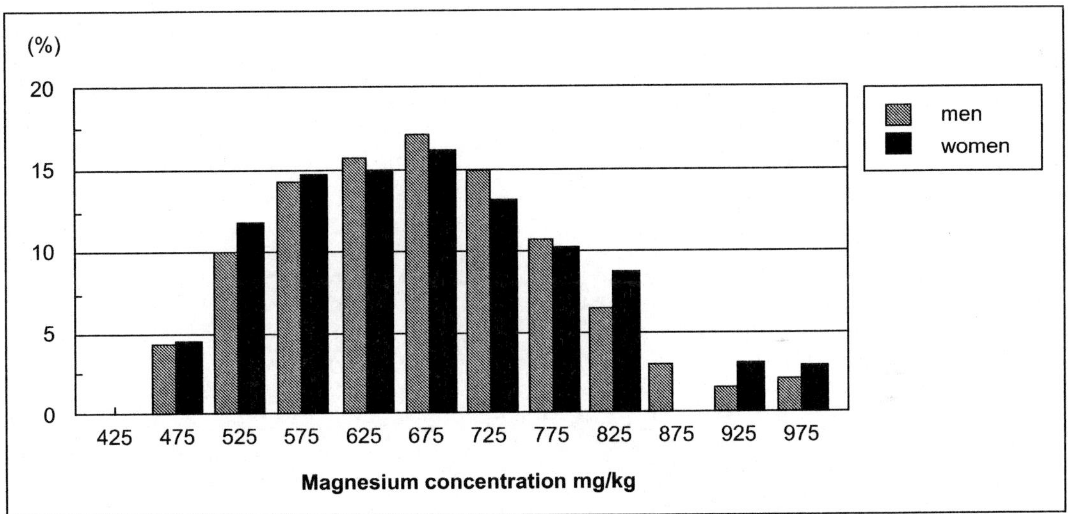

Figure 2. Distribution of the magnesium concentration in the food dry matter (average of 7 days).

Table XX. Normative requirement and recommendation of magnesium for man.

Parameter	Normative requirement					Recommendation mg/day
	Women mg/day	Men mg/day	mg/kg body weight	mg/kg DM		
				Omnivores	Vegetarians	
Mg	200	250	3.0	650	470	300; 350

supplements (Meij *et al.*, 2002; Weber and Konrad, 2002) [20, 25]. Women and men with mixed diets and without these genetic disorders are recommended to take in 300 and 350 mg/day, respectively (*Table XX*).

References

1. Aikawa JK. Magnesium. In: Merian E, ed. *Metals and Their Compounds in the Environment.* New York, Basel, Cambridge: VCH, Weinheim, 1991: 1025-34.
2. Anke M. Essential and toxic effects of macro, trace and ultratrace elements in the nutrition of man. In: Merian E, Anke M, Ihnat M, Stoeppler M, eds. *Elements and their Compounds in the Environment.* Wiley- VCH Verlag GmbH and Co. KgaA, 2004: 343-67.
3. Anke M, Dorn W, Illing-Günther H, *et al.* Die biologische Bedeutung des Vanadiums in der Nahrungskette. 5. Mitteilung: Vanadiumverzehr, Vanadiumausscheidung und Vanadiumbilanz Erwachsener in Abhängigkeit von Geschlecht, Zeit, Alter, Gewicht, Jahreszeit, Lebensraum und Leistung. *Mengen-und Spurenelemente* 1998; 18: 1028-40.
4. Anke M, Glei M, Groppel B. Magnesium in der Nahrungskette landwirtschaftlicher Nutztiere und des Menschen. *Rekasan J* 2000; 7: 13-9.
5. Anke M, Glei B, Groppel B, Rother C, Gonzales D. Mengen-, Spuren- und Ultraspurenelemente in der Nahrungskette. *Nova Acta Leopoldina* 1998; 309: 157-90.
6. Anke M, Grün M, Schneider HJ. Der Magnesiumstatus des Menschen in Abhängigkeit von Alter und Geschlecht. In: Schneider HJ, Anke M, eds. *Magnesiumstoffwechsel.* Jena: Friedrich-Schiller-Universität Jena, 1976: 36-51.
7. Anke M, Müller M, Röhrig B, *et al.* Der Chromtransfer in der Nahrungskette. 4. Mitteilung: Der Chromverzehr Erwachsener in Abhängigkeit von Zeit, Geschlecht, Alter, Körpermasse, Jahreszeit, Lebensraum, Leistung. *Mengen-und Spurenelemente* 1997; 17: 912-27.
8. Anke M, Risch M. *Haaranalyse und Spurenelementstatus.* Gustav Fischer Verlag Jena, 1979.
9. Anke M, Seifert M, Lösch E, Schmidt P. Zinc in the food chain – its biological importance. Part Five: Zinc in animal foodstuffs and beverages. *Mengen- und Spurenelemente* 2004; 22: 1825-31.
10. Arnhold W, Anke M, Glei M, *et al.* Magnesium intake of Cercopithecidae. *Trace Elements and Electrolytes* 1998; 15(1): 11-5.
11. Glei M. Magnesium in der Nahrungskette unter besonderer Berücksichtigung der Magnesiumversorgung des Menschen. *Habilitationsschrift: Biologisch- Pharmazeutische Fakultät der Friedrich-Schiller-Universität Jena, Germany.* 1995.
12. Glei M, Anke M, Arnhold W, Röhrig B. Magnesium intake and balance in adults consuming self-selected mixed or vegetarian diets. *Trace Elements and Electrolytes* 1998; 15: 111-5.
13. Glei M, Anke M, Groppel B, Müller M. Influence of geological origin of the site on the magnesium content of plant and animal foodstuffs. In: Golf S, Dralle D, Vecchiet L, eds. *Magnesium.* J. Libbey and Co, 1994: 109-13.
14. Glei M, Anke M, Röhrig B. Magnesium intake and magnesium balance of adults eating mixed or vegetarian diets. In: Fischer PWF, Abbé MRL, Cockell KA, Gibson RS, eds. *TEMA – 9.* Ottawa: NRC Research Press, 1997: 181-2.
15. Glei M, Anke M, Röhrig B, Lehmann U. Magnesium supply to adults in Germany. In: Garban Z, Dragan P, eds. *Proceedings of 2nd International Symposium on "Metal Elements in Environment, Medicine and Biology". October 27-29, Timisoara, Roumania.* 1996: 219-24.
16. Gonzalez D, Ramirez A, Hernandez M, Müller R, Anke M. Der Magnesiumverzehr erwachsener Mischköstler Mexikos. *Mengen-und Spurenelemente* 1999; 19: 130-42.
17. Harzer G, Haschke F. Micronutriants in human milk. In: Renner E, ed. *Micronutrients in Milk and Milk-based Food Products.* London and New York: Elsevier Applied Science, 1989: 125-37.
18. Iyengar GV. *Elemental Composition of Human and Animal Milk. (IAEA- TECDOC- 269).* Vienna: International Atomic Energy Agency, 1982.
19. Kumpulainen JRT. Characterization of a total diet reference material (ARC/CL HDP) for contents of essential and toxic elements. *Fresenius J Anal Chem* 1990; 338: 461-5.
20. Meij JC, Lambert PWJ, van den Heuvel N, Knoers VAM. Genetic disorders of magnesium homeostasis. *Biometals* 2002; 15: 297-307.
21. Renner E. In: *Micronutrients in Milk and Milk-based Food Products.* London and New York: Elsevier Applied Science, 1989: 125-237.
22. Shils ME. Magnesium. In: O'Dell BL, Sunde RA, eds. *Handbook of Nutritionally Essential Mineral Elements.* New York, Basel, Hong Kong: Marcel Dekker Inc, 1997: 117-52.
23. Vormann J. Magnesium. In: Merian E, Anke M, Ihnat M, Stoeppler M, eds. *Elements and their Compounds in the Environment. Vol. 2.* 2004: 587-98.
24. Vormann J, Anke M. Dietary magnesium supply, requirements and recommendations – results from duplicate and balance studies in man. *J Clin Basic Cardiol* 2002; 5: 49-53.
25. Weber S, Konrad M. Angeborene Magnesiumverlusterkrankungen. *Dtsch Arztebl* 2002; 99: 1230-8.

VIII. Magnesium and sport

Magnesium supplementation in soccer players: effects on metals related to antioxidant defense

L. Leal, E. Gómez-Trullén, M. Galvez, J.L. López-Colón*, S. Millán-Plano, J. García, J.F. Escanero

Department of Pharmacology and Physiology, Faculty of Medicine, University of Zaragoza, Spain
** Institute of Preventive Medicine of the Defence Madrid, Spain*

Abstract. The muscle during intensive exercise produces high levels of ROS. Different antioxidants has been used in order to prevent this increases. Magnesium has been also used to enhance the performance in athletes. Acute (three hours) and chronic (seven days) Mg-supplementation was supplied to soccer players with the aim to know the variations of malondialdehyde en plasma and different metals (Zn, Cu, Se and Mn) related to antioxidant defense after a ergometer bicycle test (70% of maximal intensity). The results of malondialdehyde in serum show a clear tendency ($p < 0.08$) to decrease and to increase serum Se after Mg-supplementation. The other metals analysed do not modify its levels in plasma neither after Mg-supplementation nor after the ergometer effort in all cases. These facts suggest that Mg could ameliorate the antioxidant defenses.

Free radicals, chemical species containing one or more unpaired electrons that are capable of independent existence, are produced in all living cells. Most radicals are generated *in vivo* either are reactive oxygen species (ROS) or reactive nitrogen species [1] or are derived from them. Since the 80's several papers have reported an increased production ROS correlated with different pathologies [2] and aging [3]. Dillard *et al.* [4] demonstrated for the first time, in 1978, that physical exercise can lead to an increase in lipid peroxidation. Since then, an increasing body of evidence has been accumulated to support the hypothesis that physical exercise has the potential to increase free radicals production and lead to oxidative stress. Most of the studies investigating the effects of exercise on oxidative stress are, however, focused on markers of free radical-induced tissue damage. Indications of increased damage to lipids [4-9], proteins [10] and DNA [4, 11, 12] with exercise have been well documented. Exercise-induced changes in levels of antioxidants have also been studied, but their significance to oxidative stress is harder to determine. In general, after the first reports [13, 14] of an excessive formation of ROS by the muscle during intensive physical exercise, it is being assumed that this increased production of ROS exercise-induced takes place in exercises of high intensity and has not been found on submaximal exercises [15-18].

By the other hand, an important number of intervention assays have evaluated the role of the exogen antioxidants in order to prevent the exercise-induced oxidative stress. In the same way, the use of magnesium (Mg) supplementation in athletes is being vastly employed in order to prevent the loss provoked by exercise or to correct marginal deficiencies generated by nutritional mistakes as nowadays has been suggested to increase the athletic performance [19]. In this sense, Newhouse y Finstad [20] revision about the Mg-supplementation effects in athletes reports the non-existence of these ones on performance (strength, anaerobic-lactacid, and aerobic), assuming finally that further studies in this area are justified.

In this study, it is intended to determine the effects of Mg-supplementation as short term (three hours) as long term (seven days) on ROS production and on the variations induced in the concentration in serum of the several trace elements (Se, Mn, Zn and Cu) involved in oxidative stress in athletes after a ergometer test.

Material and methods

Experimental subjects

Twelve normal subjects belonging to Real Zaragoza "C" Football Club in Spanish 3rd Division with the following anthropometric characteristics (*Table I*) were used in this study:
VO2max. was calculated indirectly following Astrand method [21].
The training programme per week is as follows: four days of two hours training each (Monday: recovery training and Wednesday, Thursday and Friday: maximal intensity training) and on Sunday competition: 90 min match.

Methodology used

The ergometer bicycle test consisted in submitting the athlete to a continuous effort of about 20 minutes with a frequency of 170 ± 3 beats per minute in 3 seasons separated by one week. The first season was for control values, the second one was performed three hours after Mg-supplementation (500 mg oral via, in a single dose as magnesium sulphate) and the last one after a week of Mg-supplementation (500 mg/day as first dose).
The ergometer test was carried out with an initial intensity of 60 watts under the value of PWC170 (cardiac intensity in watts to 170 beats), that had previously been determined by ergometer bicycle until the exhaustion carried out to calculate de VO2max. After the beginning, the power ascended (one watt each two seconds) until obtaining frequency values of 170 beats/minute. In that moment the time measurement for the test started. The achievement was about 3-7 minutes. The power was readjusted during the test and the pedal was constant (60 pedals per minute). During the test, the athletes were surveyed for controlling the frequency and the electrocardiogram.

Biochemical determinations

Before and after each test, blood was taken from an elbow's vein using EDTA as anticoagulant. After this, the blood was centrifuged (P Selecta® MEDITRONIC B-1S) at 3000 rpm for 10 minutes. The plasma was separated into aliquots of 2 ml in Eppenndorf tubes to be frozen for a temperature of -30 °C until using them.
The metals Mg, Cu and Zn were determined by atomic absorption spectrophotometry (Perkin-Elmer 1100-B cor. Norwalk CT 06856R). For Mn and Se determination an atomic absorption spectrophotometer with a graphite camera HGA 700 and an automatic shower AS-70 were used. The optic system had a simple beam and the background correction was made with a deuterium arch. For these determinations graphite tubes with pyrolytic surface and L'vov platform (Perkin-Elmer BO 137111 R) were used. The working conditions were the indicated by the manufacturer, with little modifications based on the best results obtained with diary experience.

Table I. Anthropometrical values (means ± SD) of the football players.

Anthropometrical Parameters	Means ± SD
Age (years)	17-20
Weight (Kg)	69.35 ± 5.62
Height (cm)	177.25 ± 4.11
$\mathbf{VO_{2max}}$ (L/min)	3.43 ± 0.37

Lactate determination was realized by an enzymatic method which uses the lactate oxidase to which the peroxidase is coupled to generate a dye. Finally, malondialdehyde was determined by a colorimetric method.

Statistical treatment

STAT-VIEW software was used to perform statistical treatment.

Results

Magnesium

Serum Mg values without and with Mg-supplementation of three hours and seven days do not show significant variations, neither in basal values nor in those ones obtained after the ergometer bicycle test in any of the cases performed (*Figure 1*).

Lactate

Plasmatic lactate concentrations in basal conditions increases significantly ($p < 0.01$) with the realization of a ergometer bicycle test to submaximal intensity (70%), with (acute 2.2 ± 0.3 *versus* 8.0 ± 1.7 mmol/L and chronic: 1.9 ± 0.1 *versus* 8.2 ± 1.9 mmol/L) and without (2.1 ± 0.1 *versus* 8.6 ± 1.0 mmol/L) Mg supplementation.

Malondialdehyde

The values of malondialdehyde in plasma don't variate significantly but they show a clear tendency to descend after Mg-supplementation (*Figure 2*). This decrease was not significant, probably because of the reduced number of samples.
After ergometer bicycle test to 70% of maximal intensity don't show any variation of the plasmatic values of malondialdehyde either with or without Mg-supplementation (three hours and seven days).

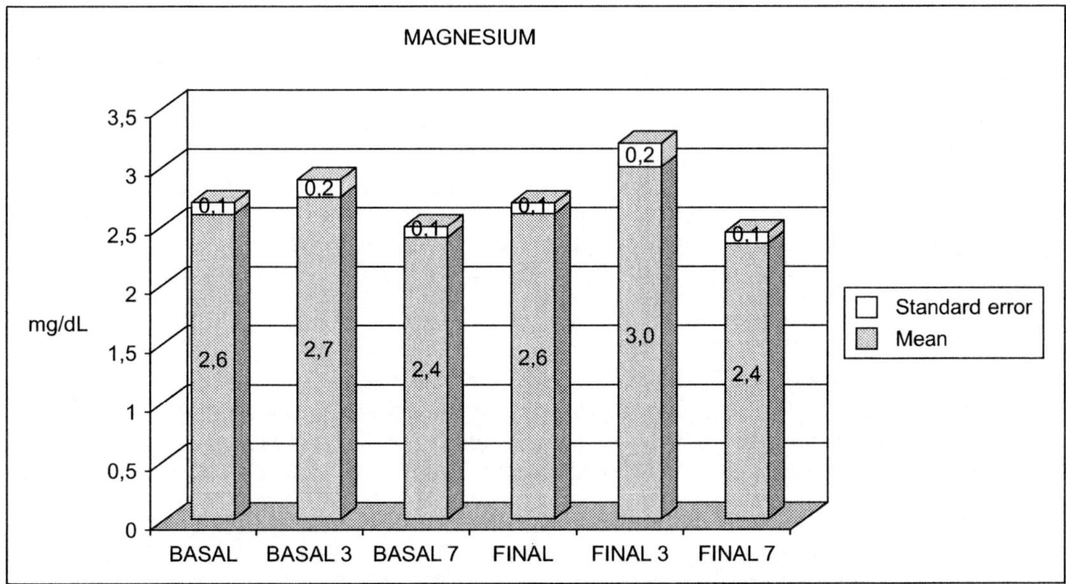

Figure 1. Serum Mg concentration (mean ± standard error) in basal conditions and after a ergometric bicycle test with and without Mg-supplementation in the indicated periods (3 hours – basal and final 3- and 7 days – basal and final 7-).

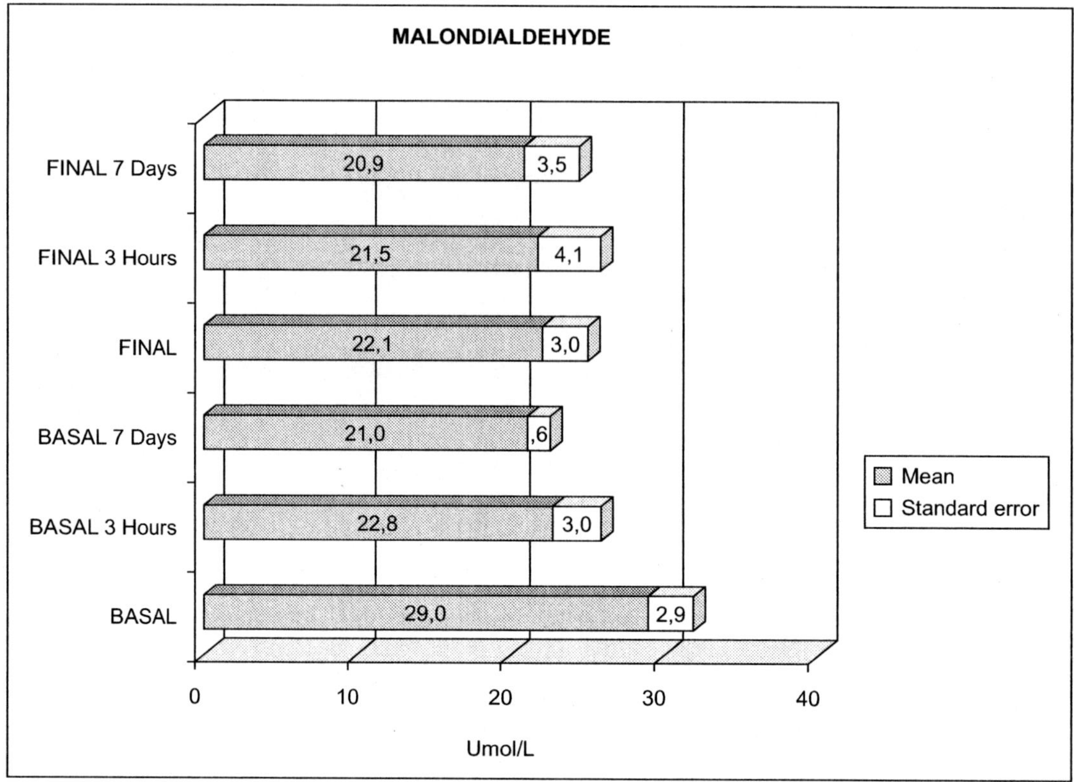

Figure 2. Serum malondialdehyde concentration (mean ± standard error), expressed as mmol/L, in basal conditions and after an ergometric bicycle test with and without Mg-supplementation in the indicated periods.

Trace elements

In *Figure 3* plasmatic values of the trace elements studied before and after ergometer bicycle test, with and without Mg-supplementation, are shown. There are not modifications either after a ergometer bicycle test or in normal conditions or after Mg-supplementation in plasma values of Cu (part A), Zn (part B) and Mn (part C). In contrast, concentrations of plasma Se (part D) show a tendency to increase, but not significantly, after Mg-supplementation; after exercise plasma Se increases without Mg-supplementation and after Mg-supplementation, either acute or chronic; in all cases these increases are not significant respect to basal values.

Discussion

The results of serum Mg after acute and chronic Mg-supplementation show no variations in relation to the values found before the supplementation and they confirm an excellent homeostatic regulation for this metal. In this line, Terblanche *et al.* [22] supplemented athletes (365 mg/day) for 4 weeks before and 6 weeks after a marathon and they reported that Mg-supplementation did not increase neither muscle or serum Mg concentrations and had no measurable effect on 42-Km marathon running performance; Weller *et al.* [23] reported that oral Mg-supplementation (500 mg Mg-oxide/day for 3 weeks) in athletes with low-normal serum Mg levels did not increase the Mg concentration in serum or any cellular compartment studied. It is important remark that the Mg clearance increased in the subjects receiving Mg-supplementation while in the placebo group the renal Mg clearance decreased. Moreover, Mg-supplementation did not affect exercise performance, neuromuscular activity, or muscle-related

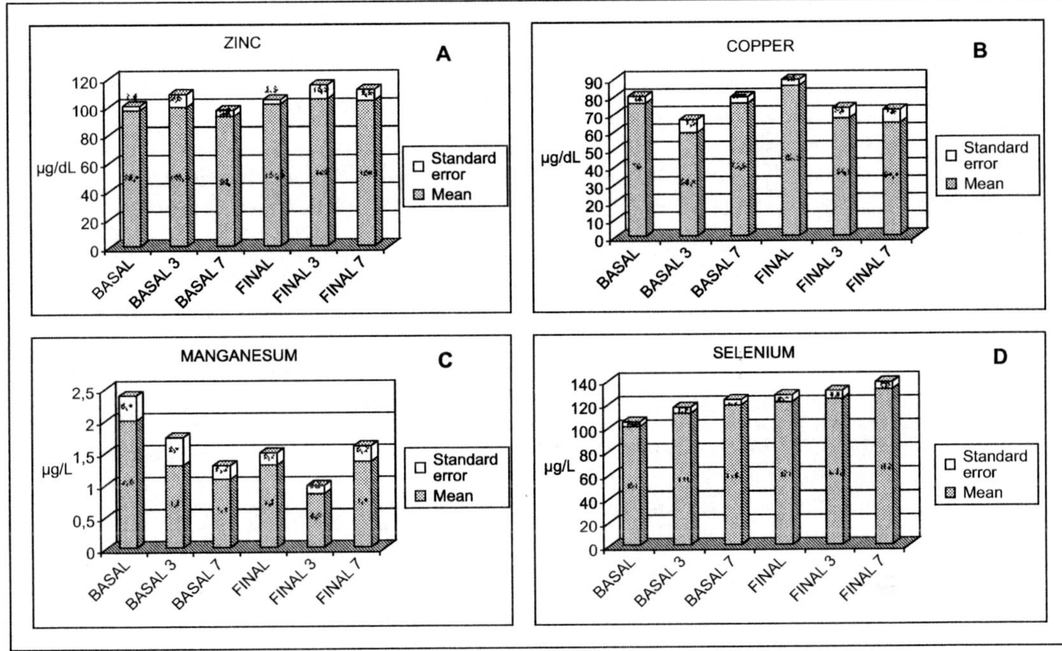

Figure 3. Serum Zn (part A), Cu (part B), Mn (part C) and Se (part D) concentrations (mean ± standard error) in basal conditions and after a ergometric bicycle test with and without Mg-supplementation in the indicated periods (3 hours – basal and final 3- and 7 days – basal and final 7).

symptoms; and finally, Finstad *et al.* [24] determined the effects of Mg-supplementation on performance and recovery in physically active women. Four weeks before the supplementation with 212 mg/day of Mg oxide, only the levels of ionic Mg increased. It is probably that the mechanisms of regulation work actively in athletes either avoiding losses generated by sport practice or eliminating the induced surcharge by the supplementations and, as a consequence, they do not generate alterations in serum total levels.

The effect of muscular exercise over Mg metabolism, after Rayssiguier revision [25], seemed to be clearly established so that sport practice of high intensity and short duration developed a hypermagnesemia, while submaximal prolongated exercise provoked a hypomagnesemia. In previous papers of our group [26, 27] was shown that in an ergometer bicycle test until exhaustion, the total serum Mg did not change while the ionic Mg decreased. The reported results in this paper do not show variations in total Mg plasmatic levels after an ergometer test (submaximal intensity) in athletes supplemented with Mg.

In spite of intense and extenuant exercise (mainly of resistance), that needs a major oxygen consumption, may provoke oxidative stress, in this study a submaximal exercise season has been chosen for several reasons: i) Moderate and regular physical exercise seems to increase the endogenous antioxidant system activity [17-19] and diminish the LDL susceptibility to oxidation [28, 29]; ii) Furthermore, the researches on general population confirm that the regular and moderated physical exercise practise helps to a health improvement so that it produces beneficious effects over cardiovascular risk factors: improves the lipidic profile [30], diminishes the blood pressure [31] and prevents the diabetes 2 [32]; and iii). Finally, Dillard *et al.* [4], in 1978, who have demonstrated for the first time that physical exercise can lead to an increase in lipid peroxidation after 60 minutes of cycling at 25-75% of VO2max.

The results showed that plasmatic malondialdehyde levels in basal conditions diminished after Mg-supplementation and, consequently, we could interpret this fact due to a protector effect of this element. In this respect, it should be reminded that Mg is considered an element of the anti-inflammatory and antioxidant systems and present a protective effect on cellular membrane [33], influencing the levels of the different modulatory factors [34].

The values of malondialdehyde did not variate after effort test neither with acute or chronic Mg-supplementation nor without it. These results agree with the ones reported by Lovlin *et al.* [7], who demonstrated that exhaustive treadmill running increased levels of malondialdehyde, whereas running at a moderate intensity (70% VO2max) failed to produce this effect, and low intensity running (40% VO2max) even decreased this marker of oxidative stress. However, we might insist as Cooper *et al.* [1], that some products of oxidative reactions may not be elevated directly after exercise, and reach their maximal levels only hours [35, 36] or even days [12] after the end of exercise. Therefore, the absence of signs of oxidative stress directly after exercise does not necessarily imply the oxidative damage.
This lack of malondialdehyde modification after the ergometer effort test must be analysed with extreme awareness. It must be set up on first place that, while oxidative stress could cause a primary decrease in antioxidants, mobilization from secondary sources, anywhere in the body, might result in an apparent increase and, in consequence, the damage can be less than expected. It has been demonstrated that GSH:GSSG ratio in blood diminishes with exercise [37, 38], whereas plasma levels of vitamins C and E tend to increase [38-41]. Therefore, although changes in oxidation state or concentration of antioxidants can point to impaired antioxidant defenses, they do not necessarily indicate tissue damage [42]. It is unclear what causes these changes, or to what extent they influence oxidative stress. A rise of antioxidant levels in plasma might enhance the antioxidant defenses in blood, but could possibly impair defenses at the sites from which they are mobilized [1].
In relation to the trace elements, Mg-supplementation did not generate any modification in basal levels of trace elements studied excepting Se. The plasmatic increases of Se after an ergometer bicycle test in athletes without Mg-supplementation and its increment in the supplemented ones, point out to a higher Se availability after Mg-supplementation. The basal serum Se tendency to increase after Mg-supplementation is reported for the first time. The exercise-induced variations of the concentrations of this metal in serum or plasma are contradictory. In this line, Fogelholm and Latineen [43] reported an increase of Se in serum after weeks of transatlantic race have while Singh *et al.* [44] have shown a significant decrease in a Navy Seals group who followed a training programme of intense physical effort and who were also supplemented with an intake of Se of 92.5 ug/day during the training week and 61.5 ug/day the week before. By contrast, Monteiro *et al.* [45] did not find these variations in aerobic training. Finally, our group [26] has reported a decrease in serum levels of Se after an ergometer bicycle test until exhaustion.
In **conclusion**, Mg-supplementation tends to decrease the plasmatic levels of malondyaldehyde and to increase those of Se. Also, Mg-supplementation does not modify any value found in basal conditions after an ergometer bicycle test of submaximal intensity. These facts suggest that Mg could enhance the antioxidant defenses. However, further research is required in order to establish firmly these conclusions.

References

1. Cooper CE, Vollaard NBJ, Choueiri T, Wilson MT. Exercise, free radicals and oxidative stress. *Biochem Soc Trans* 2002; 30(2): 280-5.

2. Cross CE. Oxygen radicals and human disease. *Ann Intern Med* 1987; 107: 526-45.

3. Harman D. Free radical theory of aging: the free radical diseases. *Age (Omaha)* 1984; 7: 111-31.

4. Dillard CJ, Litov RE, Savin WM, Dumelin EE, Tappel AL. Effects of exercise, vitamin E, and ozone on pulmonary function and lipid peroxidation. *J Appl Physiol* 1978; 45: 927-32.

5. Ashton T, Rowlands CC, Jones E, *et al.* Electron spin resonance spectroscopic detection of oxygen-centred radicals in human serum following exhaustive exercise. *Eur J Appl Physiol Occup Physiol* 1998; 77: 498-502.

6. Ashton T, Young IS, Peters JR, *et al.* Electron spin resonance spectroscopy, exercise, and oxidative stress: an ascorbic acid intervention study. *J Appl Physiol* 1999; 87: 2032-6.

7. Lovlin R, Cottle W, Pyke I, Kavanagh M, Belcastro AN. Are indices of free radical damage related to exercise intensity. *Eur J Appl Physiol Occup Physiol* 1987; 56: 313-6.

8. Niess AM, Hartmann A, Grunert-Fuchs M, Poch B, Speit G. DNA damage after exhaustive treadmill running in trained and untrained men. *Int J Sports Med* 1996; 17: 397-403.

9. Kanter MM, Lesmes GR, Kaminsky LA, La Ham-Saeger J, Nequin ND. Serum creatine kinase and lactate dehydrogenase changes following an eighty kilometer race. Relationship to lipid peroxidation. *Eur J Appl Physiol Occup Physiol* 1988; 57: 60-3.

10. Allesio HM, Hageman AE, Fulkerson BK, Ambrose J, Rice RE, Wiley RL. Generation of reactive oxygen species after exhaustive aerobic and isometric exercise. *Med Sci Sports Exerc* 2000; 32: 1576-81.

11. Poulsen HE, Loft S, Vistisen K. Extreme exercise and oxidative DNA modification. *J Sports Sci* 1996; 14: 343-6.

12. Hartmann A, Pfuhler S, Dennog C, Germadnik D, Pilger A, Speit G. Exercise-induced DNA effects in human leukocytes are not accompanied by increased formation of 8-hydroxy-2'-deoxyguanosine or induction of micronuclei. *Free Radic Biol Med* 1998; 24: 245-51.

13. Davies KJA, Quintanilha AT, Brooks GA, Packer L. Free radicals and tissue damage produced by exercise. *Biochem Biophys Res Commun* 1982; 107: 1198-205.

14. Demopoulos HB, Santomier JP, Seligman ML, Pietronigro L. Free radical pathology: rationale and toxicology of antioxidants and other supplements in sports medicine and exercise science. In: Katch FI, ed. *Sport health and nutrition.* Champaign IL: Human Kinetics, 1988.

15. Wiit EH. Exercise, oxidative damage and effects of antioxidant manipulation. *J Nutr* 1992; 122: 766-73.

16. Cao G, Chen J. Effects of dietary zinc on free radical generation, lipid peroxidation, and superoxide dismutase in trained mice. *Arch Biochim Biophys* 1991; 292(1): 147-53.

17. Goldfarb AH. Antioxidants: role of supplementation to prevent exercise-induced oxidative stress. *Med Sci Sports Exerc* 1993; 25(2): 232-6.

18. Ji Ll. Antioxidant enzyme response to exercise and aging. *Med Sci Sports Exerc* 1993; 25(2): 225-31.

19. Bohl CH, Volpe SL. Magnesium and exercise. *Crit Rev Food Sci Nutr* 2002; 42(6): 533-63.

20. Newhouse IJ, Finstad EW. The effects of magnesium supplementation on exercise performance. *Clin J Sports Med* 2000; 10: 195-200.

21. Astrand I. Aerobic work capacity in man and women with special references to age. *Acta Psysiol Scand* 1960; 49(Suppl): 169.

22. Terblanche S, Noakes TD, Dennis SC, Marais D, Eckert M. Failure of magnesium supplementation to influence marathon running performance or recovery in magnesium-replete subjects. *Int J Sport Nutr* 1992; 2(2): 154-64.

23. Weller E, Bachert P, Meinck HM, Friedmann B, Bartsch P, Mairbaurl H. Lack of effect of oral Mg-supplementation on Mg in serum, blood cells, and calf muscle. *Med Sci Sports Exerc* 1998; 30(11): 1584-91.

24. Finstad EW, Newhouse IJ, Lukaski HC, Mcauliffe JE, Stewart CR. The effects of magnesium supplementation on exercise performance. *Med Sci Sports Exerc* 2001; 33(3): 493-8.

25. Rayssiguier Y. Apport alimentaire en magnésium et exercice physique. In: Monod H, ed. *Nutrition et Sport.* Paris: Masson, 1990: 154-63.

26. Guerra M, Monje A, Perez-Beriaín RM, *et al.* Ionic magnesium and selenium in serum after a cycle-ergometric test in football-players. In: Centeno JA, Collery P, Vernet G, Finkelman RB, Gibb H, Etienne JC, eds. *Metal Ions in Biology and Medicine.* Paris: John Libbey, 2000: 501-4; (6).

27. Guerra M, Monje A, Vidal MD, *et al.* Total and ionic plasmatic magnesium assessment after a bicycle ergometric test. In: Collery P, Corbella J, Domingo JL, Etienne JC, Llobet JM, eds. *Metal Ions in Biology and Medicine.* Paris: John Libbey, 1996: 544-6; (4).

28. Williams PT, Krauss RM, Wood PD, Lindgren FT, Giotas C, Vranizan KM. Lipoproteins subfractions of runners and sedentary men. *Metabolism* 1986; 35: 45-52.

29. Sanchez-Quesada JL, Ortega H, Payes-Romero A, *et al.* LDL from aerobically-trained subjects shows higher resistence to oxidative modification than LDL from sedentary subjects. *Atherosclerosis* 1977; 132: 207-13.

30. Marrugat J, Elosua R, Covas ML, Molina L, Rubiés-Prat J. Marathon investigators. Amount and intensity of physical activity, physical fitness and serum lipid in men. *Am J Epidemiol* 1996; 143: 562-9.

31. American College of Sports Medicine. Physical activity, physical fitness and hypertension. *Med Sci Sports Exerc* 1993; 25: 1-10.

32. Helmrich SP, Ragland DR, Leung RW, Paffenbarger RS. Physical activity and reduced occurrence of non-insulin-dependent diabetes mellitus. *N Engl J Med* 1991; 325: 874-7.

33. Durlach J, Bara M. *Le magnésium en biologie et en médecine.* Paris: Editions Médicales Internationales, 2000; (2).

34. Manuel B, Keenoy Y, Moorkens G, *et al.* Magnesium and oxidative stress status in patients with chronic fatigue. In: Theophanides T, Anastassopoulou J, eds. *Magnesium: Current Status and New Developments.* Dordrecht: Kluwer Academic Publishers, 1997: 99-104.

35. Koyama K, Kaya M, Ishigaki T, *et al.* Role of xanthine oxidase in delayed lipid peroxidation in rat liver induced by acute exhausting exercise. *Eur J Appl Physiol Occup Physiol* 1999; 80: 28-33.

36. Maughan RJ, Donnelly AE, Gleeson M, Whiting PH, Walker KA, Clough PJ. Delayed-onset muscle damage and lipid peroxidation in man after a downhill run. *Muscle Nerve* 1989; 12: 332-6.

37. Dufaux B, Heine O, Kothe A, Prinz U, Rost R. Blood glutathione status following distance running. *Int J Sports Med* 1997; 18: 89-93.

38. Viguie CA, Frei B, Shigenaga MK, Ames BN, Packer L, Brooks GA. Antioxidant status and indexes of oxidative stress during consecutive days of exercise. *J Appl Physiol* 1993; 75: 566-72.

39. Gleeson M, Robertson JD, Maughan RJ. Influence of exercise on ascorbic acid status in man. *Clin Sci* 1987; 73: 501-5.

40. Duthie GG, Robertson JD, Maughan RJ, Morrice PC. Blood antioxidant status and erythrocyte lipid peroxidation following distance running. *Arch Biochem Biophys* 1990; 282: 78-83.

41. Pincemail J, Deby C, Camus G, *et al.* Tocopherol mobilization during intensive exercise. *Eur J Appl Physiol Occup Physiol* 1988; 57: 189-91.

42. Packer L. Oxidants, antioxidant nutrients and the athlete. *J Sports Sci* 1997; 15: 353-63.

43. Fogelholm GM, Latinen PK. Nutritional evaluation of a sailing crew during a transatlantic race. *Scand J Med Sci Sports* 1991; 1: 99-103.

44. Singh A, Smoak BL, Patterson KY, Le May LG, Veillon C, Deuster PA. Biochemical indices of selected trace minerals in men: Effect of stress. *Am J Clin Nutr* 1991; 53: 126-31.

45. Monteiro CP, Varela A, Pinto M, *et al.* Effect of an anaerobic training on magnesium, trace elements and antioxidant systems in a Down syndrome population. *Magnes Res* 1997; 10: 65-71.

Urea and ammonia levels in serum after an ergometer bicycle test in athletes on magnesium supplementation

L. Leal, M. Galvez, M.S. Soria, L. Roda, J. Villanueva, M. Guerra, J.F. Escanero

Department of Pharmacology and Physiology, Faculty of Medicine, University of Zaragoza, Spain

Abstract. Although Mg-supplementation is vastly used by athletes its effects on performance, recovery or overtraining are not well established. The ammonia in blood increase with the exercise and this may affect the ventilation and misadjust several functions in athletes. Twelve soccer players were supplemented with Mg and in a period of three hours and in a period of a week after it they were submitted to a ergometer bicycle test (70% maximal intensity). The basal values of ammonium decreased significantly after Mg-supplementation and the ones after ergometer bicycle test increased but not so much as when the test was performed to the same athletes without Mg-supplementation. The lactate increased similarly after exercise with and without Mg-supplementation. Other parameters analysed (magnesium, urea, enzymes) did not vary neither with Mg-supplementation nor after exercise. In brief, the results concerning with the decreases of ammonium levels in serum may be a support in order to explain the possible benefit effects of Mg-supplementation in athletes.

Magnesium (Mg) supplementation is usually applied by athletes, as much as only element as in combination with other trace elements and vitamins. Nevertheless, the benefits of this supplementation aren't well established neither on the performance nor on the recovery or on the overtraining [1].
Several studies [2-8] have shown that the physical exercise practice in humans causes an increase in the serum levels of ammonium (NH_4^+). During the same one, in muscle at work, the ADP coming from the hydrolysis of the ATP is transphosphorilated to AMP, which desamine to IMP and NH_4^+. This last reaction is catalysed by the AMP-desaminase and is the first step of the purine nucleotide cycle [9]. A part of the ammonia (NH_3) produced by the muscle in contraction is released to the blood where it is accumulated during the exercise [10-12]. The blood NH_3 has been suggested to affect the ventilation and can misadjust the motor function and the adequate working of the central nervous system [13, 14]. In that way, the blood NH_3 elimination is such important for the maintenance of the physical performance of athletes.

On the other hand, connection between the production of NH_4^+ and the glycolytic energetic metabolism has also been suggested in different papers [2, 15, 16] and because of this reason most of them usually analyses simultaneously the serum lactate and the NH_4^+ levels during the exercise.

According to all the previous data, this work has tried to verify if Mg-supplementation in athletes, as much acute (three hours) as chronic (one week), modifies the serum NH_4^+ concentrations and if the behaviour serum lactate/NH_4^+ ratio is maintained after the accomplishment of an ergometer bicycle test of submaximal intensity.

Material and methods

Experimental subjects

Twelve normal subjects belonging to Real Zaragoza "C" Football Club in Spanish 3rd Division with the following anthropometric characteristics (*Table I*) were used in this study.
VO2max. was calculated indirectly following Astrand method [17].
The training programme per week was as follows: four days of two hours training each (Monday: recovery training and Wednesday, Thursday and Friday maximum intensity training) and on Sunday competition: 90 minutes match.

Methodology used

The ergometer bicycle test consisted in submitting the athlete to a continuous effort of about 20 minutes with a frequency of 170 ± 3 beats/minute in three seasons separated by one week. The first season was for control values, the second one was after three hours Mg-supplementation (500 mg oral via, in a single dose as Mg sulphate) and the last one after a week of Mg-supplementation (500 mg/day as first dose).
The ergometer bicycle test was carried out with an initial intensity of 60 watts under the value of PWC170 (cardiac intensity in watts to 170 beats), that had previously been determined by an ergometer bicycle test until exhaustion performed in the first season to calculate the VO2max. After the beginning, the power ascended (one watt each two seconds) until reaching frequency cardiac values of 170 beats/minute. In that moment the time measurement for the test started. The achievement was about 3-7 minutes. The power was readjusted during the test and the pedal was constant (60 pedals/minute). During the test, the athlete was monitorized for controlling the cardiac frequency and the electrocardiogram.

Biochemical determinations

Before and after each test, blood was taken from an elbow's vein using EDTA as anticoagulant. Blood was taken after centrifuged (P Selecta® MEDITRONIC B-1S) at 3000 rpm for 10 minutes. The plasma was divided in aliquots of two ml in Eppenndorf tubes to be frozen for a temperature of - 30° C until using them.
Magnesium was performed by spectrophotometry of atomic absorption (Perkin-Elmer 1100-B corp. Norwalk CT 06856 R). The working conditions were those indicated by the manufacturer, with little modifications based on the best results obtained with daily experience.
For lactate determination, with enzymatic method, it was used the lactate oxidase to which the peroxidase is coupled to generate a dye. Ammonia was determined by an enzymatic kinetic assay (glutamate dehydrogenase) and the urea by another enzymatic test (urease). Finally, the enzymatic activities of total CK and LDH were performed measuring photometrically the rate of decrease in NADH (LDH) and the rate of formation of NADPH (CK).

Statistical treatment

STAT-VIEW software was used to perform statistical treatment.

Table I. Anthropometrical values (means ± SD) of the football players.

Anthropometrical Parameters	Means ± SD
Age (years)	17-20
Weight (Kg)	69.35 ± 5.62
Height (cm)	177.25 ± 4.11
$\mathbf{VO_{2max}}$ (L/min)	3.43 ± 0.37

Results

Magnesium

Serum Mg values with and without Mg-supplementation of three hours and seven days did not show significant variations, neither in basal values nor in those ones obtained after the ergometer test in any of the cases performed (*Figure 1*).

Urea

Urea levels in serum did not vary in relation to the basal values in any case (on Mg-supplementation and after ergometer bicycle test with and without Mg-suplementation) (*Figure 2A*).

Ammonia

The basal values of ammonia in serum decreased significantly ($p < 0.01$) after Mg-supplementation, so acute as chronic (*Figure 2A*). These findings have not been reproduced in posterior experiments.
The ergometer bicycle test to submaximal intensity provoked a significant increase ($p < 0.01$) in the amonia serum, in all cases (45% without Mg-supplementation and 392% and 166% after acute and chronic Mg-supplementation). However, the final values after the ergometer bicycle test decreased after Mg-supplementation in relation to those reached without one (*Figure 2B*).

Lactate and NH_4^+/lactate ratio

Plasmatic lactate concentrations in basal conditions increased significantly ($p < 0.01$) with the realization of an ergometer bicycle test to submaximal intensity (70%), with (acute 2.2 ± 0.3 *versus* 8.0 ± 1.7 and chronic: 1.9 ± 0.1 *versus* 8.2 ± 1.9) and without (2.1 ± 0.1 *versus* 8.6 ± 1.0) Mg-supplementation.

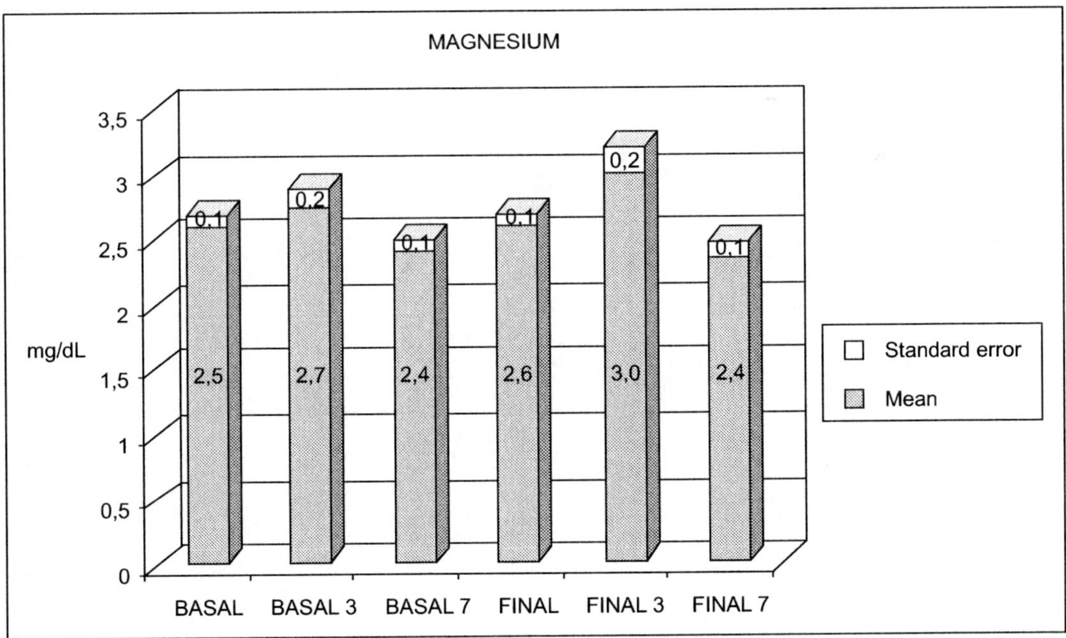

Figure 1. Serum Mg concentration (mean ± standard error) in basal conditions and after a ergometric bicycle test with and without Mg-supplementation in the indicated periods (3 hours – basal and final 3- and 7 days – basal and final 7-).

The NH_4^+/lactate ratio was higher than 60 in basal conditions decreasing to values about 13 with acute Mg-supplementation and about 25-26 after chronic one. After the ergometer bicycle test the values were 22-23 (without Mg-supplementation) *versus* 17 and 15-16 (after acute and chronic Mg supplementation). These values showed a decrease in the ratio after Mg-supplementation.

CK

Figure 3 shows the tendency to decrease in CK basal values three hours (19.5%) and a week (23.5%) with Mg-supplementation. Similar decreases were found in final values (after ergometer bicycle test). Submaximal ergometer bicycle test did not show differences respect to the basal values. In all cases existed a tendency to increase around to 15%.

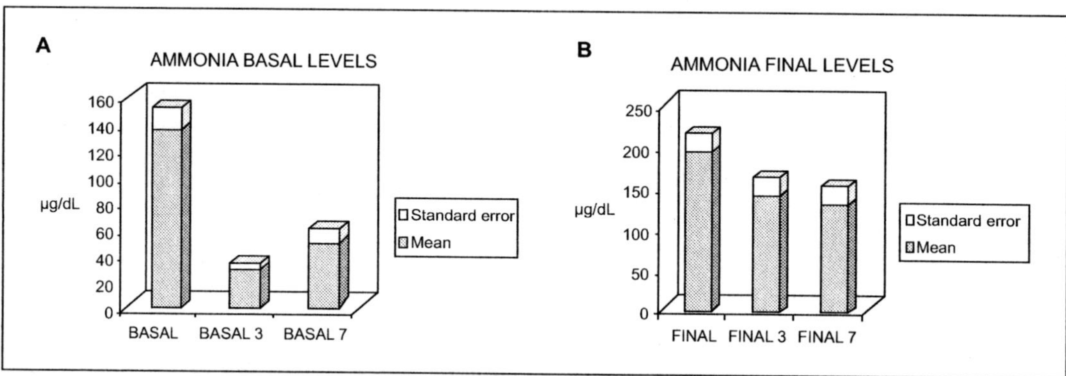

Figure 2. Serum ammonia concentration (mean ± standard error) in basal conditions (part A) and after a ergometric bicycle test (part B) with and without Mg-supplementation in the indicated periods (3 hours – basal and final 3- and 7 days – basal and final 7-).

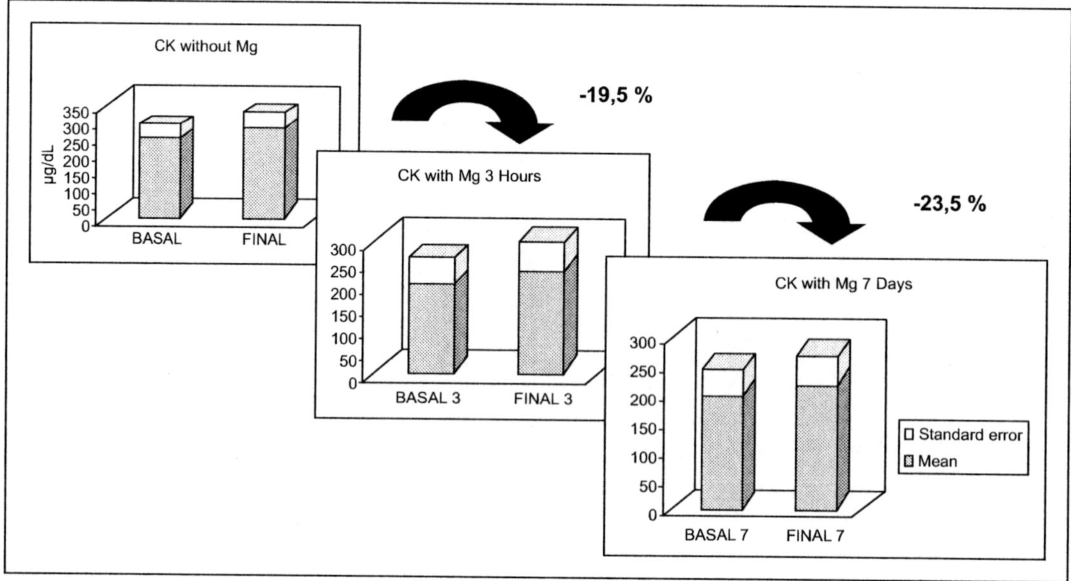

Figure 3. Serum CK activity (mean ± standard error), expressed as U/L, in basal conditions and after a ergometric bicycle test with and without Mg-supplementation in the indicated periods (3 hours – basal and final 3- and 7 days – basal and final 7-).

LDH

This enzymatic activity did not present significative changes neither Mg-supplementation nor exercise with and without Mg-supplementation.

Discussion

The results of serum Mg after acute and chronic Mg-supplementation showed no variations in relation to the values found before the supplementation and these confirmed an excellent homeostatic regulation for this metal [18]. In this line, Terblanche *et al.* [19] supplemented athletes (365 mg/day) for 4 weeks before and 6 weeks after a marathon and they reported that Mg-supplementation did not increase neither muscle nor serum Mg concentrations and had no measurable effect on 42-Km marathon running performance; Weller *et al.* [20] reported that oral Mg-supplementation (500 mg Mg-oxide/day for 3 weeks) in athletes with low-normal serum Mg levels did not increase the Mg concentration in serum or any cellular compartment studied. It is important to remark that the Mg clearance increased in the subjects receiving Mg-supplementation while in the placebo group the renal Mg clearance decreased. Moreover, Mg-supplementation did not affect exercise performance, neuromuscular activity, or muscle related symptoms; and finally, Finstad *et al.* [21] determined the effects of Mg-supplementation on performance and recovery in physically active women. Four weeks before the supplementation with 212 mg/day of Mg oxide only the levels of ionic Mg increased.
Respect the other purposes of this study, one of this consisted in analysing the effects of Mg-supplementation on the ammonia values in serum. The results found in basal values after acute and chronic Mg-supplementation were a surprise because of the decrease of serum ammonia levels may suppose [13,14]. This improvement and the smaller increase of the ammonia levels after exercise, in which to absolute values it talks about, would be a support to begin to understand the possible beneficial effects of the Mg-supplementation.
The decreases of the ammonia production and consequently the smaller levels in plasma after exercise can be explained themselves by some of the following mechanisms:

a) Inhibition or delay in the increase of the activity AMP desaminase exercise induced. Krause and Wegener [22] have suggested that in muscle the AMP desaminase activity is very low in rest and during the initial phase of the exercise, increasing in a stage of the exercise to decrease again if the exercise continues. The pH optimum for this activity is between 6.2-6.5 and has been repeatedly suggested that the acidosis increases the synthesis of IMP and, consequently, of NH_4^+. The role of Mg may regulate (inhibiting) the enzyme activity or provoking a delay in the increase stage of this activity during exercise.

b) A lesser implication of the glycolytic fibers in the contraction. The fast glycolytic muscle are the main source of blood ammonia during exercise [16]. The fact that in the exercise different types of muscular fibers can be recruited could explain that Mg would induce the implication of the no glycolytic fibers in the muscular contraction. In this sense, Wilkerson *et al.* [4] reported the existence of variations in NH_4^+ levels after an incremental load exercise test to exhaustion and after a constant load exercise test of 15 minutes at 65% VO2max, both on the ergometer bicycle and on the treadmill, concluding that the blood ammonia accumulation during exercise is critically dependent upon the test procedure.

c) Improve in the NH_4^+ elimination and reuptake by the muscle.

However, these results have not been confirmed in posterior experiments. The increased values of ammonia in basal conditions may be explained by the heterogeneous work performed during the Sunday match (time of participation, team position, intensity of the play, etc.).
In relation with the other aim of this paper: the analysis of the variations in the behaviour of the NH_4^+/lactate ratio, the reduction Mg-induced as in basal situation as after exercise may help understand the benefit of Mg-supplementation too. While ammonia decreases in basal situation and after exercise on Mg-supplementation, the levels of lactate do not vary with the supplementation and increase very similarly after exercise, independently of the Mg-supplementation. In this line, Pages *et al.* [7] analyzed the blood concentrations of ammonia and lactate in 7 volunteer male athletes before

and immediately after each segment of an endurance triathlon. Concentration of blood ammonia was increased after each of the three segments, reaching a peak after the 40 km bicycle ride. Concentrations of blood lactate were also increased over baseline. However there was a higher increase after the 1.4 km lake swim, than after the 40 km bicycle ride or after the 10 km run. These authors [7] did not find correlation between the levels of ammonia and lactate suggesting that ammonia and lactate follow different metabolic patterns. In the same way, Bouckaert and Pannier [8] reported the variations in the ammonia and lactate concentrations after an incremental load exercise test to exhaustion and after a constant load exercise test of 15 minutes at 65% VO2max, both on the ergometer bicycle and on the treadmill. During the incremental exercise test, blood ammonia levels were significantly higher on the ergometer bicycle as compared to the treadmill at the same submaximal VO2 (80% of VO2max) and at the VO2max, which was identical in the two modes of exercise. Plasma lactate levels were also significantly higher on ergometer bicycle at high submaximal exercise intensity but not at VO2max. During the constant load exercise test blood ammonia levels increased continuously and showed no differences between cycling and running, in contrast to plasma lactate accumulation, which was higher on the ergometer bicycle.

In conclusion, Mg-supplementation provokes decreases in the basal values of serum ammonia and the variations exercise-induced are lesser than in the same athletes no supplemented; respect to the NH_4^+ behaviour versus lactate must be indicated that the variations are different for these metabolites. The first facts may help understand the possible benefits of Mg-supplementation in athletes.

References

1. Newhouse IJ, Finstad EW. The effects of magnesium supplementation on exercise performance. *Clin J Sports Med* 2000; 10: 195-200.
2. Babij P, Matthews S. Rennie. MJ. Changes in blood ammonia, lactate and amino acids in relation to workload during bicycle ergometer exercise in man. *Eur J Appl Physiol Occup Physiol* 1983; 50: 405-11.
3. Dudley GA, Staron RS, Murray TF, Hagerman FC, Luginbahl A. Muscle fiber composition and blood ammonia levels after intense exercise in humans. *J Appl Physiol Respirat Environ Exercise Physiol* 1983; 54: 282-6.
4. Wilkerson JE, Batterton DL, Horvath SM. Ammonia production following maximal exercise: treadmill vs bicycle testing. *Eur J Appl Physiol Occup Physiol* 1975; 34: 169-72.
5. Wilkerson JE, Batterton DL, Horvath SM. Exercise induced changes in blood ammonia levels in humans. *Eur J Appl Physiol Occup Physiol* 1977; 37: 256-63.
6. Degoute F, Jouanel P, Filaire E. Energy demands during a judo match and recovery. *Br J Sports Med* 2003; 37(3): 245-9.
7. Pages T, Murtra B, Ibañez J, Rama R, Callis A, Palacios L. Changes in blood ammonia and lactate levels during a triathlon race. *J Sports Med Phys Fitness* 1994; 34(4): 351-6.
8. Bouckaert J, Pannier JL. Blood ammonia response to treadmill and bicycle exercise in man. *Int J Sports Med* 1995; 16(3): 141-4.
9. Andersen P, Saltin B. Maximal perfusion of skeletal muscle in man. *J Physiol* 1985; 366: 233-49.
10. Eriksson LS, Broberg S, Bjorkman O, Wahren J. Ammonia metabolism during exercise in rat skeletal muscles. *Am J Physiol* 1990; 258: E762-E766.
11. Graham TE, Bangsbo J, Gollnick PD, Juel C, Saltin B. Ammonia metabolism during intense dynamic exercise and recovery in humans. *Am J Physiol* 1989; 66: 313-7.
12. Katz A, Broberg S, Sahlin K, Wahren J. Muscle ammonia and amino acid metabolism during dynamic exercise in man. *Clin Physiol Oxf* 1986; 6: 365-79.
13. Banister EW, Cameron BJC. Exercise-induced hyperammonemia: peripheral and central effects. *Int J Sports Med* 1990; 11(suppl 2): S129-S142.
14. Wicher J, Kacemi H. Ammonia and ventilation: site and mechanism of action. *Resp Physiol* 1974; 20: 393-406.
15. Mutch BJ, Banister EW. Ammonia metabolism in exercise and fatigue. *Med Sci Sports Exerc* 1983; 15: 41-50.
16. Buono MJ, Clancy T, Cook J. Blood lactate and ammonium ions accumulation during graded exercise in humans. *J Appl Physiol* 1984; 57: 135-9.
17. Astrand I. Aerobic work capacity in man and women with special references to age. *Acta Physiol Scand* 1960; 49(Suppl): 169.
18. Leal L, Gómez-Trullen E, Galvez M, et al. Magnesium supplementation in soccer players: effects on metal related to antioxidant defence. *VIII European Magnesium Congress*, Cluj-Napoca, 2004.

19. Terblanche S, Noakes TD, Dennis SC, Marais D, Eckert M. Failure of magnesium supplementation to influence marathon running performance or recovery in magnesium-replete subjects. *Int J Sport Nutr* 1992; 2(2): 154-64.

20. Weller E, Bachert P, Meinck HM, Friedmann B, Bartsch P, Mairbaurl H. Lack of effect of oral Mg-supplementation on Mg in serum, blood cells, and calf muscle. *Med Sci Sports Exerc* 1998; 30(11): 1584-91.

21. Finstad EW, Newhouse IJ, Lukaski HC, Mcauliffe JE, Stewart CR. The effects of magnesium supplementation on exercise performance. *Med Sci Sports Exerc* 2001; 33(3): 493-8.

22. Krause U, Wegener V. Control of glicólisis in vertebrate skeletal muscle during exercise. *Am J Physiol* 1996; 270(4 Pt2): R821-R829.

IX. Posters

Comparison between the magnesium content in tap water and in natural mineral waters

O. Voroniuc[1], D. Diaconu[2], T. Navrotescu[2], A. Cojocariu[1], V. Nastase[1], R. Diaconu[2]

1. University of Medicine and Pharmacy Iaşi ; 2. Institute of Public Health Iaşi, Romania

Magnesium is an oligoelement which has an important effect on the myocardial function and the peripheral system [6, 10].
Both calcium and magnesium are involved in numerous functions and absolutely essential for the maintenance of health body. The most important biological functions of magnesium are: development of bones (about 70% of body magnesium is found in bones); the crucial part of many enzymes involved in energy production and respiration: it is an ion activating over 300 enzymes that take part in the carbohydrate, protein and fat metabolism and condition the cell and mitochondrial permeability; it can induce sleep and inhibit the neuromuscularexcitability; it stimulates the activity of some hormones (ex. vasopressin, insulin); transmission of nerve impulses and release of nerve tension; regulation of body temperature; pH balance; absorption and utilization of calcium, phosphorus, sodium, potassium, vitamins C, E, D; it is an important element which prevents the appearance and the aggravation of the atherosclerosis; it decreases the blood coagulability; it has an antiseptic effect on the musculature of the vessel walls, digestive tract and urinary system; Mg increase the antimicrobial protection by stimulating the increase of the leucocytosis when these are low, and also by stimulating the synthesis of the immunoglobulin, etc.
The importance of the Mg intake in drinking water is both quantitative and qualitative. Among all the factors studied in drinking water, the highest inverse correlation has been observed between Mg and cardiovascular mortality and morbidity, particularly fatal arrhythmic seizures. The myocardial Mg level is significantly lower in soft water areas than in hard water areas. The same is found in the heart of the subjects who have died from sudden heart attacks with or without myocardial infarcts [1, 2].
Epidemiological data in man and experimental data in rats demonstrated that intake of water containing a sufficient amount of magnesium, may prevent hypertension and nervous disturbances [12].
Drinking water contains chemically important levels of calcium (Ca^{2+}) and magnesium (Mg^{2+}). Due to the growing concern that content of drinking water may have adverse affects, the consumption of bottled mineral waters has increased.
If the lack of magnesium in the organism shows clinically, the effects may be cardiovascular, respiratory, neurological, psychiatric [8] and metabolic.

Material and methods

There is a great variation concerning the mineral content of commercially available bottled waters [7-9].
This study was performed to evaluate the levels of magnesium and other macroelements (Ca^{2+}, Na^{+}, K^{+}) in some natural mineral waters in comparison with the levels of these elements in the tap water delivered to the urban population in the territory of Moldova.

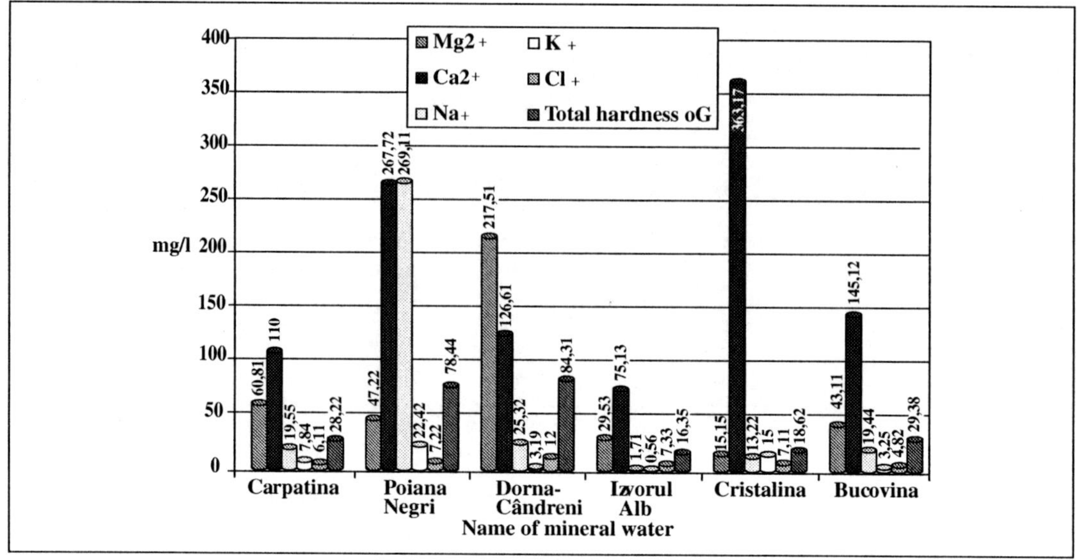

Figure 1. Chemical indicators of the mineral waters.

The Standardized methods for the determination of other chemical parameters (organic, matter, nitrate, sulphate, chloride, total hardness, hydrocarbonate) were used.

The water (tap and mineral waters) was analysed for the heavy metal contents. The microelements Cu, Cd, Pb, Ni were determined by atomic absorption spectrophotometry.

The concentration ratio of calcium and magnesium as well as of sodium and potassium in the water samples that have been analysed, were performed.

Result

Most of the commercially available mineral water are to be found in Suceava district. These waters were analysed and some of the results are presented in *Figure 1*.

Among the bottled waters that we reviewed, the magnesium content raged from 15.15-217.51 mg/l, the calcium content ranged from 75.13-363.17 mg/l *(Table I)*.

It is known that a high sodium intake and more, a high Na^+/K^+ ratio have been associated with hypertension risks [3].

The mineral water has good properties if the value of Ca^{2+}/Mg^{2+} ratio is low, and so the calcium assimilation in the body is facilitated. If this ratio is higher than optimal ratio the health effect of mineral water is decreased.

In our mineral waters the values of Ca^{2+} and Mg^{2+} concentrations ratio ranged between 0.58-24.31 and Na^+ and K^+ ratio ranged between 0.88-12.00 *(Figure 2) (Table II)*.

Seelig [4, 5] states the importance of the proportion Ca^{2+}/Mg^{2+} and considers that the optimal value should be of 2:1 to provide protection against the cardiovascular disease. A higher Ca concentration provides protection against osteoporosis, but should not enhance the heart risk or other diseases.

There are some hypotheses about the way that water content may affect health; these mostly involve either a protective action attributed to some metallic elements found in hard water or harmful effects attributed to certain metals often found in soft water [11].

Our analytical determination was performed on four metallic trace elements with toxicological effects: Pb, Cu, Ni, Cd. The heavy metal contents in the mineral water samples of Suceava area are presented in *Figure 3 (Table III)*.

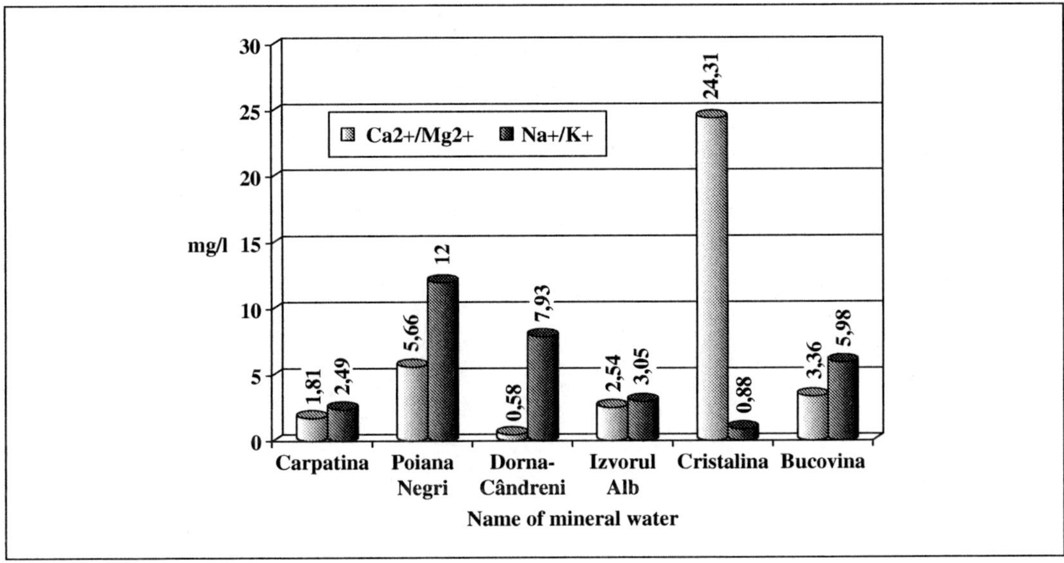

Figure 2. The values of Ca^{2+}/Mg^{2+} and Na^{+}/K^{+} ratio in the mineral waters.

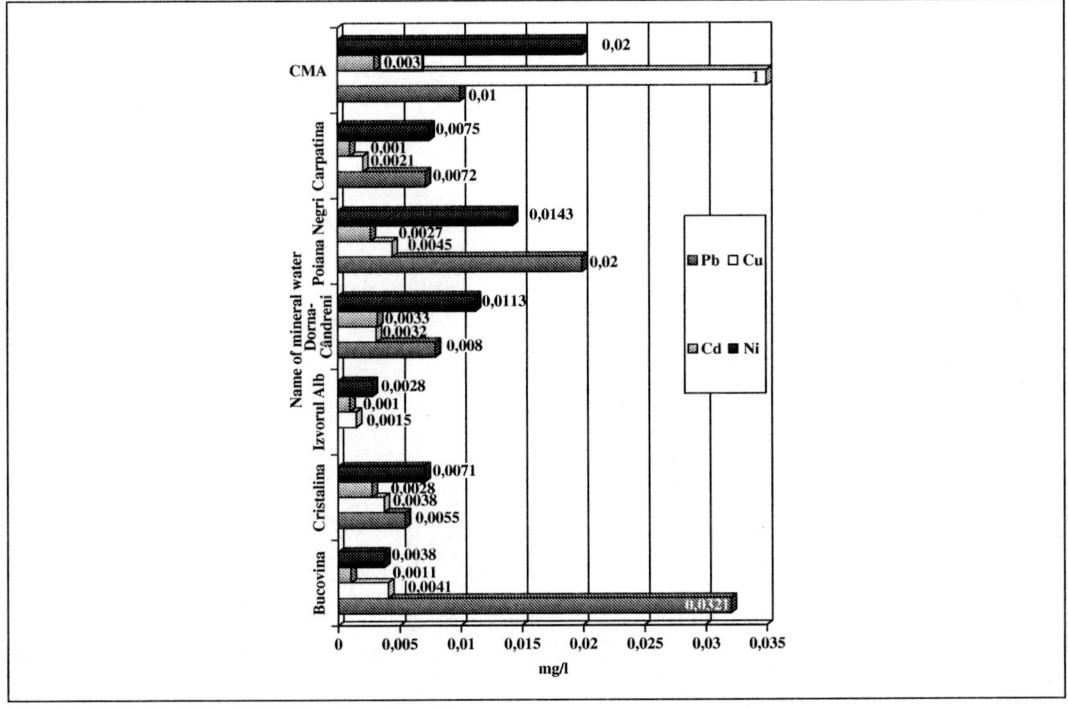

Figure 3. Selected heavy metals contents in water samples. * ND - non detec. CMA / Stas Ape minerale naturale 4450/97.

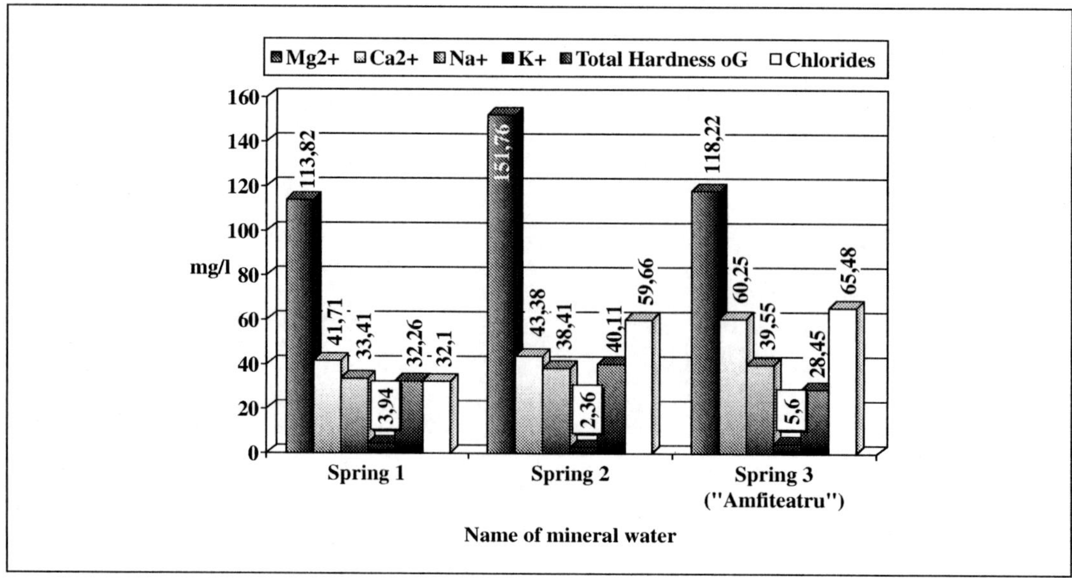

Figure 4. The mineral constituents of the springs in Iaşi city (Botanic Garden).

In Iaşi city there are seven mineral springs (in the Botanical Garden). From these mineral waters only three are used for human consumption (Springs 1, 2, 3). The mineral water "Amfiteatru" (commercially available mineral water) is the water from the spring number three. Regardless of the mineral content of these waters, we noticed the highest levels of magnesium concentration (*Figures 4, 7*) (*Table IV*). Excepting this commercially available mineral water there are in Iaşi district some springs which represent the public drinking water supply. A summary of the results concerning the water properties of the two springs in Răducăneni city revealed a high mineralization of the analysed samples (*Figures 5, 7*) (*Table V*).

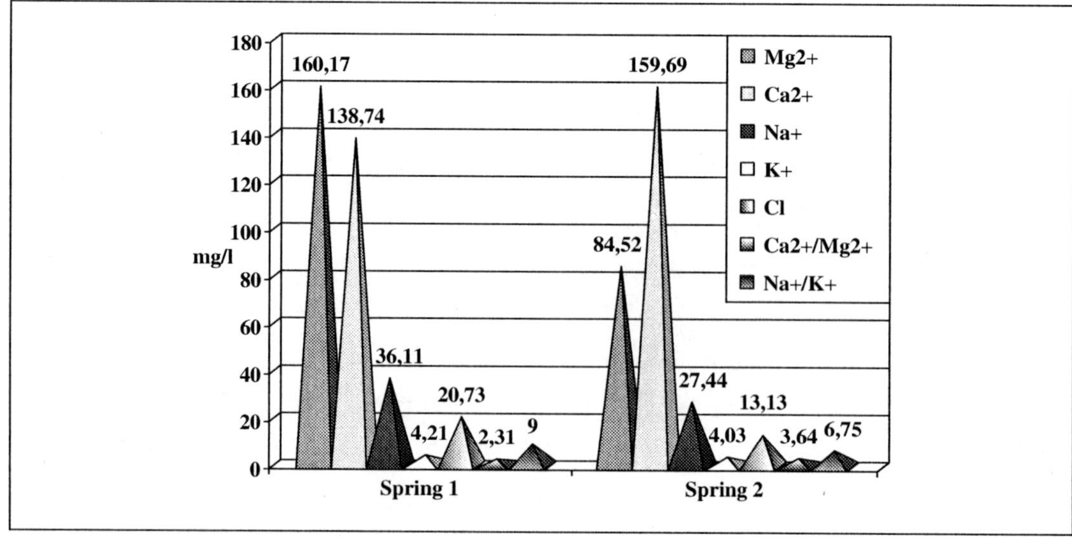

Figure 5. Water quality of the springs (Răducăneni).

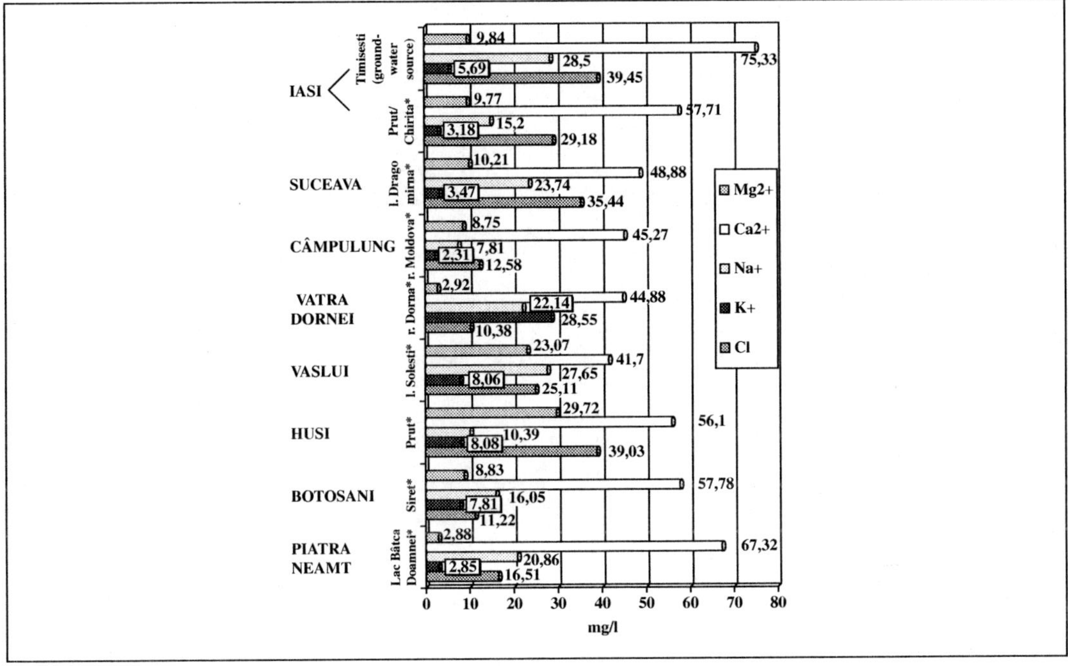

Figure 6. Some results of the chemical analysis of tap water.

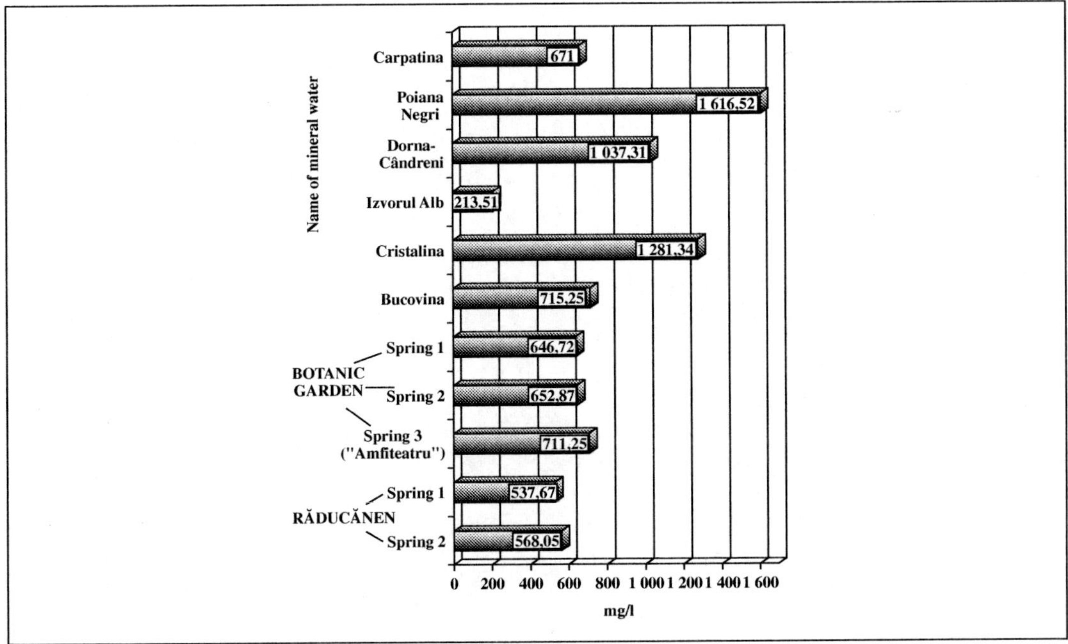

Figure 7. The concentration of hidrocarbonates in the mineral waters.

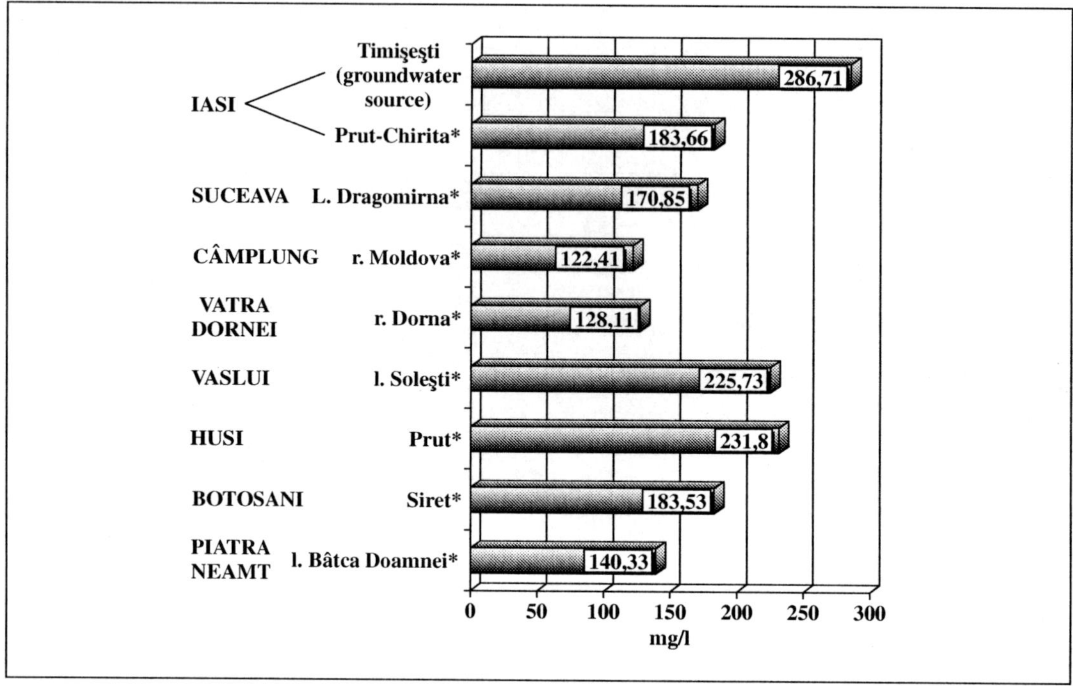

Figure 8. The concentration of the hidrocarbonates in the tap water.

In Moldova Territory the public drinking water supplies are derived from both surface and groundwater. The mineral content of underground water being influenced by the nature of the geological structure from water sources area, by the extent of pollution and by other hydrogeographical factors. In our study we compared the level of magnesium and that of the other chemical constituents of tap water with the levels registered in the mineral water samples. For this purpose, we analysed the drinking water delivered to the population in the districts of Iaşi, Suceava, Vaslui, Botoşani, Piatra Neamţ (*Figures 6, 8*).

The results from the chemical analysis showed that most of the samples had the low levels of inorganic constituents, including magnesium (*Table VI*).

Conclusions

Water is one of the main dietary components. Its quality plays an important role for the health of the population. In some areas the quality of drinking water is unsui and mineral waters frequently are used in place of tap water.

Table I. Chemical indicators of the mineral waters.

Names of mineral waters	Mg^{2+}	Ca^{2+}	Na^{+}	K^{+}	HCO_3^-	Cl	Total hardness ^0G
			mg/l				
Carpatina [1]	60.81	110.00	19.55	7.84	671.00	6.11	28.22
Poiana Negri [2]	117.22	267.72	269.11	22.42	1616.52	7.22	78.44
Dorna-Cândreni [3]	217.51	126.61	25.32	3.19	1037.31	12.00	84.31
Izvorul Alb [4]	29.53	75.13	1.71	0.56	213.51	7.33	16.35
Cristalina [5]	15.15	363.17	13.22	15.00	1281.34	7.11	18.62
Bucovina [6]	43.11	145.12	19.44	3.25	715.25	4.82	29.38

Table II. The values of Ca^{2+}/Mg^{2+} and Na^{+}/K^{+} ratio in the mineral waters.

Names of mineral waters	Ca^{2+}/Mg^{2+}	Na^{+}/K^{+}
Carpatina [1]	1.81	2.49
Poiana Negri [2]	5.66	12.00
Dorna-Cândreni [3]	0.58	7.93
Izvorul Alb [4]	2.54	3.05
Cristalina [5]	24.31	0.88
Bucovina [6]	3.36	5.98

Table III. Selected heavy metal contents in water samples.

Samples	Pb	Cu	Cd	Ni
		mg/l		
Carpatina [1]	0.0072	0.0021	0.001	0.0075
Poiana Negri [2]	0.0200	0.0045	0.0027	0.0143
Dorna-Cândreni [3]	0.0080	0.0032	0.0033	0.0113
Izvorul Alb [4]	ND*	0.0015	0.001	0.0028
Cristalina [5]	0.0055	0.0038	0.0028	0.0071
Bucovina [6]	0.0321	0.0041	0.0011	0.0038

Table IV. The mineral constituents of the springs in Iaşi city (the Botanical Garden).

Parameters	Spring 1	Spring 2	Spring 3 ("Amfiteatru")
Mg^{2+} (mg/l)	113.82	151.76	118.22
Ca^{2+} (mg/l)	41.71	43.38	60.25
Na^{+} (mg/l)	33.41	38.41	39.55
K^{+} (mg/l)	3.94	2.36	5.6
Total hardness (^{0}G)	32.26	40.11	28.45
Chlorides (mg/l)	32.10	59.66	65.48
Hydrocarbonates (HCO_3^-) (mg/l)	646.72	652.87	711.25

Table V. Water quality of the springs (Răducăneni).

Springs	Mg^{2+}	Ca^{2+}	Na^{+}	K^{+}	HCO_3^-	Cl	Ca^{2+}/Mg^{2+}	Na^{+}/K^{+}
				mg/l				
1	160.17	138.74	36.11	4.21	537.67	20.73	2.31	9.00
2	84.52	159.69	27.44	4.03	568.05	13.13	3.64	6.75

Our study revealed the following aspects:

– the mineral natural waters represente an important source of magnesium for the human body, with a beneficial influence on health;

– these mineral waters have a high mineralization level;

– a variation in the magnesium content of the mineral waters has been registered< in the most of samples we noticed a high magnesium concentration;

– with regard to the Ca^{2+}/Mg^{2+} ratio, there was registered a high value for the mineral water named "Cristalina" (Suceava district);

Table VI. Some results of the chemical analysis of tap water.

Cities	Nature of source	Mg^{2+}	Ca^{2+}	Na^{+}	K^{+}	HCO_3^-	Cl
				mg/l			
Iaşi	Timişeşti	9.84	75.33	28.50	5.69	286.71	39.45
	Prut-Chiriţa	9.77	57.71	15.20	3.18	183.66	29.18
Suceava	L. Dragomirna	10.21	48.88	23.74	3.47	170.85	35.44
Câmpulung	r. Moldova	8.75	45.27	7.81	2.31	122.41	12.58
Vatra Dornei	r. Dorna	2.92	44.88	22.14	28.55	128.11	10.38
Vaslui	Soleşti	23.07	41.70	27.65	8.06	225.73	25.11
Huşi	Prut	29.72	56.10	10.39	8.08	231.80	39.03
Botoşani	Siret	8.83	57.78	16.05	7.81	183.53	11.22
Piatra Neamţ	Lac Bâtca Doamnei	2.88	67.32	20.86	2.85	140.33	16.51

– comparison of the level of magnesium in mineral waters and tap water in some areas of Moldova territory, revealed the lowered concentrations in tap water;

– distribution of heavy metals with toxicological properties in the mineral water samples do not suggest outrunning of the admissible values in the sanitary standards.

The regular consumption of the mineral waters with the best content of magnesium is advisable, because this can ensure an important quantity (until 50%) of necessary of magnesium on a daily basis, with good results on health.

References

1. Durlach J, Bara M, Guiet-Bara A. Magnesium level in drinking water: its importance in cardiovascular risk. In: Itokawa I, Durlach J, eds. *Magnesium in health and Disease.* London: John Libbey, 1989: 173-82.
2. Durlach J. In: *Clinical forms of primary magnesium deficiency. Modern life-styles, lower energy intake and micronutrient status.* Berlin: Spring, 1991: 156-67.
3. Rylander R, Rubennowitz E. Magnesium and calcium in drinking water and cardiovascular mortality. *Scan Y Wor Environ Health* 1991; 17: 91-4.
4. Seelig MS. Magnesium (Mg) treatment or supplementation is often needed when calcium (Ca) intake is increased. *Blaine Journal* 1998.
5. Seelig MS. Epidemiology of water magnesium; evidence of contributions to health. Proceedings of Mg Symposium, Vichy, France 2000.
6. Topalov V, Kovacevic D. Magnesium cardiology. *Med Pregl* 2000; 53: 319-24.
7. Evandri MG, Bolle P. Pharmaco-toxicological screening of commercially available Italian natural mineral waters. *Farmaco* 2001; 5: 475-82.
8. Nichifor M, *et al.* In: *Changes in Mg2+ and other cations plasmatic concentration in patients with depression. Abstract book, 7-th European Magnesium Congress, Spain.* 2001: 34.
9. Galan P, Arnaud MJ. Contribution of mineral waters to dietary calcium and magnesium intake in France adult population. *J Am Diet Assoc* 2002: 1658-62.
10. Marque S, Jacqmin Gadda H. Cardiovascular mortality and calcium and magnesium in drinking water: an ecological study in elderly people. *Eur J Epidemiol* 2003; 18: 305-9.
11. Sabatier M, Arnaud J, Kastenmayer P, Rytz A, Barclay D. Meal effect on magnesium bioavailability from mineral water in healthy women. *Am J Clin Nutr* 2003; 75(1): 65-71.
12. Nerbrand C, Agrens L, Lenner RA. The influence of calcium and magnesium in drinking water and diet on cardiovascular risk factors in individuals living in hard and soft water areas with differences in cardiovascular mortality. *BMC Public Health* 2003; 18: 20-1.

Effect of treatment with magnesium orotate in patients with chronic heart failure

D. Gaita, S. Mancas, M. Iurciuc, C.A. Sarau, B. Mut, C. Dina, A. Ionac, D. Cozma, D. Lighezan, A. Avram, I. Branea, P. Armean, S.I. Dragulescu

Institute of Cardiology Timisoara, Clinic of Cardiac Rehabilitation, Bd. CD Loga nr.49, 300020 Timisoara, Romania

Abstract. The benefit of the treatment with magnesium orotate (magnerot) was assessed in a randomised, single blind and placebo controlled study. Respecting the inclusion criteria were selected 32 patients with ischemic chronic heart failure in early postoperative period after CABG. The main improvements induced by magnesium orotate are the increase in exercise capacity (distance ambulated during 6 minutes walk test and ergospirometric parameters) and the reduction of ventricular premature beats. The treatment was well tolerated and the adverse reactions were not significant. The study strongly suggests the benefit of magnesium orotate added to classical antiischemic therapy in the complex management of coronary patients after CABG.

Orotic acid, a naturally occurring substance, is a key intermediate in the biosynthetic pathway of the pyrimidines. Previous heart investigations suggest that orotate can protect recently infarcted against a further ischemic stress and may be beneficial in certain types of experimental cardiomyopathy [1]. The experimental studies demonstrate that orotic acid acts to maintain the cardiac pool of high-energy phosphate rather than stimulate ribonucleic acid production and protein synthesis [2]. The beneficial effect of magnesium orotate (an orotate salt) in the treatment of cardiomyopathic hamsters seems to be greater than the effect of orotate itself. The results showed that magnesium orotate reduced the severity of microscopic damage in the myocardium and skeletal muscle. Furthermore, magnesium orotate tended to reduce mortality compared with untreated hamsters [3]. The benefit of treatment with magnesium orotate in patients with coronary artery disease was demonstrated in a double-blind, prospective, placebo - controlled study, but with a small number of patients [4]. Four weeks of therapy with magnesium orotate improved exercise duration, reduced ventricular ischemia and increased ejection fraction.

Until now the benefit of treatment with magnesium orotate in the early postoperative period of patients with cardiac insufficiency and coronary artery by-pass grafts was not demonstrated.

The heart failure patients have some particularities in the early postoperative period after coronary artery by-pass grafts: hemodynamic instability, myocardial ischemia, reperfusion syndrome and intensive diuretic therapy. In this situation magnesium orotate seems to be, at least theoretically, a rational choice of treatment.

Methods

The study was performed on 32 patients, with ischemic chronic heart failure. Inclusion criteria were: age 40-75, coronary artery disease (high degree stenosis of at least 2 main vessels proved by coronary angiography with or without a history of myocardial infarction), relatively stable patients in the early period (24-48 hrs) after coronary artery by-pass grafting, left ventricular ejection fraction 25-35%, treated with standard medication.
Exclusion criteria were: hemodynamically significant valvulopathy, cardiac pacemaker and treatment with magnesium orotate more than one month during the last three months.

Study design

The study was performed in the Institute of Cardiology of Timisoara, respecting the Good Clinical Practice Guidelines. The study was approved by the Ethical Committee of the Institute of Cardiology of Timisoara. Before inclusion, patients signed the informed consent. The study was randomised, single blinded and placebo controlled. The tablets (placebo and magnesium orotate), identically shaped, were offered by Worwag Pharma. After inclusion patients were treated t.i.d. with 1000 mg magnesium orotate (magnerot - Worwag Pharma) for Group A or placebo for Group B for 2 months.
The endpoints were: functional capacity (Minnesota Living with Heart Failure Questionnaire and Dyspnea Fatigue Index), exercise capacity (6-minute walk test and ergospirometric parameters), left ventricular systolic function during rest (2D echocardiography), left ventricular wall motions (2D echocardiography) and arrhythmia (Holter monitoring).

Quality of life

We used "Minnesota Living with Heart Failure Questionnaire" (21 questions for each a scale of 0 to 5: 0 - normal function, 5 - very much altered function) and "Dyspnea Fatigue Index": (DFI, 3 questions, for each a scale of 0 to 4: 0 – very much altered function, 4 - normal function). The patients fulfilled the questionnaires before and two months after operation.

Exercise capacity

All patients performed a 6 minute walk test on a 47 m long corridor, one week and two months after the operation.
One month after the operation the patients underwent a maximal exercise test using an ergospirometer JAEGER Oxycon Delta. Before tests the ergospirometer was calibrated at the atmospheric pressure and at the temperature and humidity of the effort test laboratory of the Cardiac Rehabilitation Clinic. We monitored maximal oxygen uptake (VO_2max, l $O_2 min^{-1}$), carbon dioxide production (VCO_2, l $CO_2 min^{-1}$), minute ventilation (VE, l min^{-1}), respiratory rate (RR, resp min^{-1}), heart rate (HR, beat min^{-1}), respiratory exchange ratio (RER=VCO2/VO2), pulse-oxygen (VCO_2/HR, ml O_2 beat min^{-1}), anaerobic threshold (AT, ml O_2 min^{-1}), respiratory equivalent of oxygen (VE/VO_2) and carbon dioxide (VE/VCO_2) [5]. The effort test was performed on bicycle (Schiller) with increasing work with 20 W every 2 minutes. Electrocardiogram and blood pressure were monitored during the effort test.

Echocardiography

The ejection fraction of the left ventricle was determined using Simpson's rule (apical view) before and two months after operation. The result represents the average of at least two successive measurements of the cardiac cycles [6].
Segmental wall motion was assessed using the 16-segment model before and two months after operation. Each segment was semi-quantitatively graded as follows: 1 = normal, 2 = hypokinetic, marked reduction of endocardial motion and thickening: 3 = akinetic, absence of motion and thickening and 4 = dyskinetic paradoxal wall motion [7].
Wall motion score index was obtained by dividing the sum of individual segmental scores by the total number of interpretable segments. We used echocardiographs: Ving Med Sonolayer 800 and Toshiba Sonolayer SSH 140A.

Table I. Baseline characteristics and medication.

Baseline characteristics		
	Group A (magnerot) n = 16	**Group B (placebo)** n = 16
(years)	53 ± 6	51 ± 8
(M/F)	13/3	12/4
height (cm)	174 ± 6	173± 5
weight (kg)	76 ± 8	74 ± 10
NYHA (II/III)	9/7	8/8
number of grafts (2/3/4)	3/9/4	3/8/5
symptoms duration (years)	5.5 ± 1.4	5.6 ± 1.7
EF (%)	31 ± 3	30 ± 4
Medication		
Diuretics	16(100%)	16(100%)
ACE inhibitors	5(31%)	6(37%)
Digoxin	14(87%)	13(81%)
Nitrate	3(18%)	4(25%)
Beta-blocker	5(31%)	7(43%)-
Calcium Channel Blocker	3(18%)	2(12%).
Amiodarone	4(25%)	3(18%)
Aspirin	5(31%)	6(37%)

Holter monitoring

The assessment of cardiac arrhythmia was performed one week and two months after operation using Simens Siestole System.

Statistical analysis

Statistical analyses were performed with SPSS for Windows. The data were presented as mean ± standard deviation. The results in the same group and between groups were analyzed with ANOVA and were compared using Student t test; p value < 0.05 was considered statistically significant. After inclusion all the patients were addressed to the rehabilitation program and were controlled twice a week in the ambulatory service of Cardiac Rehabilitation Clinic.

Results

Patients

According to inclusion criteria, the patients were selected early after operation (24-48 hours). The initial characteristics of patients are presented in *Table I.*

There were no statistically significant differences between groups regarding baseline characteristics. Two patients in Group A and one patient in Group B did not complete the study for subjective reasons. The results of tests before and after operation are presented in *Table II.*

In both groups we observed a significant improvement of the functional capacity after operation. Even the decrease of Minnesota score was greater in Group A (13% vs. 38%); the difference was not statistically significant.

Table II. Quality of life evaluation.

	Group A (magnerot)			Group B (placebo)		
	baseline	**final**	**p**	**baseline**	**final**	**p**
Minnesota	63 ± 6	41 ± 8	< 0.001	59 ± 7	43 ± 6	< 0.001
DFI	7.3 ± 1.4	8.4 ± 1.5	< 0.05	7.2 ± 1.3	8.5 ± 1.2	< 0.05

Exercise capacity

The results of the 6 minutes walk test (one week and 2 months after operation) are presented in *Table III*. Even baseline parameters were comparable between groups, the distance ambulated during 6 minutes walk test was greater in Group A. The ergospirometric data are presented in *Table IV*. Even if the rehabilitation program was similar in both groups, we noticed a significant improvement of exercise capacity in Group A ($p<0.05$ for exercise duration of the test. VO_2. VE/ VO_2 and $p<0.01$ for VO_2 max, AT, VO_2/HR).

Echocardiography

The echocardiographic measurements (before and after the operation) are presented in *Table V*. Left ventricular ejection fraction was not changed significantly during the follow up period. Even if the wall motion score was ameliorated after operation, the improvement was not statistically significant between groups.

Table III. Functional capacity evaluation – 6 minute walk test (m).

	Baseline	**Final**	**p value**
Group A (magnerot)	341 ± 21	425 ± 32*	< 0,01
Group B (placebo)	351 ± 25	398 ± 28	< 0.05

*p < 0.05 when we compare the final results of both groups.

Table IV. Ergospirometric parameters.

	Group A (magnerot)	**Group B (placebo)**	**p value**
exercise duration (min)	12.6 ± 1.4	11.3 ± 1.4	< 0.05
VO_2max (ml $kg^{-1}min^{-1}$)	18.9 ± 2.1	15.6 ± 2.4	< 0.01
AT (ml $kg^{-1}min^{-1}$)	14.2 ± 1.8	11.1 ± 1.8	< 0.01
VCO_2(ml $kg^{-1}min^{-1}$)	20.6 ± 1.9	17.1 ± 1.6	< 0.05
VCO_2/ VO_2	1.09 ± 0.02	1.10 ± 0.02	ns
HR max (beat min^{-1})	138 ± 12	139 ± 16	ns
VE($lmin^{-1}$)	53 ± 4	51 ± 5	ns
VE / VO_2	38 ± 3	46 ± 4	< 0.05
VO_2/HR	10.4 ± 1.5	8.3 ± 1.2	< 0.01

Table V. Ecocardiographic data.

	Group A (magnerot)		Group B (placebo)	
	Baseline	**final**	**Baseline**	**final**
LVEF(%)	31 ± 3	33 ± 4	30 ± 4	31 ± 5
Wall motion score	2.1 ± 0.4	1.9 ± 0.5	2.2 ± 0.5	2 ± 0.3

Table VI. Holter monitoring parameters.

	Group A (magnerot)		Group B (placebo)	
	Baseline	**final**	**Baseline**	**final**
ST	8	5	10	6
SE	2.1 ± 0.3	1.9 ± 0.4	2.6 ± 0.5	2.3 ± 0.3
AF	12	7	13	5
VT	2	0	2	1
VPB	1.9 ± 0.2	0.7 ± 0.3	1.6 ± 0.3	1.5 ± 0.2

ST - paroxistical supraventricular tachycardia (number of episodes) - sum. SE - supraventricular extrasistolia (percent from total number of beats) - mean. AF - atrial fibrillation or flutter (number of episodes) - sum. VT - ventricular tachycardia (number of episodes) - sum. VPB - ventricular premature beats (percent from total number of beats) - mean.

Holter monitoring

The results regarding supraventricular and ventricular arrhythmia are presented in *Table VI*.

We noticed in both groups a significant decrease of supraventricular tachycardia and atrial fibrillation or flutter.

The decrease of ventricular premature beats was observed only in Group A (1.9±0.2 to 0.7±0.3, $p<0.05$).

Adverse reactions

The incidence of adverse reactions regarding the study medication, are comparable between groups (*Table VII*). The intensity of adverse reactions was low, disappeared after the symptomatic treatment and it was not necessary to stop the study medication.

Discussion

Magnesium is well known for its favorable effect on reducing coronary resistance, influencing myocardial reperfusion, inhibiting platelets aggregation and suppressing life threatening ventricular arrhythmia [8-11]. The studies which used magnesium in acute myocardial ischemia (especially acute myocardial infarction) demonstrated the reduction of general mortality with 20-25%, of cardiac mortality with 20% and of early evolution to heart failure with 25% [12-15].

However, orotic acid can improve the energy status of the recently infarcted myocardium [15].

Moreover, the combination of magnesium with orotic acid (magnesium orotate, a salt of orotic acid) may reduce the severity of chronic myocardial dysfunction and structural damage of the failing heart and can improve exercise tolerance in patients with coronary artery disease [16].

Having in mind these results it is not surprising to see the benefit of the treatment with magnesium orotate (magnerot) in patients with ischemic heart failure studied in early postoperative period after coronary artery by-pass graft.

Good selection of patients for operation, the choice of optimal timing for performing cardiac intervention and the inclusion in a tailored rehabilitation program was a guarantee for the improvement of exercise capacity, quality of life and functional capacity of the operated heart (control group).

Table VII. Adverse events.

	Group A (magnerot)	Group B (placebo)
Nausea	3	1
GastricPain	2	1
Headache	2	3
Heartburn	3	2

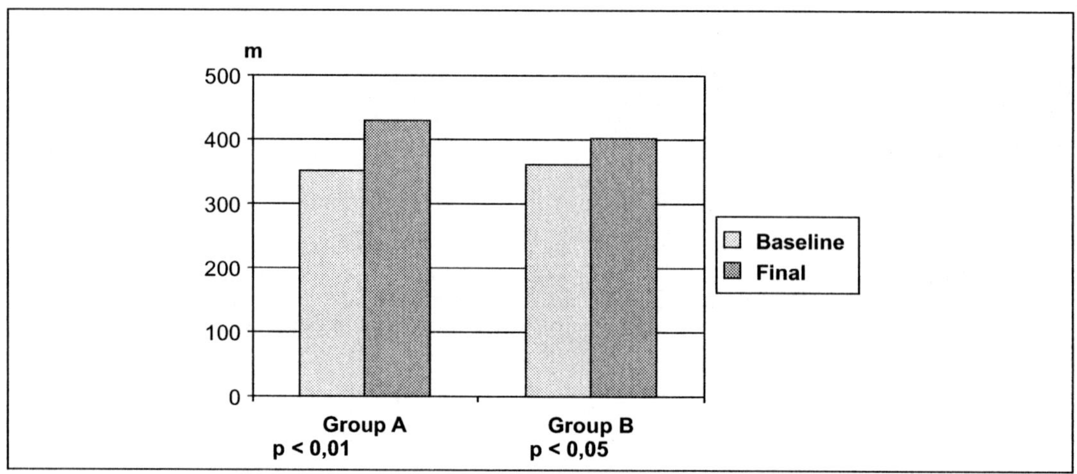

Figure 1. Distance ambulated during 6-minute walk test (m).

The fragile equilibrium of hemodynamics, early postoperative myocardial ischemia, reperfusion syndrome and the ionic imbalance induced by high dose of diuretics claim a protective treatment. Magnesium orotate seems to be one of the solutions for these patients. In our study the main improvements induced by magnesium orotate **(magnerot)** are the increase in exercise capacity (distance ambulated during 6-minute walk test and ergospirometric parameters) and the reduction of ventricular premature beats (*Figures 1, 2*).

Magnesium orotate was well tolerated and the adverse reactions were not significant.

In conclusion, this study, even if made on a small number of patients, confirmed the initial hypothesis that magnesium orotate can lead to improvement of exercise tolerance and left ventricular function of patients after coronary artery by-pass grafting. It should be mentioned that jn comparison to other antiischemic drugs the benefit of administration of magnesium orotate is not followed by adverse reactions. The study strongly suggests the benefit of magnesium orotate added to classical antiischemic therapy in the complex management of this special category of coronary patients.

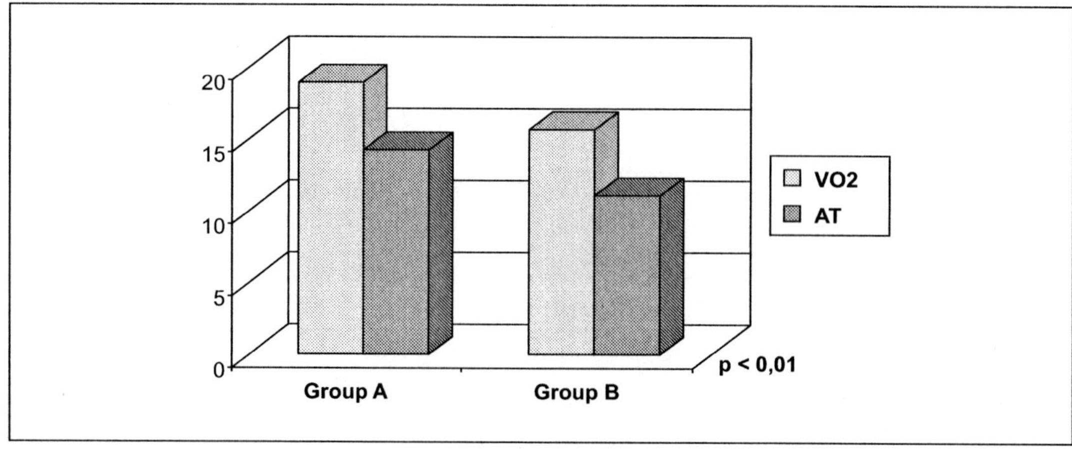

Figure 2. Ergospirometric parameters.

Further studies are necessary to demonstrate all the positive effects of magnesium orotate upon patients with coronary artery disease with or without heart failure. Furthermore it is necessary to introduce the hard end-points (morbidity and mortality), together with the surrogate end-points (hospitalization, cost/effective ratio) and to compare the long-term follow up results.

Acknowledgements

Authors would like to thank Worwag Pharma GmbH&Co. KG Germany (Romanian representative – Mrs. Norina Gavan), which supported this study. Authors also thank the Chief nurses Mrs. Daniela Demea, Lucia Cosma and Stela Reghis for their help in conducing this study.

References

1. Rosenfeldt FL. Metabolic supplementation with orotic acid and magnesium orotate. *Cardiovasc Drugs Ther* 1998; 12: 147-52.
2. Rosenfeldt FL, R1chalds SM, Lin S, *et al.* Mechanism of cardioprotective effect of orotic acid. *Cardiovasc Drugs Ther* 1998; 12: 159-71.
3. Jasmin G, Proschek L. Effect of orotic and magnesium orotate upon the development and progression of the UM-X7.J hamster hereditary cardiomyopathy. *Cardiovasc Drugs Ther* 1998; 12: 189-97.
4. Geiss KR, Stergiou N, Jester H, *et al.* Effects of magnesium owtale on exercise tolerance in patients with chronic heart disease. *Cardiovasc Drugs Ther* 1998; 12: 153-7.
5. Branea I, Gaita D, Mancas S, *et al.* Benefit of long-term exercise training in patients with left ventricular dysfunction concerning survival and clinical events. *Canad J Cardiol* 1997; 136: 166-7.
6. Dragulescu SI, Pescariu S. In: *Ghid de ecocardiografie.* Timisoara: Ed Brumar, 1998: 56-62.
7. Shiller NB, Shah PM, Crawford M, *et al.* Recommendations for quantitation of the left ventricle by two-dimesional echocardiography (American Society Of Echocardiography Committee on Standards, Subcommittee on Quantitation of Two-Dimensional Echocardiograms). *J Am Soc Echocardiogr* 1989; 2: 358-67.
8. Noods KL. Possible pharmacologial actions of magnesium in acute myocardial infarction. *Br J Clin Pharmacol* 1991; 32: 3-10.
9. Schechter M, Kalinsky E, Rabinowitz B. The rationale of magnesium supplementation in acute myocardial infarction: a review of literature. *Arch Intern Med* 1992; 15: 2189-96.
10. Arsenian MA. Magnesium and cardiovascular disease. *Prog Cardiovasc Dis* 1993; 35: 271-3.
11. Vigorito C, Giordano A. Hemodynamic effects of magnesium sulfate on the normal human heart. *Am J Cardiol* 1991; 167: 1435-7.
12. Herzog WR, Schlossborg ML. Timing of magnesium therapy affects experimenta infarctsize. *Circulation* 1995; 92: 2622-6.
13. Antman EM. Randomized trials of magnesium for acute myocardial infarction: big numbers do not tell the whole story. *Am J Cardiol* 1995; 75: 391-3.
14. Woods KL, Fletcher S. Intravenous magnesium sulfate in suspected acute myocardial infarction: results of the second Leicester Intravenous Magnesium. Intervention Trial (LIMIT-2). *Lancet* 339:1553–8.
15. Rossi A, Olivares J. Basis of pyrimidine nucleotide metabolism in the myocardites. *Cardiovasc Drug Ther* 1998; 12: 171-9.
16. Golf SW, Bender S, Grottner J. On the significance of magnesium in extreme physical stress. *Cardiovasc Drug Ther* 1998; 12: J97-J200.

Hypomagnesemia in patients with diabetes and ischemic heart disease

M. Onaca, A.P. Babes, M. Motocu, L. Demian, A. Onaca, N. Negrutiu

University of Oradea, Faculty of Medicine, Clinical Hospital of Oradea, Department of Diabetes, Nutritional and Metabolic Diseases, Romania

Magnesium represents the second most important intracellular cation, its importance in physiological growing and metabolic processes being well known. Magnesium plays its role as a cofactor in many enzymatic reactions [1–3].

Involving of the magnesium ion in the lipidic metabolism is related to the role plated in the appearance of atherogenesis [4,5].

The presence of a certain antagonism between cholesterol blood level and magnesium blood level was proved (magnesium deficit and hypercholesterolemia). The following changes are noted at cellular level and lipid metabolism due to magnesium deficit:

1. Increase of plasma cholesterol;
2. Changes of plasma lipoproteins level;
3. Changes of low-density lipoproteins (LDL) metabolism;
4. Stimulating of smooth muscle cells migrating towards coronary arteries intimae;
5. Stimulating of calcium ion penetration into endothelial arterial cells followed by the increase of cytosolic calcium concentration;
6. Altering of linoleic and arachidonic acids conversion.

Patients with diabetes mellitus often present low magnesium levels-plasmatic and or intracellular, and its degree is related to metabolic alterations [6].

Magnesium deficit is interrelated with the appearance of ischemic heart disease at diabetic patients [7, 8].

Background and aims

The study below is proposed to show the presence of hypomagnesemia at patients with diabetes who have macro-angiopathic abnormalities like the ischemic heart disease, the role played by the Magnesium ion in atherogenesis being well known.

Material and method

During the study, a group of 70 patients with diabetes was studied, 60 patients with the type 2 and 10 with type 1, all of them being under the evidence of the Diabetes Centre of Oradea. The study had in view the degree of the metabolic control, the presence of the ischemic heart disease, and the serum, urinary and erythrocytic Magnesium levels. The group of witnesses included 30 patients with neither diabetes nor abnormal oral glucose tolerance test.

The patients included in the study were divided into 3 categories considering:

– the degree of metabolic control (good, moderate and insufficient) relayed to glicemic values a jeun, 2 hours after meals and the Hb A1c. Due to atomic spectrophotometric method of absorption, the following laboratory findings were dosed:

– plasma magnesium (normal ranges 1.6-2.2 mg/dl);
– erythrocytic magnesium (normal ranges 5,2+/-0,34 mEq/l);
– urinary magnesium (normal ranges about 100 mg/24 hours).

The presence of the ischemic heart disease was studied and connections were made with plasma magnesium and with the degree of metabolic control.

Results

Gender repartition of the patients included in the study was the following: *Figure 1.*
The patients' average of age was the following (*Figure 2*):
– Control group: 49.3 years;
– Type 1 Diabetes: 45.5 years;
– Type 2 Diabetes: 51.6 years.

The average of the disease evolution is shown in *Figure 3*
The degree of metabolic control was appreciated as good, moderate or insufficient, depending on the glycemic values a jeun, 2 hours after meals and the Hb A1c values .The majority of the patients included in the study were classified as having a moderate metabolic control: 58.3% of them were diagnosed with type 2 diabetes mellitus, while 40% were with type 1. See *Table I.*

The average values of plasma, erythrocytic and urinary Magnesium

Plasma Magnesium values were on the average lower at type 1 diabetes patients who were also diagnosed with ischemic heart disease (1.25+/-0.2 mg/dl vs. 1.4+/-0.2mg/dl). Regarding Magnesium contained in the red cells, the lowest average values were found at patients with type 1 diabetes with ischemic heart disease (4.9+/-0.6 mEq/l vs. 1.45+/-0.2mEq/l).Urinary Magnesium was found in average on highest level at type 2 diabetes patients with ischemic heart disease (115.3+/-11.7mg/24h vs. 109.7+/-12.3 mg/24h) especially at those patients with an insufficient metabolic control. See *Table II.*
Plasma Magnesium was found at lowest level at type 1 DM with insufficient metabolic control (*Table III, Figure 4*).
The diminution prevalence of plasma magnesium under normal ranges was 73% at type 2 diabetes patients and 60% at type1.

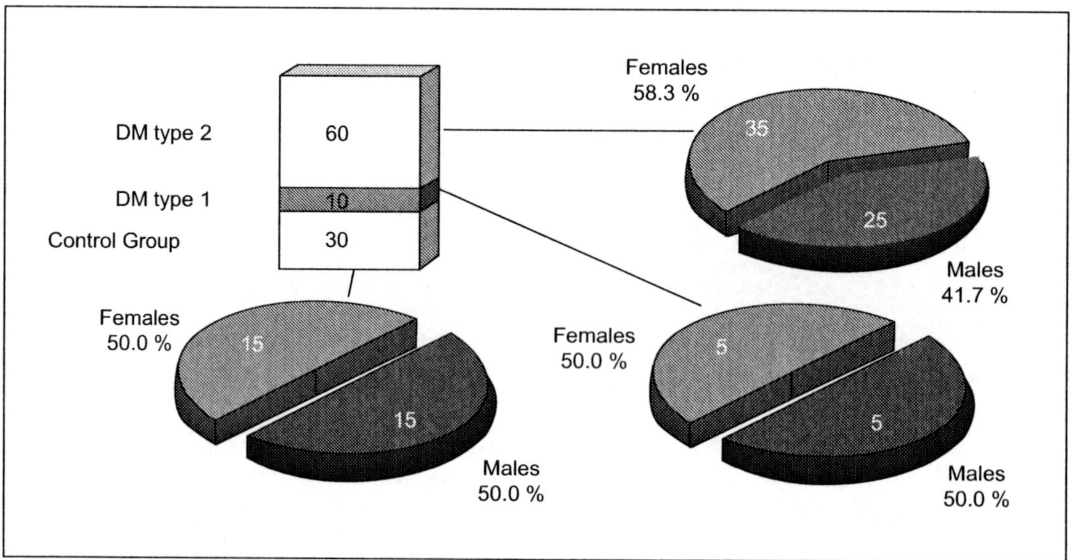

Figure 1. Gender repartition of the patients.

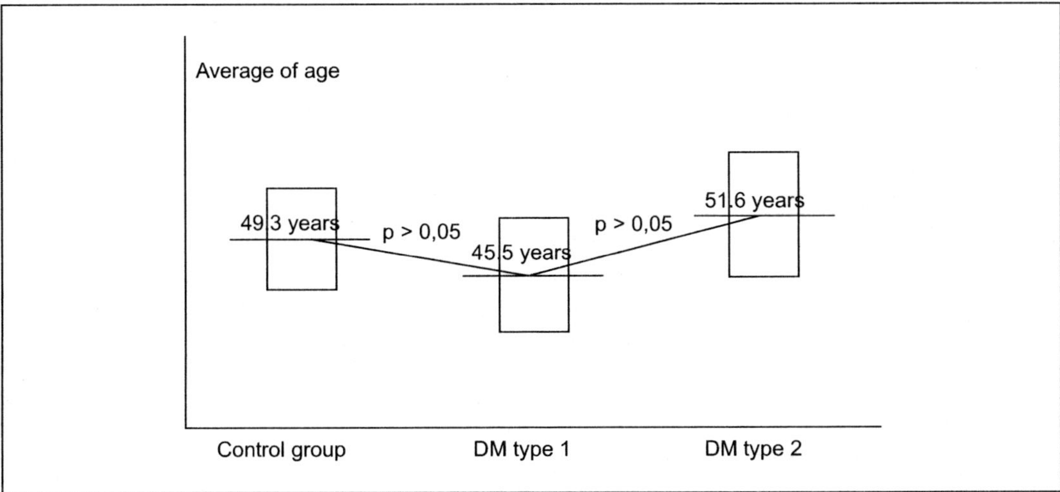

Figure 2. The patients' average of age.

Erythrocytic Mg-its values were noted as being correlated to the presence of ischemic heart disease, and with the degree of metabolic control (worse metabolic control, lower level of erythrocytic Mg) (*Table IV, Figure 5*).

The prevalence of low levels erythrocytic magnesium was 50% at type 1 and 58% at type 2 diabetes patient.

Urinary Magnesium-its lowest levels were found at type 2 diabetes patients who were also diagnosed with metabolic disorders (*Table V, Figure 6*).

Urinary magnesium values above normal ranges were found at 40% of the type 1 and at 80% of type 2 diabetic patients.

The prevalence of ischemic heart disease is shown in the following graphic (*Figure 7*).

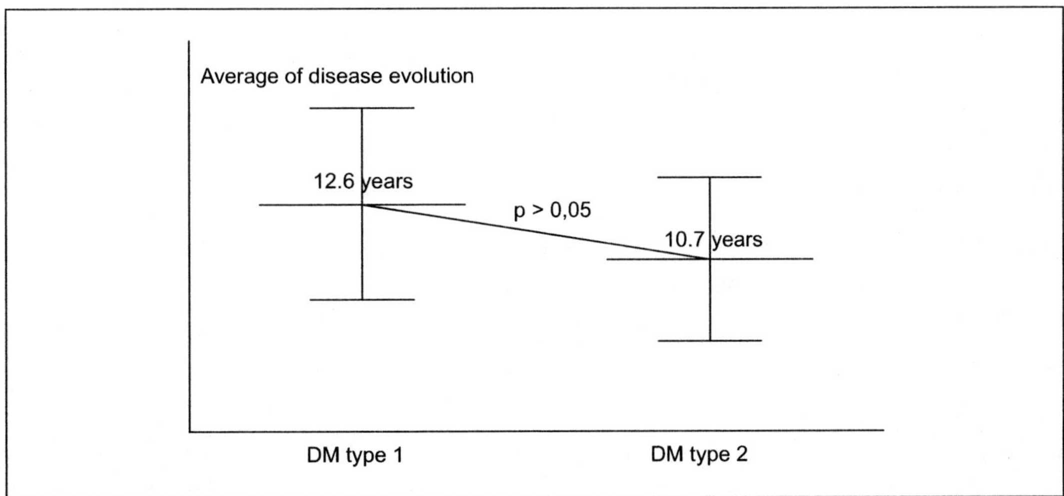

Figure 3. The average of the disease evolution.

Table I. The degree of metabolic control.

	Metabolic control							
	Good		Moderate		Insufficient		Total	
	Nr.	%	Nr.	%	Nr.	%	Nr.	%
TYPE 1 DM	**5**	**50,0**	**4**	**40,0**	**1**	**10,0**	**10**	
Without IHD	3	60,0	3	75,0	-	-	6	60,0
With IHD	2	40,0	1	25,0	1	100,0	4	40,0
TYPE 2DM	**20**	**33,3**	**35**	**58,3**	**5**	**8,3**	**60**	
Without IHD	10	50	13	37,2	2	40,0	25	41,7
With IHD	10	50	22	62,8	3	80,0	35	58,3
Control group							**30**	
Without IHD							22	73,3
With IHD							8	26,7

Prevalence of IHD in correlation with Mg

The prevalence of ischemic heart disease was statistically significantly ($p < 0.05$) correlated to the diminution of plasma and erythrocytic Magnesium, as well as with the increase of urinary Magnesium (*Table VI*) (*Figures 8– 10*).

Conclusions

– The patients with type 2 of diabetes were found with low values for the serum and erythrocytic Magnesium.

– Urinary Mg values were directly interrelated to the degree of the metabolic disorders, fact that explains why glycosuria is responsible for the increase of magnesiuria.

Table II. The average values of plasma, urinary and erythrocytic Magnesium.

	Plasma Mg mg/dl	Erythrocytic Mg mEq/l	Urinary Mg mg/24h
Without IHD			
Type 1 DM	1,4 ± 0,2	5,1 ± 0,4	108,0 ± 12,5
Type 2 DM	1,5 ± 0,3	5,1 ± 0,3	109,7 ± 12,3
Control group	2,1 ± 0,4	5,2 ± 0,6	101,2 ± 11,3
With IHD			
Type 1 DM	**1,25 ± 0,2**	**4,9 ± 0,5**	113,0 ± 11,6
Type 2 DM	1,3 ± 0,2	5,0 ± 0,5	**115,3 ± 11,7**
Control group	1,7 ± 0,3	5,1 ± 0,6	100,8 ± 10,2

Table III. The average values of plasma Magnesium.

Plasma Mg	Metabolic control		
mg/dl	Good	Moderate	Insufficient
Without IHD			
Type 1 DM	1,5 ± 0,3	1,4 ± 0,2	-
Type 2 DM	1,5 ± 0,2	1,2 ± 0,2	1,2 ± 0,2
With IHD			
Type 1 DM	1,25 ± 0,1	1,15 ± 0,4	**1,0 ± 0,2**
Type 2 DM	1,3 ± 0,2	1,1 ± 0,2	1,1 ± 0,4

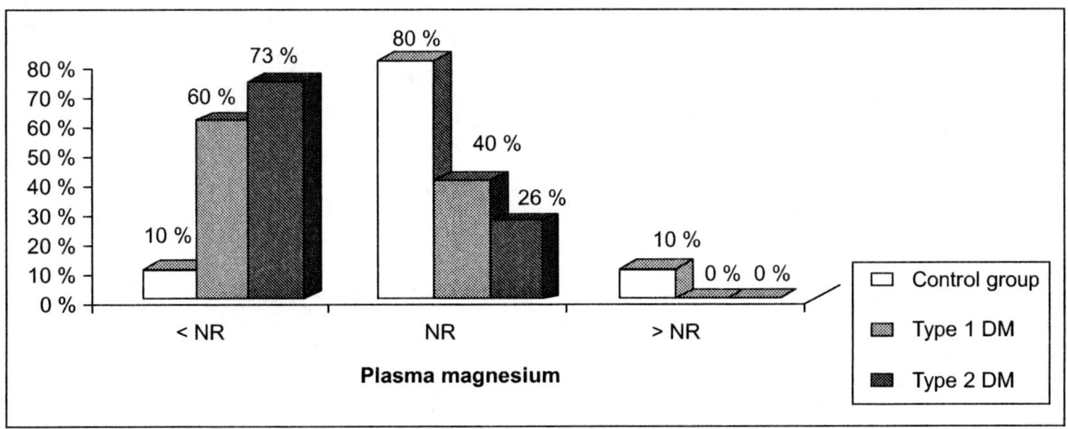

Figure 4. Plasma Magnesium changes at diabetic versus nondiabetic patients.

Table IV. The average values of erythrocytic Magnesium.

Erythrocytic Mg	**Metabolic Control**		
mEq/l	**Good**	**Moderate**	**Insufficient**
Without IHD			
Type 1 DM	5,3 ± 0,5	5,0 ± 0,4	-
Type 2 DM	5,4 ± 0,6	5,1 ± 0,6	4,5 ± 0,5
With IHD			
Type 1 DM	5,2 ± 0,6	4,9 ± 0,0	4,5 ± 0,0
Type 2 DM	5,0 ± 0,6	4,8 ± 0,5	4,5 ± 0,5

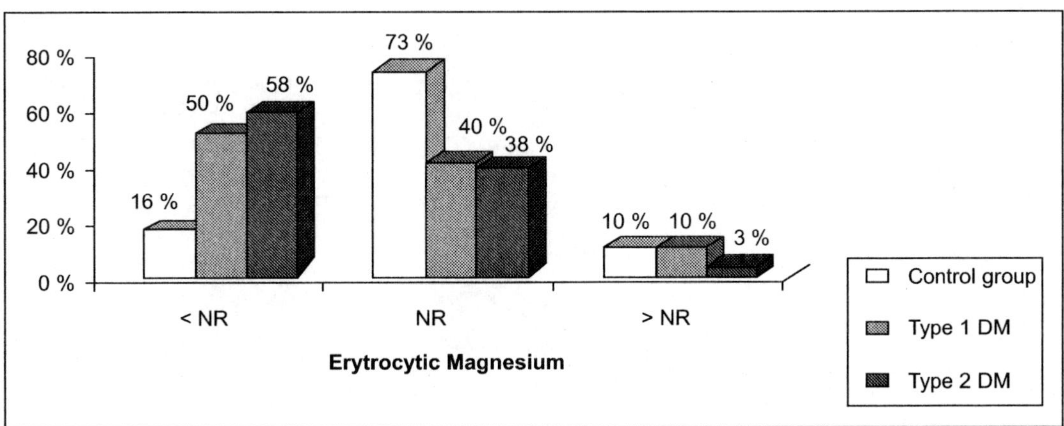

Figure 5. Erythrocytic Magnesium changes at the diabetic versus nondiabetic patients.

Table V. The average of urinary Magnesium.

Urinary Mg mg/24h	Metabolic control Good	Moderate	Insufficient
Without IHD			
Type 1 DM	104,7 ± 12,6	110,3 ± 14,4	-
Type 2 DM	99,5 ± 11,9	117,4 ± 14,1	129,0 ± 15,8
With IHD			
Type 1 DM	110,0 ± 10,7	120,0 ± 8,5	124,0 ± 10
Type 2 DM	106,9 ± 11,7	118,2 ± 13,5	**128,3 ± 14,6**

– The prevalence of ischemic heart disease was statistically significantly correlated to the diminution of plasma and erythrocytic magnesium, as well as with the increase of urinary magnesium.

– The correlation between Mg homeostasis and the presence of ischemic heart disease at diabetic patients is an extra- proof to the well-known implication of Magnesium in atherogenesis and in inducing the coronary spasm as well.

– The appearance of macroangiopathic complications at diabetic patients who are diagnosed with Mg deficit involves the necessity of supplying the patients' therapy with Mg.

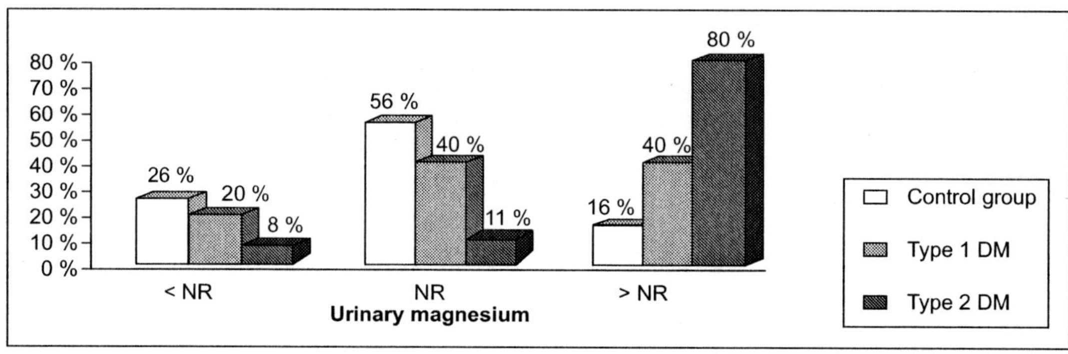

Figure 6. Urinary Magnesium changes at the diabetic versus nondiabetic patients.

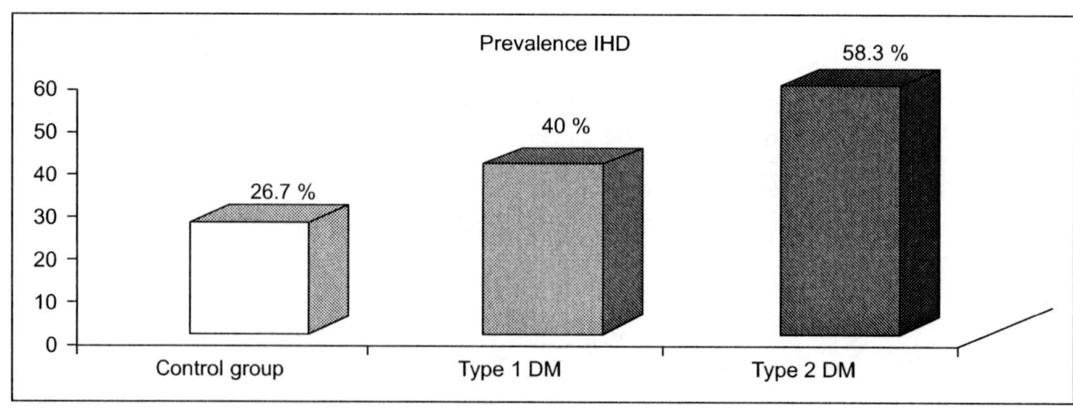

Figure 7. Prevalence of ischemic heart disease.

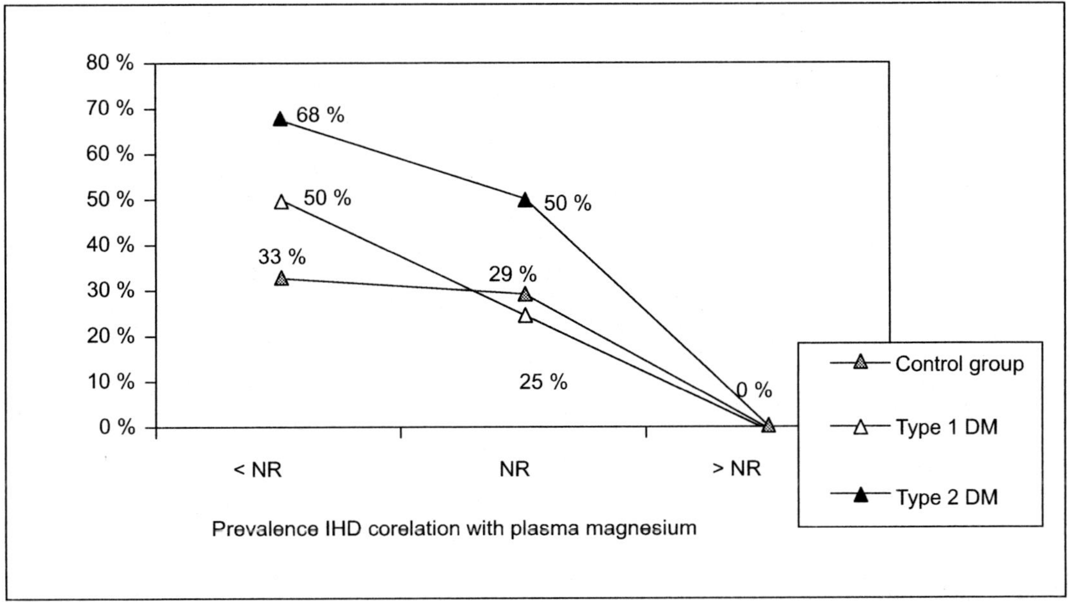

Figure 8. Prevalence IHD corelation with plasma magnesium.

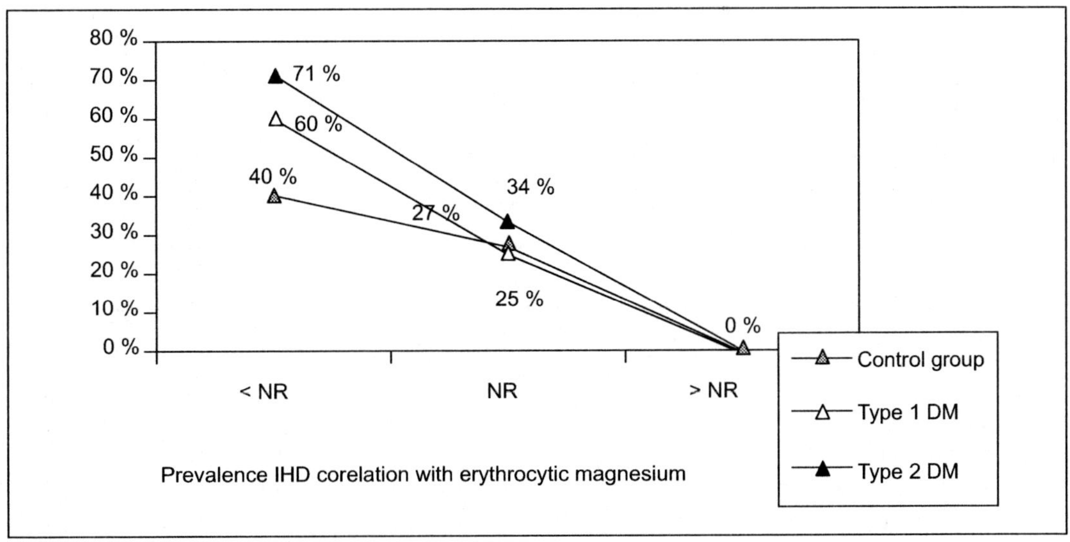

Figure 9. Prevalence IHD corelation with erythrocytic magnesium.

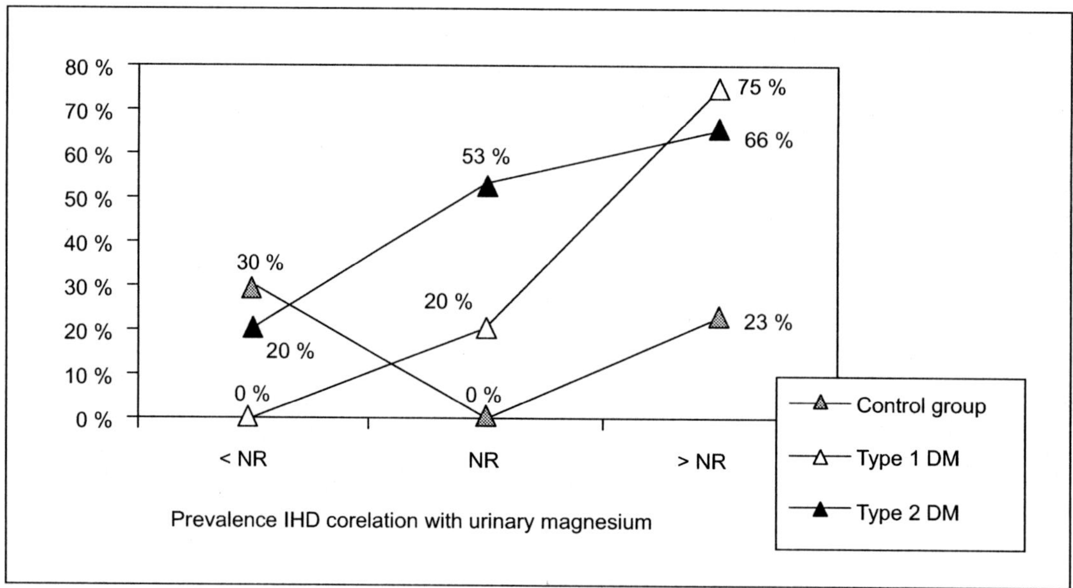

Figure 10. Prevalence IHD corelation with urinary magnesium.

Table VI. Magnesium correlation with IHD at diabetic and nondiabetic patients.

	Plasma Mg			**Erythrocytic Mg**			**/Urinary Mg**		
	<NR	**NR**	**>NR**	**<NR**	**NR**	**>NR**	**<NR**	**NT**	**>LN**
Control group	3 (10,0%)	24 (80,0%)	3 (10,0%)	5 (16,6%)	22 (73,3%)	3 (10%)	8 (26,6%)	17 (56,6%)	5 (16,6%)
IHD	1 (33,3%)	7 (29,2%)	-	2 (40%)	6 (27,3%)	-	4 (30,8%)	-	4 (23,5%)
TYPE 1 DM	6 (60,0%)	4 (40,0%)	-	5 (50,0%)	4 (40,0%)	1 (10,0%)	2 (20,0%)	4 (40,0%)	4 (40,0%)
IHD	3 (50,0%)	1 (25,0%)	-	3 (60%)	1 (25%)	-	-	1 (25%)	3 (75,0%)
TYPE 2 DM	44 (73,3%)	16 (26,7%)	-	35 (58,3%)	23 (38,3%)	2 (3,33%)	5 (8,33%)	7 (11,6%)	48 (80,0%)
IHD	30 (68,2%)	8 (50,0%)	-	25 (71,4%)	8 (34,7%)	-	1 (20%)	4 (57,2%)	32 (66,7%)
Total DM	50 (71,4%)	20 (28,6%)	-	40 (57,2%)	27 (38,6%)	3 (4,3%)	7 (10,0%)	11 (15,7%)	52 (74,2%)
IHD	33 (66%)	9 (45%)	-	28 (70,0%)	22 (81%)	-	1 (14,3%)	5 (45,4%)	35 (67,3%)

References

1. Durlach J. *Le magnesium en pratique clinique.* Paris: Bailliere JB, EM Inter, 1985.
2. Miu N, Dragotoiu G. In: *Magneziul în biologia şi patologia umană.* 2000: 8-12; (Cluj-Napoca:).
3. Mărgineanu O, Miu N. *Oligomineralele în biologie şi patologie.* Cluj-Napoca: Ed Dacia, 1984.
4. Paccalin J. Le magnesium en therapie. *Le Gazette medicale* 1993; 100(28): 41-5.
5. Lehr D. Magnesium and cardiac necrosis. *Magnesium Bull* 1981; 3: 178-91.
6. Durlach J, Durlach V. *Magnesium* 1984; 3: 109-31.
7. Wen-Gang Z, *et al.* Hypomagnesemia and heart complications in diabetes mellitus. *Chin Med J (Engl)* 1987; 100: 719-22.
8. Arsenian MA. Magnesium in cardiovascular disease. *Progres in cardiovascular disease* 1993; XXXV(4): 271-310.

Magnesium supplementation therapy in type 2 diabetes patients

M. Onaca, A.P. Babes, M. Motocu, L. Demian, A. Onaca, N. Negrutiu

University of Oradea, Faculty of Medicine, Clinical Hospital of Oradea, Department of Diabetes, Nutritional and Metabolic Diseases, Romania

Magnesium is a very important cation, which allows the good performance of the human body in all the physiological functions. Magnesium is almost as important as sodium is that it is involved in various enzymatic reactions and different metabolic processes [1–3]. Magnesium is essential in preserving the functional capacity and structure of mytochondrias, in developing all the enzymatic reactions catabolised by ATP, as well as in the function of the enzimes transference of phosphats proteins and nucleic acids syntehetizing. Magnesium plays its role in the membrane permeability (especially mythocondrials) and in neuromuscular excitability, level in which this ion, besides calcium, play its role in neuromuscular excitability regulation. Magnesium ions adjust metabolic reactions, they are needed in processes as anaerobic glycosilation oxidatting phosphorylation activating alkaline and acid phosphatases, pyrophosphatases and ATP-ase. Synthesis of the second cyclic AMP messanger is addicted to magnesium-related adenylcyclase. In each ATP-related enzymatic reaction, magnesium acts like a catalyser [4–6].

Therefore, as a synthesis regarding the previous ideas we speak about 3 directions of magnesium action [1,3,4]:

- it is essential for the macroergic constituents (ATP, GTP, UTP), activity and synthesis;
– synthesis of the hydrogen and electrons transporters (phosphopiridinnucleotides, flavonnucleotides, co-enzyme A, all of them relayed on magnesium);
– activating a big amount of enzymatic reactions are magnesium-addicted.

The cells contain big magnesium amounts, with a variety between 25-35 mEq/l (depending on the cell type), situated especially into the nucleus and mytochondrias, bound to nucleic acids. Magnesium plasma concentration is: 1,6-2,2 mEq/l (0,62-1,45 mmol/l) which is related to the 14-33mg/l. About 75% of the plasma magnesium circulates as ionisated form while the rest of 25% is bound to albumins and globulines magnesium is prominently excreted though kidneys , the urinary excreted magnesium ranges about 6-20 mEq daily (3-10 mmol daily). Diabetes patients often present a diminution regarding plasma and cell magnesium, which becomes more evident at those patients who are affected by metabolic disorders.

Purpose of study

The current study emphasized the existence of hypomagnesemia at type 2 diabetes patients and the effects of magnesium therapy at these patients. Starting with the well-know implications of hypomagnesemia in the appearance of the late complications of diabetes (macro and microangiopathy) ADA considered the opportunity of magnesium supplementing therapy at type 2 diabetes patients [7-11].

Material and method

A number of 40 type 2 diabetes patient, were included in the study (15 of them treated with insulin and 25 treated with oral agents), all of them were registered at the Diabetes Clinical Centre of Oradea.

The values of plasmatic and urinary magnesium , as well as before meal plasma glucose, cholesterol, blood pressure were measured at the beginning and at the end of study.
The patients age ranged between 40-65 years. The duration of their diseases ranged between 5- 15 year. Plasma magnesium was monitored at the beginning and at the end of the study. Other monitored parameters were: before meal plasma glucose, glycosilated hemoglobin, blood cholesterol and blood pressure. For a period of 3 months, through the administering of Magnerot 3x1 pill/day, magnesium supplementing therapy was applied, even at those patients who ranged normal values for plasma magnesium. The idea relied in the fact that normal plasma magnesium ranges do not exclude a possible existing magnesium deficit at cellular level.

Study results

Plasma magnesium ranges: At the beginning of the study (the so-called To moment), a number of 29 patients (representing 72,5%) showed low levels of plasmatic magnesium (A range about 1,4 +/- 0,2 mEq/l), these values being as low as the level of the metabolic control. At the end of the average plasmatic magnesium values was noticed to 1,6+/- 0,5 mEq/l. only a number of 9 patent (22,5%) showed low plasma magnesium levels (*Figure 1*).

Eliminated urinary magnesium

A number of 32 patients showed increased levels of eliminated urinary magnesium, during To the average values were between 120+/-2 mg/24h, while at T1, 115+/-2 mg/24h.

Blood pressure values

At To, a number of 30 patients were diagnosed with arterial hypertension and followed the corresponding treatment. The average values of their blood pressure ranged between 155+/-4 mmHg (systolic pressure) and 95+/-5mmHg (diastolic pressure) in To moment and T1 moment:138+/-5 mmHg at 85+/-5 mmHg (*Figure 2*).
The average value of plasma cholesterol was at To between 280+/-4 mg% and at T1 between 245+/-5 mg% (*Figure 3*).

Conclusion

Patients with type 2 diabetes showed low plasma magnesium levels and high urinary magnesium level, these changes depending on the metabolic control. After administering for a 3 months period of MAGNEROT 3x1 pill daily plasma magnesium levels increased, while the urinary ones did not suffer major changes, urinary excretion being under the influence of glicosuria.

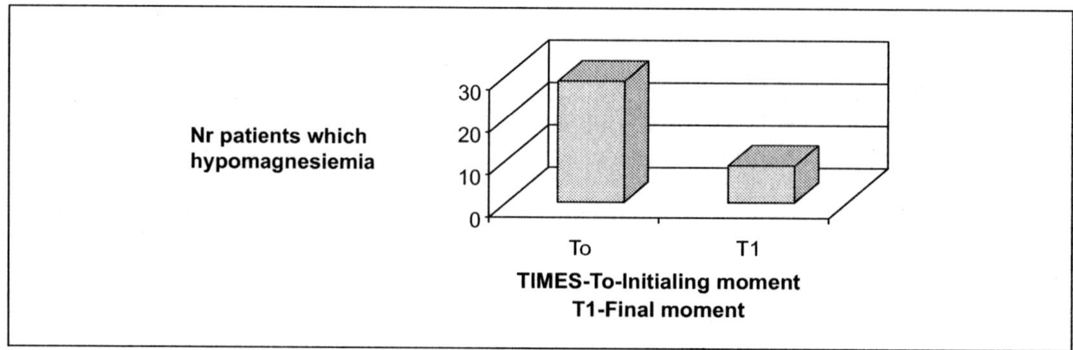

Figure 1. The presence of hypomagnesemia at diabetes patients included in the study.

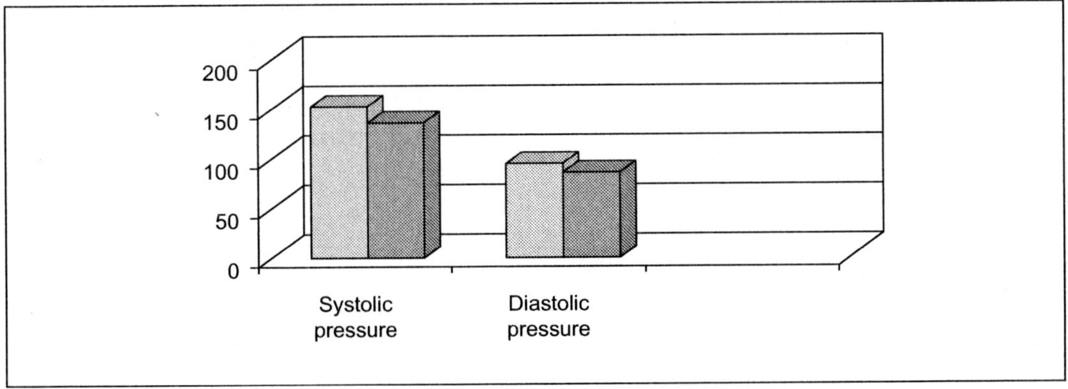

Figure 2. Blood pressure values at patients included in the study.

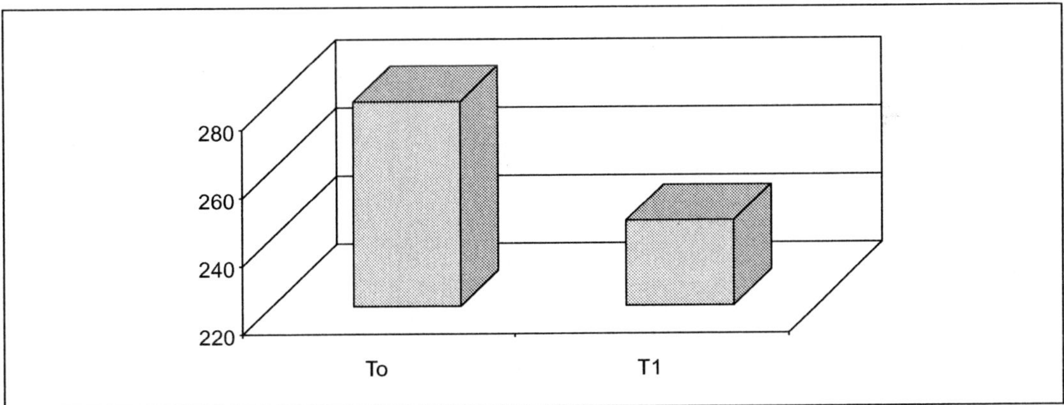

Figure 3. Plasma cholesterol of the patients in the study.

After the administration of magnesium, an improvement of blood pressure values as well an increase of hypotensor treatment efficiency were noted. Magnesium therapy in hypertension treatment may be explained, due to the following mechanisms:

- decrease of arteries basal tonus;
- antistressing effect;
- diminishing of catecholamines release.

Plasma cholesterol range diminished after MAGNEROT administration, due to the following changes produced by magnesium, regarding to the magnesium metabolism:

- diminishing of plasma cholesterol concentration;
- modifying of lipoproteins concentration.

The present study certifies once again the importance of magnesium and the need for substituting therapy, especially at diabetes patients, who present a factory metabolic control is. The need for the supplementing is justified in the appearance of the late diabetes complications.

References

1. Durlach J, Durlach V. *Magnesium* 1984; 3: 109-31.

2. Durlach J, Pedrerey C. *Magnesium Rev Franc Endocrinol Clin* 1977; 18(2): 123-7.

3. Gontea I. *Alimentaţia raţională a omului*. Bucureşti: Ed. Didactică şi Pedagogică, 1970.

4. Aikawa JK. The biochemical and cellular fonction of Mg. Intern Symp On Mg. *Deficit in Human Pathol* 1971: 39-54.

5. Gunther T. Biochemistry and pathobiochemistry of magnesium. *Magnesium Bull* 1981; 3: 91-101.

6. Margineanu O, Miu N. *Oligoelementele in biologie si patologie*. Cluj Napoca: Ed Dacia, 1984.

7. Lima M, Cruz T, Carreiro Pousada J, *et al.* The effect of magnesium supplementation in increasing doses on the control of type 2 diabetes. *Diabetes Care* 1998; 21: 682-6.

8. Paolisso G, Sgambato S, Gambardella A, *et al.* Daily magnesium supplements improve glucose handling in elderly subjects. *Am J Clin Nutr* 1992; 55: 1161-7.

9. Smellie WS, O'Reilly D, St J, Martin BJ, Santamaria J. Magnesium replacement and glucose tolerance in elderly subjects. *Am J Clin Nutr* 1993; 57: 594-5.

10. Sjorgren A, Floren CH, Nilsson A. Oral administration of magnesium hydroxide to subjects with insulin dependent diabetes mellitus. *Magnesium* 1988; 121: 16-20.

11. Valk HW, Verkaaik R, van Rijn HJM, *et al.* Oral magnesium supplementation in insulin-requiring type 2 diabetic patients. *Diabet Med* 1998; 15: 503-7.

The pattern of magnesium and calcium metabolism in spasmophylia, hyperthyroidism and osteoporosis

I. Duncea, C.E. Georgescu, A. Paul, G. Hazi, G. Drăgotoiu, L. Gozariu

Clinic of Endocrinology, University of Medicine and Pharmacy « Iuliu Haţieganu » Cluj-Napoca, Cluj-Napoca, Romania

The concomitant measurement of both serum and urinary magnesium (Mg) and calcium (Ca) allows a better assessment of the body pool of these cations. This may be of significant importance in some endocrine diseases such as spasmophylia, hyperthyroidism or osteoporosis, known to involve alterations of the magnesium and calcium homeostasis. Information about the metabolic balance of Mg and Ca in the human body can be obtained through various methods. However, a complete evaluation often needs expensive and laborious techniques.

Material and methods

In the present work, serum and twenty-four hours-urinary magnesium and calcium were assessed in a total of 367 women admitted at the Clinic of Endocrinology, Cluj-Napoca. Of the 367 subjects, 139 women were asymptomatic and were considered as controls, 143 women presented with various signs of spasmophylia, 49 women were on antithyroid medication for hyperthyroidism and 36 women had osteoporosis, as assessed by osteodensitometry.

The levels of Mg and Ca in both serum and 24-hours urine were assessed using an EEL atomic absorption spectrophotometer. Data were expressed as mean ± SD. The differences between groups were analyzed using the Student's *t* two-tailed test.

Results

Despite mean serum levels of magnesium and calcium within the normal range, patients with spasmophylia had significantly lower serum concentrations of calcium as compared to controls ($p < 0.01$, *Table I*). In addition, the concentrations of both urinary magnesium and calcium were markedly diminished in this group of patients in comparison to healthy controls ($p < 0.0001$, *Table I*).

No differences in serum levels of magnesium or calcium were observed in our study groups between controls and patients treated for hyperthyroidism. However, women on antithyroid therapy also had lower concentrations of urinary magnesium and calcium compared to healthy subjects ($p < 0.0001$, *Table I*). Women with postmenopausal osteoporosis had significantly lower serum and urinary calcium levels in comparison to controls ($p < 0.01$, *Table I*). A tendency towards lower circulating magnesium levels was noticed in this group of patients but the difference was not statistically significant as compared to healthy controls. However, postmenopausal women had a significantly reduced excretion of magnesium in the urine ($p < 0.01$, *Table I*).

Eventually, the groups of patients suffering from spasmophylia and hyperthyroidism were subdivided in two subgroups each: a subgroup of women with normal ovarian activity and a subgroup of postmenopausal women. The previous comparisons were applied on the subgroups. The results are presented in *Tables II and III*.

Table 1. Mean concentrations of serum and daily urinary Ca and Mg of patients with various endocrine diseases (spasmophylia, hyperthyroidism and postmenopausal osteoporosis) and controls.

Study Group	Serum Ca (mg/dl)	Serum Mg (mg/dl)	Urinary Ca (mg/24 ore)	Urinary Mg (mg/24 ore)
Controls (n = 139)	9.35	2.01	158.37	85.87
Spasmophylia (n = 143)	9.19**	1.94	103.91***	60.24***
Hyperthyroidism (n = 49)	9.31	1.98	126.47***	67.94***
Osteoporosis (n = 36)	9.21**	2.00	130.08**	70.31**

**$p < 0.01$; ***$p < 0.001$.

As shown, the status of the ovarian function appears not to influence significantly the ions' metabolism (*Tables II and III*).

Discussion

The simultaneous determination of Mg and Ca ions from both serum and from 24-hours urine allows the evaluation of the homeostasis of these cations in disorders such as spasmophylia, hyperthyroidism and osteoporosis.

In spasmophylia, the assessment of urinary Mg and Ca is useful to establish the low pool level of the ion in the body. Although the values of the total circulating Ca and Mg do not allow the detection of this disorder in some patients, our study clearly shows that patients with spasmophylia may exhibit lower mean serum Ca and urinary Ca and Mg levels than controls. Of the calciotropic hormones, the parathyroid hormone (PTH) increases intestinal absorption of Mg and facilitates renal reabsorption of the ion [1], thus hypoparathyroidism is associated with a decrease of the Mg body pool. Assessment of 25 (OH) vitaminD_3 concentrations may be of interest in patients with spasmophylia. A mild 25 (OH) vitaminD_3 deficiency may contribute to a decreased intestinal absorption of both cations. Other studies found no relationship between Mg absorption and the vitamin D status [2]. Thyroid hormones have a minor role in the regulation of Mg metabolism in euthyroid subjects. On the other hand, thyrotoxicosis

Table 2. Effects of ovarian function on serum and daily urinary Ca and Mg levels of patients with spasmophylia.

Ovarian status	Serum Ca (mg/dl)	Serum Mg (mg/dl)	Urinary Ca (mg/24h)	Urinary Mg (mg/24h)
Premenopausal (n = 94)	9.19	1.92	99.95	62.31
Postmenopausal (n = 49)	9.19	1.96	111.51	56.27

Table 3. Effects of ovarian function on serum and daily urinary Ca and Mg levels of patients treated for hyperthyroidism.

Ovarian status	Serum Ca	Serum Mg	Urinary Ca	Urinary Mg
Premenopausal (n = 25)	9.30	2.01	112.20	62.80
Postmenopausal (n = 24)	9.31	1.95	141.33	73.29

is associated with profound disturbances of Mg metabolism such as increased urinary Mg excretion and low serum, skeletal and intracellular Mg concentrations, leading to a negative cationic balance [3]. Antithyroid therapy normalized Mg concentrations in our hyperthyroid patients. The presence of low urinary Mg and Ca in subjects with treated hyperthyroidism may reflect a low dietary intake of these ions. Previous studies have shown that low serum Mg concentrations may persist in hyperthyroid subjects even after therapy [3] and may be explained, at least in part, by the low Mg intake that characterizes the diet in most countries from Europe. Dietary Mg supplements of 15 mmol may normalise circulating Mg levels after 100 days in euthyroid subjects [4]. Both Ca and Mg play important roles in bone metabolism [5]. A low body pool of Mg may be associated with low BMD values and increased risk of osteoporosis [6–8]. To conclude, the study data confirm the importance of the simultaneous measurements of magnesium and calcium in blood and urine in order to define precisely the alterations of both cations associated with some endocrine diseases. It appears that ovarian activity exerts no significant influence on changes of the Mg and Ca metabolic balance that may be detected in these endocrine diseases.

References

1. Broadus AE. Physiological functions of calcium, magnesium, and phosphorus and mineral ion balance. In: Murray JF, ed. *Primer on the metabolic bone diseases and disorders of mineral metabolism*. New York: Raven Press, 1993: 41-5.

2. Lemann J. Intestinal absorption of calcium, magnesium and phosphorus. In: Murray JF, ed. *Primer on the metabolic bone diseases and disorders of mineral metabolism*. New York: Raven Press, 1993: 46-9.

3. Pepene C, Duncea I, Dragotoiu G, Orbai P, Hazi G. Dinamica magneziului seric si urinar sub terapie cu antitiroidiene si glucocorticoizi in tirotoxicoza. *Sibiul Medical* 2002; XIII(4): 476-8.

4. Golf SW. Biochemistry of Magnesium in Man. In: Golf SW, Dralle D, Vecciet L, eds. *Magnesium*. 1994: 31-41.

5. Duncea I, Pepene C. Implicatiile deficitului de magneziu in osteoporoza – update. *Sibiul Medical* 2002; XIII(3): 289-91.

6. Rude RK, Gruber HE, Wei LY, Frausto A, Mills BG. Magnesium deficiency: effect on bone and mineral metabolism in the mouse. *Calcif Tissue Int* 2003; 72: 32-41.

7. Tucker KL, Hannan MT, Chen H, Cupples LA, Wilson PW, Kiel DP. Potassium, magnesium, and fruit and vegetable intakes are associated with greater bone mineral density in elderly men and women. *Am J Clin Nutr* 1999; 69: 727-36.

8. Gur A, Colpan L, Nas K, *et al*. The role of trace minerals in the pathogenesis of postmenopausal osteoporosis and a new effect of calcitonin. *J Bone Miner Metab* 2002; 20: 39-43.

The relationship of bone mineral density to magnesium intake and metabolism in postmenopausal women

C.E. Georgescu, I. Duncea, P. Orbai, G. Hazi, G. Drăgotoiu, L. Gozariu

Clinic of Endocrinology, University of Medicine and Pharmacy "Iuliu Haţieganu" Cluj-Napoca, Cluj-Napoca, Romania

Magnesium (Mg) influences both mineral and bone matrix metabolism by effects on hormones and other factors that regulate the skeletal and mineral metabolism, as well as by direct effects on bone cells. *In vitro* data have shown that Mg increases the activity of osteoblasts, resulting in enhanced type I procollagen synthesis and accelerated alkaline phosphatase activity and favours the effects of $1{,}25(OH)_2D_3$ on bone. Thus, Mg deficiency may lead to a decrease in the activity of mature osteoblasts and bone formation [1]. Disorders in which Mg depletion is common associate a high incidence of osteoporosis in both animals and humans. In the rat, prolonged Mg deficiency is associated with a loss of trabecular tissue and reduced bone rigidity [2]. Likewise, the Mg-deficient diet results in the mouse in impaired bone growth, decreased osteoblast number and loss of trabecular bone with enhanced cytokine (IL-1, TNFα) activity in bone [3]. On the other hand, in animal studies, long-term excessive Mg intake may be deleterious for bone [4]. The purpose of our research was to analyze the relationship of several parameters reflecting Mg metabolism and Mg intake with bone mass in postmenopausal women.

Material and methods

The present study was carried out in 71 postmenopausal women (mean age 62 ± 9 years, range 31-78 years), outpatients at the Clinic of Endocrinology Cluj-Napoca. Bone mineral density (BMD) of the lumbar spine (L_1-L_4) and the hip were measured by dual X-ray absorptiometry (DPX-NT, GE, USA). In addition, several parameters of Mg metabolism were assessed. Serum and 24-hours urinary magnesium levels were determined using an EEL atomic-absorption spectrophotometer; the Mg content in red blood cells was assessed in 30 patients after lysis of red blood cells by several cycles of freezing and thawing. The dietary Mg intake was quantified using a specific questionnaire and scored. Data were expressed as mean ± SD. Pearson's correlation coefficients were used to describe the associations between BMD and parameters of Mg metabolism. A p-value of ≤ 0.05 was considered statistically significant.

Results

Of the 71 patients, 67 (80%) women had a low or medium Mg intake, whereas only 14 (20%) reported a high Mg dietary intake (*Fig. 1*). About half of the postmenopausal women entering the present study were osteopenic (44%), as determined by DXA, according to WHO criteria (-1SD ≤ Tscore < -2.5SD), 34% were diagnosed with osteoporosis (T score ≤ -2.5 SD) and 22.5% of postmenopausal women had normal BMD (T score > -1SD).

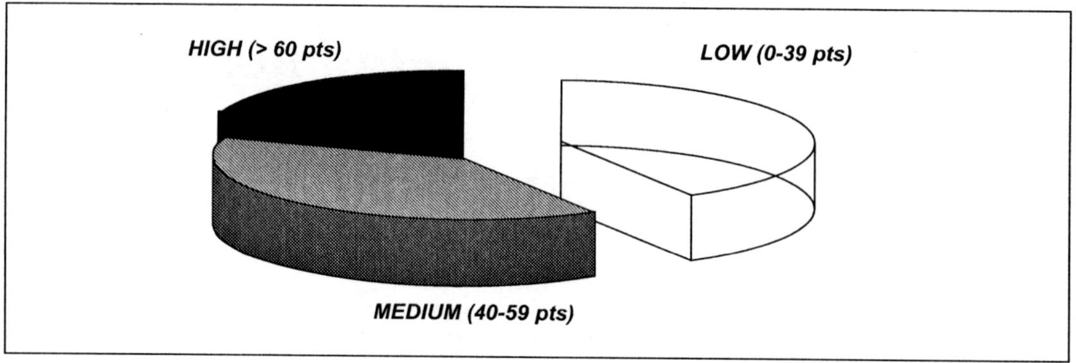

Figure 1. Results of a questionnaire on average Mg intake in 71 postmenopausal Romanian women, aged 62 ± 9 years.

A tendency towards lower serum Mg concentrations (mean 1.98 ± 0.12 mg/dl) was seen in patients with low Mg intake (r = 0.20, p = 0.08, *Fig. 2*). No significant association was described between daily urinary Mg levels (mean 79.6 ± 0.29 mg/d) and Mg intake (r = 0.03, p = 0.76). Serum or urinary Mg levels were not associated with the concentration of Mg in red blood cells (r = 0.15, p = 0.41; r = 0.31, p = 0.09).

The content of Mg in red blood cells (mean 4.64 ± 0.13 mg/dl) was not related to the dietary Mg score (r = 0.19, p = 0.32). As shown in *Table II*, a weak positive correlation was noted between circulating Mg levels and lumbar spine BMD (r = 0.19, p = 0.05) and a tendency towards a positive correlation was noted between serum Mg levels and the total hip BMD (r = 0.18, p = 0.06). Neither lumbar spine nor hip BMD was associated to urinary Mg levels in our study group (*Table II*). However, women with a low Mg intake had lower lumbar spine and femoral neck BMD values, and the correlation between Mg intake and BMD was significant at the hip (r = 0.36, p = 0.01, *Table II*) (*Table I*).

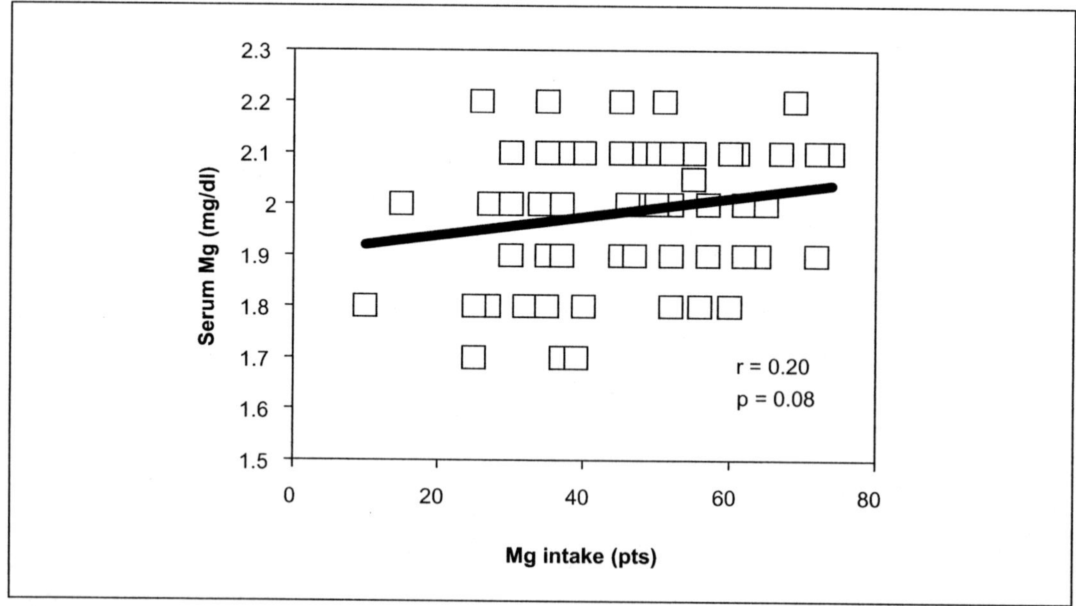

Figure 2. Relationship between serum Mg and average Mg intake in 71 postmenopausal Romanian women.

Table I. The dietary Mg intake was defined according to this questionnaire. The intake of following sorts of food at least 3 times/week was quantified with 2, 5 or 10 points as mentioned.

10 points	5 points	2 points
Chocolate	Milk	Mustard
Cacao	Pork meat	Currants
Coffee	Cheese	Fresh carrots
Nuts	Chestnuts	Fresh cauliflower
Peanuts	Pumpkin and sunflower seeds	Fresh celery
Cereals	White bred	Olives
Rye bred	Bananas	
Peas, beans	Ham	
Pasta	Fish	
Rice	Potatoes	
	Spinach	

Discussion

Data from the present study show that postmenopausal women with a low dietary Mg intake exhibit lower hip BMD values. Few studies have addressed the issue of Mg metabolism or Mg intake and bone mass. Diet supplementation with Mg suppresses bone turnover in young healthy adult males [5]. In members from the original Framingham Heart Study, Mg intake was associated over a 4 years survey with greater BMD at the hip for both men and women and with greater BMD at the forearm for men [6]. Data were similar for other alkaline-producing dietary components. Other studies have reported lower serum Mg concentrations in postmenopausal women with osteoporosis as compared to women without osteoporosis [7].

A decreased intestinal responsiveness to calcitriol has been shown in the ovariectomized rat model, thus leading to decreased absorption of calcium and, possibly, magnesium in the gut. In our study, a weak association was seen between serum Mg levels and lumbar BMD but not hip BMD. Dietary Mg supplements for two years in osteoporotic women were followed by an increase in BMD [8]. The positive significant correlation between Mg intake and hip BMD observed by us may suggest that the dietary Mg intake is linked to BMD in postmenopausal women; this observation may predict a high risk of osteoporosis in women with a long-term low Mg intake.

Table II. Pearson's correlation coefficients between various parameters of Mg metabolism and BMD at the lumbar spine and the hip in postmenopausal women. No relationship between bone mass and urinary Mg concentrations was seen in our study. Serum Mg concentrations were weakly associated to lumbar spine BMD. In the subset of women in whom the content of Mg within red blood cells was assessed, no association between intracellular Mg levels and BMD was observed. However, hip bone mass appeared to be related to the dietary Mg intake of postmenopausal women. *p = 0.05, **p = 0.01.

Parameter of Mg metabolism	L_1-L_4 BMD (g/cm^2)	Femoral neck BMD (g/cm^2)	Wards triangle BMD (g/cm^2)	Total hip BMD (g/cm^2)
Serum Mg (n = 71)	r = 0.19* p = 0.05	r = 0.11 p = 0.31	r = 0.15 p = 0.16	r = 0.18 p = 0.06
Urinary Mg (n = 71)	r = -0.16 p = 0.17	r = -0.14 p = 0.21	r = -0.12 p = 0.30	r = -0.001 p = 0.98
Red blood cells Mg (n = 30)	r = 0.26 P = 0.16	r = 0.04 p = 0.81	r = 0.10 p = 0.57	r = 0.02 p = 0.89
Dietary Mg intake (n = 71)	r = 0.22 p = 0.06	r = 0.21 p = 0.07	r = 0.19 p = 0.09	r = 0.36** p = 0.01

References

1. Duncea I, Pepene C. Implicatiile deficitului de magneziu in osteoporoza – update. *Sibiul Medical* 2002; XIII(3): 289-91.
2. Stending-Lindberg G, Koeller W, Bauer A, Rob PM. Experimentally induced prolonged magnesium deficiency causes osteoporosis in the rat. *Eur J Intern Med* 2004; 15: 97-107.
3. Rude RK, Gruber HE, Wei LY, Frausto A, Mills BG. Magnesium deficiency: effect on bone and mineral metabolism in the mouse. *Calcif Tissue Int* 2003; 72: 32-41.
4. Riond JL, Hartmann P, Steiner P, *et al.* Long-term excessive magnesium supplementation is deleterious whereas suboptimal supply is beneficial for bones in rats. *Magnes Res* 2000; 13: 249-64.
5. Dimai HP, Porta S, Wirnsberger G, *et al.* Daily oral magnesium supplementation supresses bone turnover in young adult males. *J Clin Endocrinol Metab* 1998; 83: 2742-8.
6. Tucker KL, Hannan MT, Chen H, Cupples LA, Wilson PW, Kiel DP. Potassium, magnesium, and fruit and vegetable intakes are associated with greater bone mineral density in elderly men and women. *Am J Clin Nutr* 1999; 69: 727-36.
7. Gur A, Colpan L, Nas K, *et al.* The role of trace minerals in the pathogenesis of postmenopausal osteoporosis and a new effect of calcitonin. *J Bone Miner Metab* 2002; 20: 39-43.
8. Stending-Lindberg G, Tepper R, Leichter I. Trabecuklar bone density in a two-year controlled trial of peroral magnesium in osteoporosis. *Magnes Res* 1993; 6: 155-63.

Effect of magnesium sulphate in parenteral administration in acute reserpine induced ulcer in rats

V. Sandor[1], G. Dragotoiu[2], V. Cristea[3], C. Muresan[1], L.T. Krausz[1], C.C. Luca[1]

1. Department of Pharmacology and Toxicology; 2. Department of Endocrinology; 3. Department of Immunopathology, "Iuliu Hatieganu" University of Medicine and Pharmacy, Cluj-Napoca, Romania

Abstract. Magnesium compounds administered by oral route are among the most efficient antacids. In order to study the effects of these drugs administered parenterally in gastric ulcer, an acute model of gastric ulcer was used. The experiments were performed on Wistar-Bratislava male rats weighing 160-190 g. The gastric ulcerations were induced by reserpine i.p. 5.0 mg/kg with the animals being killed 8 h after its administration. The controls received saline solution i.p. and the treated group was injected with magnesium sulphate i.p., two doses of 100 mg/kg, 30 minutes before and 4 hours after reserpine. The incidence, number and severity of the ulcerations were determined. The statistic analysis of the results revealed the aggravating effect of magnesium sulphate in reserpine induced ulcerations in rats.

Keywords: magnesium, reserpine, ulcerations, rat

Magnesium compounds administered by oral route are among the most frequently and efficient antacid drugs used [1, 2]. Their antiulcer effects were revealed by numerous types of experimental ulcer [3-5]. The action of the magnesium compounds on gastroduodenal pathology in parenteral administration is less known [6]. Under this aspect the effects of magnesium sulphate were followed up on an acute model of gastric ulcer in rats.

Materials and methods

The experiments were carried out on Wistar-Bratislava albino male rats with a body weight of 160-190 g obtained from the Biobase of "Iuliu Hatieganu" University of Medicine and Pharmacy in Cluj-Napoca. The animals were fed with a standard diet and kept at a constant temperature and natural light - dark regimen.

The gastric ulcerations were induced by reserpine. Two groups of rats were randomly formed and these were placed in cages with wide space wire floors to avoid coprophagia. The food was withdrawn 12 hours before the production of the gastric ulcerations. After this period all the animals were injected intraperitoneally (i.p.) with reserpine 5 mg/kg and were kept at the same fasting regimen with water ad libitum. The first group (10 rats) besides reserpine, received saline solution (i.p.). The second group (11 rats) was injected i.p. with magnesium sulphate 200 mg/kg, 30 minutes before and 4 hours after reserpine administration. Eight hours after reserpine administration the animals were killed and the stomachs were removed. These were opened on the great curvature, mildly washed with saline solution and examined under a binocular 5 x magnifying lens. The general aspect of the mucosa, the presence of the macroscopic hemorrhages and the number and severity of the gastric lesions were assessed [7].

Statistical analysis

The incidence of the gastric ulcerations and of the massive hemorrhages was calculated as percentage for each type of lesion. The comparison between the two groups was made using the Fischer exact test of probability. The number of gastric ulcerations was evaluated in a differentiated manner, depending on the size and location of the lesion. Thus, the following were determined: the total number of gastric ulcerations, the number of lesions with a length > 1 mm, the number of confluent ulcerations as well as the number of lesions situated in the gastric antrum. The results were expressed as the mean and standard error ($\bar{x} \pm s.e.$). The comparison of the results of the two groups was made using the bilateral Student's t-test and Mann - Whitney non-parametric test. This last test was used because the dispersion of the values was particularly wide. The null hypothesis was rejected at values of $p < 0.05$ [8]. The severity of the lesions of the mucosa was appreciated on a non-parametric scale from 0 to 4, calculating the ulcer index (UI). This index was analyzed statistically both by Student's t-test and by a particular parameter, the protection ratio (Rp%), calculated according to the formula:

$$Rp\% = \frac{\text{Control UI - treated UI}}{\text{Control UI}}$$

Rp% was considered statistically significant at absolute values > 33% [7].
In each group the parametric Bravais correlation coefficient between the ulcerogenesis indexes was calculated; the statistical significance was accepted at values of $p < 0.05$.

Drugs used

1. Reserpine phosphate (Raunervil®, Sicomed S.A., 2 ml vials, 0.25% solution)
2. Magnesium sulphate (Sicomed S.A., Romania, 20% solution). The dose administered was 200 mg/kg/1 ml.

Ethical norms

The experiment was made in accordance with the Helsinki Declaration on the experiments carried out on animals.

Results

The body weight of the two groups of rats was homogeneous. The incidence of the gastric lesions and of the hemorrhages did not differ statistically in the two groups (*Table I*). The number of gastric ulcerations, calculated in a differentiated manner, was higher in the group treated with magnesium sulphate as compared with the control group (*Table II*). The statistic significance was blurred by the very wide distribution of the results. Thus, for the total number of ulcerations, the mentioned significance was obvious in Mann - Whitney test, but it was at the limit in Student's t-test. For the large ulcerations the two tests showed a slight disagreement. The comparison of the severity of the gastric lesions (UI) was not significant in the parametric and non-parametric tests, but Rp% showed a significant aggravation (-35.065%). Positive correlations were noted between the ulcerogenic indexes for total number of gastric ulcerations, the number of lesions with a length > 1 mm, and UI, which confirmed the validity of the way of assessment (*Table III*).

Table I. Incidence of gastric ulcerations and hemorrhagia.

	Groups	Ulcers				Hemorrhagia
		Total	> 1 mm	Confluent	Antral	
I	Control	100%	100%	30%	60%	20%
II	Magnesium	100%	100%	36,364%	45,45%	9,09%
p	Fischer test	1	1	1	0.67	0.5865

Table II. Number and severity of gastric ulcerations.

	Groups	Number of ulcerations ($\bar{x} \pm s.e.$)				Ulcer index (UI)
		Total	> 1 mm	Confluent	Antral	
I	Control	19.5 ± 6.103	9 ± 2.196	0.4 ± 0.221	1.9 ± 0.605	1.75 ± 0.281
II	Magnesium	46.091 ± 11.535	18 ± 3.628	0.818 ± 0.464	0.909 ± 0.368	2.364 ± 0.262
p	"t" Student	0.063	**0.052**	0.44	0.169	0.126
	Mann -Whitney	**0.05**	0.06	> 0.1	> 0.1	> 0.1

Table III. Parametric coefficients of correlations (Bravais).

	Groups	Total ulcerations vs.				Ulceration > 1 mm vs.			Confluent vs.		Antral vs.
		> 1 mm	Confluent	Antral	UI	Confluent	Antral	UI	Antral	UI	UI
I	Control	**0.009**	0.910	0.550	**0.021**	0.100	0.332	**0.001**	0.193	0.110	0.495
II	Magnesium	**0.001**	0.753	0.661	**0.001**	0.375	0.705	**0.001**	**0.038**	0.188	0.342

Discussions

The incidence of the gastric lesions was similar in the two groups. In exchange, the increase in the number of ulcerations and in the degree of their severity following i.p. administration of magnesium sulphate revealed a tendency to aggravate the ulcerogenic effects of reserpine.

However, most data reported in the literature showed the antiulcer effects of magnesium compounds in experimental models. These actions were noted in rats following p.o. administration of magnesium antacids in gastric ulcerations produced by ethanol [4], of magnesium aspartate HCl in the model by immobilization [3], and of i.p. magnesium sulphate in PAF (platelet-activating factor)-induced ulcerations [6]. Magnesium deficiency obtained by diet in rats [9] or by genetic selection in mice [10] increased the susceptibility of the animals to the ulcerogenic effect of stress. Magnesium administration in these conditions exerts clear gastroprotective effects [3]. These findings can be correlated with the effects of persistent hypomagnesemia on gastroduodenal mucosa. In rats magnesium deficiency reduces the number of cells and increases serotonin level in the gastric mucosa [11]. The mastocytes in the duodenal submucosa proliferate and present degranulation processes simultaneously with the rise of the plasma concentration of histamine. The chromatic characteristics of the mastocytes are suggestive for the alteration of histamine storage and release capacity [12].

The gastroprotection characteristic for magnesium compounds can first be related to their action on the gastric secretion. Magnesium sulphate administered i.v. in healthy human volunteers did not influence the submaximal secretion to pentagastrin [13], but diminished one-half maximal secretion of gastrin [14]. In patients with duodenal ulcer, various magnesium compounds administered by oral or i.v. route did not modify the basal or histamine stimulated gastric acid secretion [15]. The stimulation of the gastric secretion and potentiation of gastrin on the same secretion by calcium, are inhibited by magnesium sulphate administered i.v. [13] In exchange, magnesium sulphate did not influence the release of gastrin through calcium [13].

The essential intervention of magnesium on gastric acid secretion seems to be exerted by the modulation of the effects of calcium on the coupling of the stimulus with the secretion [16]. In isolated gastric parietal cells of rats, Mg^{2+} in increased extracellular concentrations lowered the basal and carbachol stimulated acid secretion; it diminished the effects of histamine of increase of cAMP and of gastric secretion. At the same time the intracellular concentration of free calcium and the availability of the intracellular calcium stores were reduced [17].

On the other hand, by a different mode of action, magnesium facilitated the calcium mechanisms of gastric secretion. Magnesium, like other cations (Ca^{2+} and Gd^{3+}) activated the stomach calcium-sensing receptor (SCAR). SCAR is located on the basolateral membrane of the gastric parietal cells and by stimulation increases the intracellular concentration of calcium and the gastric secretion [18].

Independently of calcium mechanisms, magnesium can influence the gastric secretion by means of ADP-ribosylation factor 6 (ARF6). A protein belonging to the family of monomeric GTP-ases, ARF6 regulates the membrane traffic in the gastric parietal cell. Magnesium stimulates the GTP-ase activity of ARF6 and allows its continuous recycling, contributing to the activation of the proton pump [19].
Even if the data presented above are conclusive for the gastroprotective effect of magnesium compounds, several different findings also exist. Thus, supplementation with magnesium sulphate of an ulcerogenic diet does not modify the rate of occurrence of the gastric lesions in growing pigs [20]. In the same sense we can mention the interaction between magnesium hydroxide and ibuprofen in buffered tablets. This association increases significantly the risk of occurrence of the gastric erosions in healthy volunteers [21].
In this context our results are not in a total disagreement with the available reference sources. For the elucidation of the aspects related to the intervention of magnesium in ulcerogenesis, in our laboratory the experiments were continued on more ulcer models. The data obtained by the use of a wide range of magnesium sulphate doses are being processed.

Conclusions

1. In acute administration, reserpine produces gastric ulceration with an incidence of 100%.
2. Magnesium sulphate injected i.p. in a dose of 100 mg/kg does not change the incidence of the ulcerations and of gastric hemorrhages produced by reserpine.
3. Magnesium sulphate increases the number and severity of reserpine induced ulcerations in rats.

References

1. Hoogerwerf WA, Pasricha PJ. Agents used for control of gastric acidity and treatment of peptic ulcers and gastroesophageal reflux disease. In: Hardman JG, Limbird LE, Goodman Gilman A, eds. *Goodman & Gilman's the pharmacological basis of therapeutics. 10th Ed.* NewYork: The McGraw-Hill Companies, Inc, 2001: 1005-20.
2. Maton PN, Burton ME. Antacids revisited: a review of their clinical pharmacology and recommended therapeutic use. *Drugs* 1999; 57: 855-70.
3. Mesmer M, Fischer G, Classen HG. Stress-Abschirmung durch Magnesium. *Arzneimittelforschung* 1981; 31: 389-91.
4. Nosal'ova V, Babul'ova A, Bauer V. Gastric transmucosal potential difference: effect of antisecretory and gastroprotective drugs. *Gen Pharmacol* 1998; 30: 325-9.
5. Yu BP, Sun J, Li MQ, Luo HS, Yu JP. Preventive effect of hydrotalcite on gastric mucosal injury in rats induced by taurocholate. *World J Gastroenterol* 2003; 9: 1427-30.
6. Nechifor M, Neughebauer BI, Adomnici M, Teslaru E, Filip C, Negru A. The influence of some cations on PAF-induced gastric mucosal damage in rats. *Rom J Physiol* 1993; 30(3-4): 151-4.
7. Sandor V, Cuparencu B. Analysis of the mechanism of the protective activity of some sympathomimetic amines in experimental ulcers. *Pharmacology* 1977; 15: 208-17.
8. Trîmbiţaş RT. *Metode Statistice.* Cluj-Napoca: Presa Universitară Clujeană, 2000.
9. Mangler B, Fischer G, Classen HG. The influence of magnesium deficiency on the development of gastric ulcers in rats. *Magnesium Bull* 1982; 4: 9-12.
10. Henrotte JG, Aymard N, Allix M, Boulu RG. Effect of pyridoxine and magnesium on stress-induced gastric ulcers in mice selected for low or high blood magnesium levels. *Ann Nutr Metab* 1995; 39: 285-90.
11. Artizzu M. Cell population kinetics and biochemical changes in the rat stomach during magnesium deficiency. *Biochem Exp Biol* 1979; 15: 217-22.
12. Kraeuter SL, Schwartz R. Blood and mast cell histamine levels in magnesium-deficient rats. *J Nutr* 1980; 110: 851-8.
13. Christiansen J, Rehfeld JF, Stadil F. Interaction of calcium and magnesium on gastric acid secretion and serum gastrin concentration in man. *Gastroenterology* 1975; 68: 1140-3.
14. Christiansen J, Rehfeld JF, Kirkegaard P. Interaction of calcium, magnesium, and gastrin on gastric acid secretion. *Gastroenterology* 1979; 76: 57-61.
15. Puscas I, Nadaban F, Voicu L, *et al.* Magnesium and gastric acid secretion. *Acta Hepatogastroenterol (Stuttg)* 1977; 24: 288-92.
16. Mooren FC, Stoll R, Spyrou E, Beil W, Domschke W. Stimulus-secretion coupling in rat parietal cells is affected by extracellular magnesium. *Biochem Biophys Res Commun* 1994; 204: 512-8.

17. Mooren FC, Geada MM, Singh J, Stoll R, Beil W, Domschke W. Effects of extracellular Mg2+ concentration on intracellular signalling and acid secretion in rat gastric parietal cells. *Biochim Biophys Acta* 1997; 135: 279-88.

18. Geibel JP, Wagner CA, Caroppo R, *et al.* The stomach divalent ion-sensing receptor scar is a modulator of gastric acid secretion. *J Biol Chem* 2001; 276: 39549-52.

19. Matsukawa J, Nakayama K, Nagao T, Ichijo H, Urushidani T. Role of ADP-ribosylation factor 6 (ARF6) in gastric acid secretion. *J Biol Chem* 2003; 278: 36470-5.

20. Wesoloski GD, Jensen AH, Ladwig VD, Gosser H. Effects of different concentrations of corn oil, magnesium sulfate, and oat hulls in a porcine ulcerogenic ration. *Am J Vet Res* 1975; 36: 773-5.

21. Maenpaa J, Tarpila A, Jouhikainen T, *et al.* Magnesium hydroxide in ibuprofen tablet reduces the gastric mucosal tolerability of ibuprofen. *J Clin Gastroenterol* 2004; 38: 41-5.

Magnesium and oxidative stress during pregnancy and postpartum

V. Papadopol, I. Palamaru, G. Mancas, G. Albu

Institute of Public Health, Iasi, Romania, Department of Food Hygiene and Nutrition, Victor Babes 14, 700465 Iaşi, Romania

As literature shows, there may exist a subclinical deficiency of magnesium (Mg) even in normal pregnancy because of the increased Mg requirement and renal losses [1]. Mg deficiency seems to negatively influence metabolic homeostasis resulting in multiple disturbances.

On the other hand, lipid peroxidation products are significantly raised in normal pregnancy as compared with nonpregnant women. Thus, is possible that pregnancy is a stimulus for lipid peroxidation [2].

The link between Mg deficiency and oxidative stress is unclear in human, although, in experimental animals, severe Mg deficiency has been induced modifications in physico-chemical characteristics of lipoproteins among which the enhanced susceptibility to oxidative stress [3]. In addition to releasing free radicals, lack of Mg can also negatively influence the status of some antioxidative enzymes in the body. Otherwise, the protective properties of various antioxidant drugs and nutrients show that free radicals are involved in the injury process of Mg deficiency [4].

The study aimed to find out the relationship between the Mg status and oxidative stress markers during pregnancy, when the homeostatic mechanisms are disturbed, and about two months post partum.

Subjects. Methods

A group of 294 healthy, randomly selected pregnant women were investigated in the first half of pregnancy (10.2 ± 3.6 weeks). They were between 18 and 39 years aged (28.2 ± 4.3 years) at the beginning of the study and they were sent by their general practitioners for the routine analyses. From the same pregnant women group, 124 women in the second half of pregnancy (31.5 ± 3.8 weeks) were retested and 46 from these post partum (8.2 ± 2.5 weeks) were again retested. Morning fasting blood samples were collected by venipuncture. Serum Mg was determined by colorimetric method with xilylidyl blue [5] and erythrocyte Mg by atomic absorption spectrophotometry [6]. Serum lipid were assessed by colorimetric method with phospho-vanillin. The next oxidative stress markers were determined also by colorimetric methods: plasma lipid peroxides [7], total blood glutathione (GSH) [8], and erythrocyte superoxid dismutase (SOD) [9].

Statistical processing was performed by Epi Info program and included: Student "t" test, "chi square" test for frequencies comparison and Pearson correlation index (r).

Results and discussion

Serum and erythrocyte Mg values significantly decreased in the second stage of pregnancy compared with the first stage and postpartum ($p < 0.001$ in all cases) (*Table I*).

Lipid peroxides concomitantly varied with serum lipid, being significantly increased in the second half of pregnancy compared with the first half of pregnancy ($p < 0.001$) and postpartum period ($p < 0.05$).

Contrary, total blood GSH significantly decreased in the second stage of pregnancy versus the first stage of pregnancy ($p < 0.005$) or post partum period ($p < 0.001$).

Table I. Mean values of the biological indicators.

	First half of pregnancy	Second half of pregnancy	Post partum	Normal range
Serum Mg (mmol/L)	0.81 ± 0.04	0.78 ± 0.04 a: p < 0.001 b: p < 0.001	0.80 ± 0.04	0.77-1.00
Erythrocyte Mg (mmol/L)	1.90 ± 0.07	1.81 ± 0.15 a: p < 0.001 b: p < 0.001	1.92 ± 0.20	2.00 – 3.00
Total lipids (g/L)	6.37 ± 0.64	7.87 ± 0.90 a: p < 0.001 b: p < 0.001	7.05 ± 0.75	5.00 – 8.00
Plasma peroxides (μmol/L)	9.47 ± 2.08	10.36 ± 1.85 a: p < 0.001 b: p < 0.05	9.67 + 2.16	< = 3.3
Total blood GSH (mmol/L)	1.05 ± 0.19	0.99 ± 0.17 a: p < 0.005 b: p < 0.001	1.20 ± 0.16	0.97 – 1.30
Erythrocyte SOD (U/Ht)	7.01 ± 0.71	6.91 ± 0.65 b: p < 0.05	7.16 ± 0.69	6.50 – 7.00

a: compared with the first half of pregnancy; b: compared with the post partum period.

Erythrocyte SOD was significantly lower in the second stage only versus post partum period ($p < 0.05$). Our results emphasize that Mg levels, especially erythrocyte Mg level were generally low in our group of pregnant women-mothers as mean values (under lower cut-off points for erythrocyte Mg) or better the deficits frequencies show (*Table II*).

Regression analysis used for the pregnant women emphasized a negative correlation between gestational age and serum Mg ($r = -0.27$; $n = 372$; $p < 0.001$), erythrocyte Mg ($r = -0.19$; $n = 358$; $p < 0.001$) and plasma GSH ($r = -0.12$; $n = 361$; $p < 0.05$) and a positive correlation with lipid peroxides ($r = +0.21$; $n = 354$; $p < 0.002$).

Serum and erythrocyte Mg did not correlate with oxidative stress markers (*Table III*).

Conclusions

Our study emphasizes significant lower Mg levels in the second half of pregnancy versus the first one in a normal pregnant women group and postpartum. At the same time, significant variations of the oxidative stress markers were noticed. Even the two groups of indicators are not correlated, the findings of our study may be used for a specific therapy in order to avoid the disturbances associated with Mg deficits and antioxidant defence deficit.

Table II. Frenquencies of magnesium deficits (%).

	First half of pregnancy	Second half of pregnancy	Post partum
Serum Mg < 0.77 mmol/L	18.0	39.5 a: p = 0.000 b: p = 0.002	15.2
Erythrocyte Mg < 2 mmol/L	65.4	84.7 a: p = 0.000 b: p = 0.000	53.5

a: compared with the first half of pregnancy; b: compared with the post partum period.

Table III. Relationship between magnesium status and oxidative stress markers.

	r	n	p
Serum Mg – Lipid peroxides	- 0.06	352	> 0.05
Serum Mg – GSH	+ 0.06	358	> 0.05
Serum Mg – SOD	- 0.03	319	> 0.05
Erythrocyte Mg - Lipid peroxides	- 0.06	343	> 0.05
Erythrocyte Mg – GSH	- 0.09	347	> 0.05
Erythrocyte Mg – SOD	- 0.08	312	> 0.05

References

1. Spatling L, Disch G, Classen H-G. Magnesium in pregnant women and the newborn. *Magnesium Res* 1989; 2(4): 271-80.
2. Hubel CA. Oxidative stress in the pathogenesis of preeclampsia. *Proc Soc Exp Biol Med* 1999; 222: 222-35.
3. Keenoy BM, Moorkens G, Vertommen J, Noe M, Neve J, Leeuw ID. Magnesium status and parameters of the oxidant-antioxidant balance in patients with chronic fatigue: Effects of supplementation with magnesium. *J Am Coll Nutr* 2000; 19(3): 374-82.
4. Rayssiguier Y, Gueux E, Bussiere L, Durlach J, Mazur A. Dietary magnesium affects susceptibility of lipoproteins and tissues to peroxidation in rats. *J Am Coll Nutr* 1993; 12(2): 133-7.
5. Manta I, Cucuianu M, Benga G, Hodârnău A. *Metode biochimice în laboratorul clinic.* Cluj-Napoca: Dacia, 1976: 239-41.
6. Sherwood RA, Rock BF, Stewart A, Saxton RS. Magnesium and the premenstrual syndrome. *Ann Clin Biochem* 1986; 23: 667-70.
7. Yagi K. Lipid peroxides and human diseases. *Chem Phys Lip* 1986; 45: 337-51.
8. Parinandi NL, Thompson EW, Schmid HHD. Diabetic heart and kidney exibit increased resistance to lipid peroxidation. *Biochim Biophys Acta* 1990; 1047: 63-9.
9. Minami M, Yoshikawa H. Simplified assay method of superoxide dismutase activity for clinical use. *Clin Chim Acta* 1979; 92(3): 337-42.

Serum calcium and magnesium levels in urban landscape workers

F. Gradinariu, C. Croitoru, B. Scutaru, V. Hurduc, M. Margineanu

Institute of Public Health, Occupational Medicine Department, Str. Dr. Victor Babes 14, 700465 Iasi, Romania

Abstract. A cross-sectional epidemiological study was made upon a group of urban landscape workers in order to determine the influence of the workplace conditions upon their health status. In this paper we analyze the relationship between serum essential cations Ca and Mg and the pathology recorded in these workers, trying to identify in what extent it would be a professionally-induced condition. In the investigated group consisting of 54 subjects, (55.5% male), with an average age of 37.4±10.9 years and an average length of service in this workplace of 8.6±7.9 years. 27.7% of the subjects had hypocalcemia and 9.2% of them had hypomagnesemia. 7.4% of them had clinical signs of spasmophilia, without any biochemical change. Total and reduced glutathione were decreased in greater extent in low cation serum subgroup than in total exposed. Low serum Mg levels (1.8-2.1 mg/dl) correlated positively ($r = 0.57$ $p < 0.01$) with inhibited blood SOD activities (49-106 U/ml), suggesting a link with the impairment of the oxidant defence system. In spite of the apparent environmental non-risky area where these subjects are working, their health status is damaged. The work conditions and lifestyle are both acting upon their health quality. That is why it is more difficult to quantify the extent of occupational factors effects.

Aim

Just like street vendors and traffic policemen, landscape workers are spending their working time in areas with intense traffic, exposed to street hazards including fuel vapors, street dusts, engine combustion gases, sulphur and nitrogen oxides, noise. That is why it seemed reasonable to investigate their health status in order to determine the influence of the workplace conditions upon it. This paper analyzes only the relationship between serum essential cations Ca and Mg and the pathology recorded in these workers, trying to identify in what extent it would be a professionally-induced condition.

Material and methods

The investigated group consisted of 54 subjects, (55.5% male), with an average age of 37.4 ± 10.9 years and an average length of service in this workplace of 8.6 ± 7.9 years. The investigative protocol included: clinical examination, EKG, electroneuromyography, respiratory functional tests, biochemical and hematological tests. Ca and Mg serum levels were determined by commercial kits provided by Nobis Labordiagnostica SRL Cluj-Napoca (Romania). Whole blood superoxide dismutase was assayed by xanthine oxidase method with kits supplied from Randox Laboratories Ltd. (Antrim, UK).

Risk evaluation was made by noxious substances assays in the workplace air and also by biological exposure markers. The results were compared to those of a matched-control group. Statistical analysis was performed using Student's test.

Results and discussion

Environmental air measurements

Particulate-bound HPAs in urban air ranged between 80-1250 ng benzo(*a*)pyrene/m^3 compared to the levels < 1 ng benzo(*a*)pyrene/m^3 determined near a coal-fired power plant. The amount of total suspended particulate matter (SPM) in residential air varied between 0.26-0.95 mg/m^3 and between 4.7-32 mg/m^3 in industrial area. Nevertheless, SEM (silver coverage electron microscopy) analysis revealed that urban air SPM contains mainly respirable particles with dimensions < 3μ which are completely retained by respiratory tract [1]. These findings confirm the speculations according to which air pollution in high traffic areas is more intense than in industrial environments [2,3].

Clinical investigation

In exposed group the most frequent diseases were the cardiovascular and endocrine disorders mainly spasmophylia (*Table 1*). This could be a consequence of work conditions as well as of lifestyle and nutrition. The high rate of women in the studied group (44.5%) and the great percent of hypocalcemia detected (27.7%) might be attributed to diselectrolitic troubles experienced by these female subjects. The cardiovascular diseases were represented mainly by circulatory troubles like varicose veins and piles. These might be consequences of prolonged orthostatic and vicious work postures doubled by heavy physical work. If we take into account the meteorological conditions, these employees working often in cold, rainy and windy weather, it is not surprising the high prevalence of musculoskeletal diseases of 29.6% and of polyneuropathy cases 25.9%. Digestive diseases (25.9%) might have multifactorial causes, including occupational ones. The permanent contact with the soil and different types of plants, especially flowers, pollen, spores mushrooms and dust, explain the high rate of sensitisations manifested as allergies (27.5%), conjunctivitis (14.8%) or dermatological syndromes (5.5%).

Blood cation measurements

In the exposed group we detected 27.7% subjects with hypocalcemia, 9.2% with hypomagnesemia and 7.4% with clinical signs of spasmophylia, without any biochemical changes. Control group had no abnormal serum cations level.

In low serum cations subgroup the most frequent disorders are the endocrine ones (63.1%) accounted to the two essential cations deficit. Or it is well-known the tremendous importance of the Mg/Ca balance upon whole endocrine equilibrium [4]. Polyneuropathy, detected in 26.3% of cases might be also related to Ca and Mg deficits. The 15.8% digestive disorders are represented by biliary diseases. Deficits in tissular Mg level are also responsible for biliary motility troubles [4]. Anemia and allergies are present each in 10.5% of these subjects. These conditions are both worsened by magnesium deficit.

Table 1. Morbidity pattern (% of each group).

Crt. No.	Disease	Total exposed N = 54	Low level cations group N = 19	Control group N = 50
1	Cardiovascular diseases	31.4	0	12.5
2	Endocrine Disease (spasmophylia)	31.4	63.1	2.3
3	Musculoskeletal diseases	29.6	0	5.7
4	Allergy	27.5	10.5	1.8
5	Digestive disease	25.9	15.8	6.4
6	Polyneuropathy	25.9	26.3	0
7	Conjunctivitis	14.8	0	0
8	Metabolic and nutrition disease	11.1	6.8	5.7
9	Dermatological disease	5.5	0	0
10	Anemy	3.7	10.5	0
11	Healthy	18.5	15.8	43.8

Biological exposure markers

Whole blood total and reduced glutathione (GSSG and GSH) are illustrative biological markers of exposure to environmental chemicals being involved in hepatic detoxication. Their values were reduced in much more subjects from low serum cation group (42% respectively 72%) compared to total exposed group (39%, respectively 26%), suggesting a much more intense impairment of liver metabolizing capacity.

Reduced glutathione is involved in multiple stages of the antioxidant defence. First of all, damaging free radicals are primarily quenched by glutathione (GSH) with subsequent formation of harmful glutathionyl radical [5]. Secondly it is the substrate of enzymes such as glutathione peroxidase and glutathione reductase with crucial role in enhancing the scavenging of reactive oxygen species and free radicals induced by electrophilic xenobiotics [6]. It has been demonstrated in many animal and *in vitro* experimental models, that glutathione depletion might inhibit the activity of antioxidant protecting enzymes [7,8]. Ghutathione deficit might also affect the mechanism of protection against chemical injury, mediated via antioxidant-dependent induction of detoxifying enzymes, including glutathione-S-transferases [9].

Oxidative stress markers

Low serum Mg levels (1.8-2.1 mg/dl) correlated positively ($r = 0.57$ $p < 0.01$) with inhibited blood SOD activities (49-106 U/ml), suggesting a link with the impairment of the oxidant defence system. The normal SOD activity of red blood cells ranges between 164-240 U/ml in healthy subjects (*Fig. 1*).

The Mg involvement in antioxidant system defence is well documented by in vitro studies [10] where it was shown that Mg deficiency affected the intracellular responses of endothelial cells by stimulating and then inhibiting protecting enzymes activity. In animal models hypomagnesemia induced gradually decreases of erythrocyte SOD activity [11]. In alloxanic diabetic rats Mg supplements could in part restore the antioxidant parameters, decreasing the oxidative stress [12]. As long as magnesium imbalance might influence antioxidant defence system and can interfere indirectly even with xenobiotic detoxifying mechanisms, it is obvious that all measures to avoid the deficit of this cation, are to be taken as soon as possible. We recommend monitoring of Ca and Mg levels in working population. Dietary supplements of magnesium should be recommended to all workers which are exposed to certain risks at their workplace.

Conclusions

1. In spite of the apparent environmental non-risky area where these subjects are working, their health status is damaged. The work conditions and lifestyle are both acting upon their health quality. That is why it is more difficult to quantify the extent of occupational factors effects.

2. Due to the involvement of magnesium in protective and indirectly in metabolizing pathways of organism's homeostasis, we recommend its monitoring and correction in selected groups of working population.

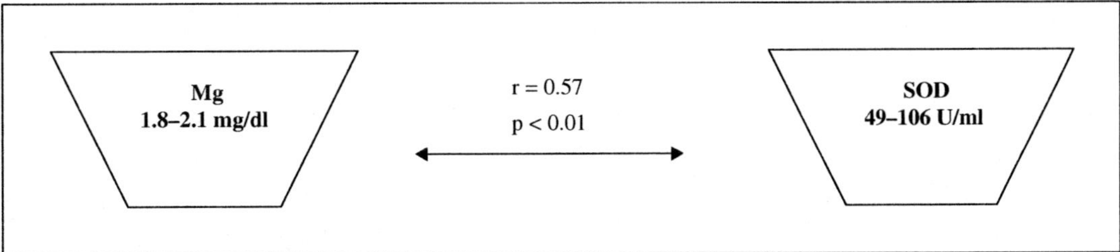

Figure 1. The correlation between serum magnesium and red blood cell SOD activity.

References

1. Branisteanu R, Aiking H. Exposure to polycyclic aromatic hydrocarbons in occupational versus urban environmental air. *Int Arch Occup Environ Health* 1998; 71: 533-6.

2. Ando M, Katagiri K, Tamura K, *et al.* Indoor and outdoor air pollution in Tokio and Beijing supercities. *Atmos Environ* 1996; 30: 695-702.

3. Kireev GV, Dimant IN, Oskin VI, Tatarski VP. Evaluation of the level of cancerogenic waste in energetics and calculation of its effects on air pollution. *Gig Sanit* 1994; 7: 5-7.

4. Miu N, Dragotoiu G. *Magneziul in biologia si patologia umana.* Cluj-Napoca: Casa Cartii de Stiinta, 2000.

5. Borisenko GG, Martin I, Zhao Q, Amoscato AA, Kagan VE. Nitroxides scavenge myeloperoxidase-catalyzed thiyl radicals in model systems and in cells. *Am Chem Soc* 2004; 126(30): 9221-32.

6. Halliwell B. Antioxidants in human health and disease. *Annu Rev Nutr* 1996; 16: 33-50.

7. Klouche K, Morena M, Canaud B, Descomps B, Beraud JJ, Cristol JP. Mechanism of in vitro heme-induced LDL oxidation: effects of antioxidants. *Eur J Clin Invest* 2004; 34(9): 619-25.

8. Nemeth K, Mezes M, Gaal T, Bartos A, Balogh K, Husveth F. Effect of supplementation with methionine and different fat sources on the glutathione redox system of growing chickens. *Acta Vet Hung* 2004; 52(3): 369-78.

9. Iqbal M, Okada S. Induction of NAD(P)H:quinone reductase by probucol: a possible mechanism for protection against chemical carcinogenesis and toxicity. *Pharmacol Toxicol* 2003; 93(6): 259-63.

10. Zhou Q, Olinescu RM, Kummerow FA. Influence of low magnesium concentrations in the medium on the antioxidant system in cultured human arterial endothelial cells. *Magnes Res* 1999; 12(1): 19-29.

11. Kuzniar A, Mitura P, Kurys P, Szymonik-Lesiuk S, Florianczyk B, Stryjecka-Zimmer M. The influence of hypomagnesemia on erythrocyte antioxidant enzyme defence system in mice. *Biometals* 2003; 16(2): 349-57.

12. Hans CP, Chaudhary DP, Bansal DD. Effect of magnesium supplementation on oxidative stress in alloxanic diabetic rats. *Magnes Res* 2003; 16(1): 13-9.

Hypomagnesemia and occupational risk in ceramic industry

F. Gradinariu, B. Scutaru, V. Hurduc, V. Cazuc, C. Croitoru, M. Ghitescu, A. Maftei, M. Margineanu

Institute of Public Health, Occupational Medicine Department,. 14 Dr. Victor Babes Str, 700465 Iasi, Romania

Abstract. We evaluated the occupational risk in ceramic industry by investigating the health status in relationship with the workplace conditions trying to understand whether hypomagnaesemia is a cause or a consequence of the pathology encountered. The investigated group consisted of 81 workers (9% women), with an average age of 38.35 ± 9.45 years and with an average length of service in a ceramic factory of 16.37 ± 9.93 years. 15% of the group had serum Mg level of 1.8 mg/dl, 4% of the subjects had hypocalcaemia, and 4% had both Ca and Mg low levels. The abnormal values of CO biological exposure marker carboxyhemoglobin (COHb) correlated with low serum magnesium ($r = -0.71$, $p < 0.05$), suggesting a link of this deficit with working conditions. Mg deficiency plays an important role in the etiology of work-related diseases found in the studied group.

Ceramic industry is known as a heavy industrial branch where noise impact, heat stress, physical work overload and air dusts represent real and serious occupational hazards. That is why we investigated the health status of workers in a ceramic brick manufacture from the cooking and drying sections. We also evaluated work conditions trying to find the relationship between them and the changes in blood biochemical pattern with special reference to the Ca/Mg balance.

Material and methods

The survey was made by a cross-sectional epidemiological study. The investigated group consisted of 81 workers (9% women), with an average age of 38.35 ± 9.45 years and with an average length of service in a ceramic factory of 16.37 ± 9.93 years. They were investigated by a complex protocol, including clinical, biochemical and hematological investigations, EKG, pulmonary functional and audiometric tests. The results were compared to a matched-control group. Serum Ca and Mg levels were assayed by commercial kits supplied by Nobis Labordiagnostica SRL Cluj-Napoca (Romania). Particulate matter and carbon monoxide in workplace air were determined by standard methods. Noise measurements were performed with Quest sonometer model 2900 (Quest Technologies, USA). Microclimate assessment included air temperature, velocity and relative humidity. End-shift blood carboxyhemoglobin (COHb), the carbon monoxide biological exposure marker, was determined with $PdCl_2$ by Richterich method [1].

Results and discussions

Workplace air measurements

Total dusts levels were 13.3 mg/m^3 (TWA for 8 hours) versus 10 mg/m^3 the EU TWA [2]. We had to compare our results with the EU standards, because our legislation has omitted to indicate threshold limits for total dusts. Carbon monoxide average level detected near the cooking furnaces and in the vicinity of drying devices was 19 mg/m^3, very near to the ceiling of the Romanian TWA which is 20 mg/m^3 [3]. Daily individual noise exposure was 93.3 dB(A), exceeding the Romanian upper limit of 87 dB(A) [3]. In brick section microclimate measurements showed ambient temperature of 32°C, a relative humidity of 79.4% and air velocity of 0.708 m/s. These data describe a warm, wet and poorly ventilated indoor environment.

Clinical investigations

The clinical investigation revealed a complex morbidity in the studied group (*Table I*). Cardiovascular diseases were found with a prevalence of 49.4%. This frequency might be connected to the work conditions: heavy work in noise and high temperature and also to the high ratio of men (91%) and smokers (72.8%) in the group. The next frequent encountered diseases were the musculoskeletal disorders with a percent of 39.5%, generated by the heavy-duty physical work of these employees. On the third place were the endocrine disorders with a frequency of 37%. They were represented by obesity, diabetes mellitus, goiter and chronic tetany, diseases etiologically bound to magnesium deficiency. Digestive disorders present in 33% of the group might be attributed partially to stress and also to lifestyle. Work stress, is coming not only from specific duties, responsibilities and relationships or from three shifts working program. It might originate also from noise and heat. High temperature is causing intense transpiration and consequently minerals loss. Stress by itself whatever its origin, is a source of hypomagnesemia. [4].

We assume that the high rate of respiratory and auditory diseases is a direct consequence of work conditions as showed dusts and noise levels in workplace measurements.
In the control group the morbidity was moderate, as showed in *Table I*.

The morbidity pattern of deficient cations group reflected the one of the whole exposed group, but on the first place there are the endocrine diseases. We can speculate that this parallel morbidity seems to suggest that low cation level might be occupationally induced or at least related.

Blood cations measurements

In the exposed group we found Ca and Mg serum abnormalities such as: 15% of the subjects had low serum Mg level of 1.8 mg/dl; 4% of the total group had hypocalcemia and other 4% had both Ca and Mg low levels. We found also a group of 4 subjects (6%) showing clinical signs of chronic spasmophylia, but having normal serum content of both cations. All subjects in control group had normal serum cations level.

Table I. The morbidity pattern in exposed, low level cations and control group (% of each group).

Crt. No.	Disease	Exposed group N = 81	Subgroup with low cations level N = 15	Control group N = 79
1	Cardiovascular diseases	49.4	33.3	7.2
2	Musculoskeletal disorders	39.5	33.3	8.7
3	Endocrine and metabolic diseases	37.0	67.0	4.3
4	Digestive diseases	33.3	33.3	16.0
5	Respiratory diseases	27.0	22.0	6.1
6	Visual disorders	27.0	13.3	5.0
7	Auditory disorders	16.0	13.3	0
8	Dermatologic disorders	10.0	7.0	3.2
9	Healthy	13.5	0	53.6

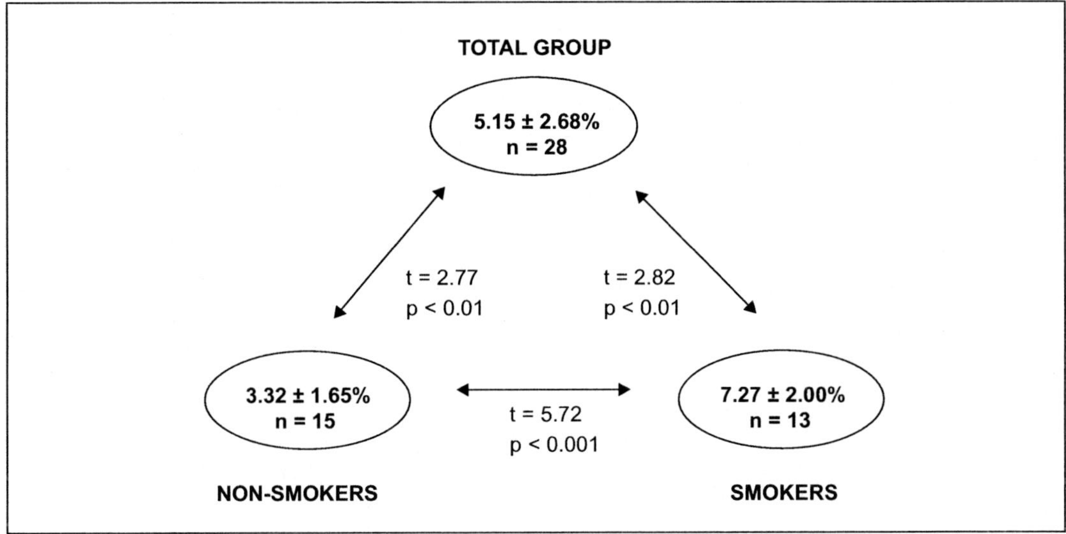

Figure 1. Statistical differences between COHb values concerning smoking habit.

Biological exposure marker

In a subgroup of 28 subjects of the exposed group, working in the vicinity of cooking furnaces, we determined end-shift blood carboxyhemoglobin (COHb) levels. Its values are expressed as percentage of total individual hemoglobin. The average value of the total group was 5.15±2.68%, significantly higher than that of the non-smokers 3.32±1.65% ($p < 0.01$, $t = 2.77$), and significantly lower than the one of smokers 7.27±2.00% ($p < 0.01$, $t = 2.82$). The statistical difference between smokers and non-smokers average value was even much more evident ($p < 0.001$, $t = 5.74$). It is considered that a COHb value above 5% in non-smokers and 10% in smokers is evidence of occupational exposure to CO [3,5]. The maximum COHb values found were: 7.71% in non-smokers and 10.80% in smokers. In control non-smokers we found a COHb average value of 1.22±0.59%.

Occupational absorption of CO induces first mild hypoxemia. This condition might determine itself the decrease of serum Ca and Mg. Human volunteers exposed to high altitude hypoxic conditions elicited such changes [6]. Work overload is another feature of daily activity of these workers. It was shown that even in well-trained individuals, short supramaximal exercise can induce hypoxemia [7] and intense physical exercise can determine significant serum Mg decreases [8]. Consequently intense physical work might be as well, a cause of hypoxemia and serum Mg loss (*Figure 1*).

We observed that the abnormal high carboxyhemoglobin values found in 9 subjects were all associated with low Mg levels between 1.8-2.1 mg/dl. Between these two markers there was a negative correlation ($r = -0.71$, $p < 0.05$). The significance is rather weak because of the small number of cases. The association between an abnormal high exposure marker and an abnormal low blood Mg levels might be another argument that hypomagnesemia is a consequence of work conditions (*Figure 2*).

Analyzing the relationships between the occupational risks and the morbidity recorded and its reflection in the blood biochemical pattern, it is difficult to postulate whether hypomagnesemia is an etiological factor or a consequence of the pathology encountered. As long as this pathology could by itself determine the blood cations imbalance, it looks like a vicious circle of cause-effect relationships. That is why we recommend Mg-enriched diet as a prophylactic protective measure in these workers. We also recommend blood Ca and Mg monitoring in medical periodic examination. This might permit applying the magnesium dietary supplements in the right moment.

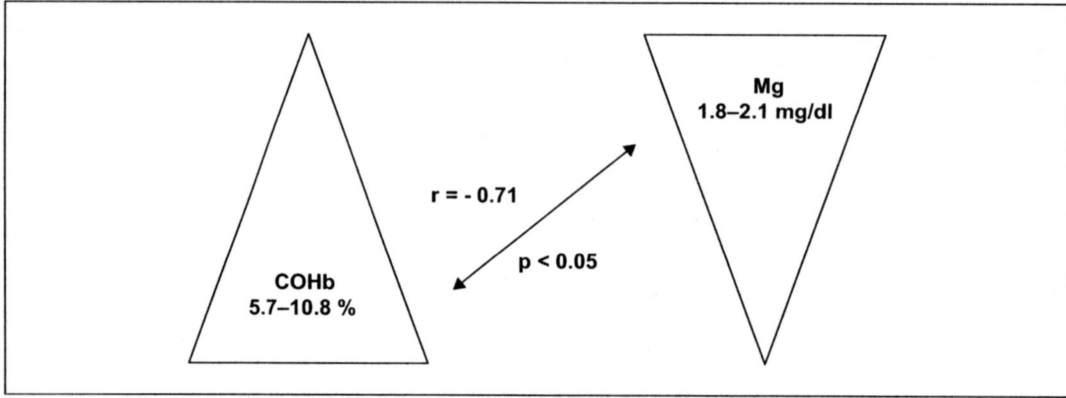

Figure 2. Correlation between COHb and low Mg level.

Conclusions

1. Mg deficiency was more frequent than hypocalcaemia; Mg deficiency might play an important role in the etiology of work-related diseases diagnosed in the studied group.
2. The correlation found between the biological exposure marker carboxyhemoglobin and the low serum Mg level suggests that hypomagnesemia is related to work conditions.
3. Further technical measures are necessary in the workplace to diminish occupational risks in order to avoid any harmful effect upon employees' health.
4. We recommend serum Mg and Ca monitoring in order to apply dietary supplements.

References

1. Rictherich R. *Chimie clinique théorique et pratique.* Paris: Ed. Doin, 1967.
2. BIA. *Report Grenzwerteliste 2/2003, Sichereit und Gesundheitsschutz bei der Arbeit.* Sankt Augustin: Ed. HVBG, 2003.
3. MMSS&MSF. *Norme generale de protectie a muncii.* Oltenita: Ed. Institutul National de Cercetare-Dezvoltare Bucuresti, GLG-Print, 2002.
4. Miu N, Dragotoiu G. *Magneziul in biologia si patologia umana.* Cluj-Napoca: Casa Cartii de Stiinta, 2000.
5. Thomas L. *Labor und Diagnose.* Marburg: Die Medizinische Verlagsgessellschaft, 1988.
6. Chatterji JC, Ohri VC, Chadha KS, *et al.* Serum and urinary cation changes on acute induction to high altitude (3200 and 3,771 m). *Aviat Space Environ Med* 1982; 53(6): 576-9.
7. Roberts D, Smith DJ. Erythropoietin concentration and arterial haemoglobin saturation with supramaximal exercise. *J Sports Sci* 1999; 17(6): 485-93.
8. Inaba R, Lanphere C, Fujita S, *et al.* Changes of peripheral blood parameters in aerobic dance instructors after aerobic dance instruction. *Nippon Eiseigaku Zasshi* 1991; 46(4): 883-9.

Effects of diets with different magnesium content on homocysteinemia in young and old rats

V. Gaume[1,2], C. Demougeot[2], M. Adrian[2], S. Devaux[2], P. Laurant[2], F. Mougin[1,2], A. Berthelot[2]

1. UFR STAPS, 31 rue de l'Epitaphe, F-25000 Besançon, France; 2. Laboratoire de Physiologie et de Pharmacologie-Nutrition Préventive et Expérimentale, EA 3921, Faculté de Médecine-Pharmacie, F-25030 Besançon, France

Abstract. A considerable number of experimental, epidemiological and clinical studies have pointed out the role of abnormal dietary deficiency of magnesium (Mg) in the etiology of cardiovascular pathology. Up to now, the underlying mechanisms involved in the deleterious effects of Mg deficiency are not fully understood. Many studies have identified a mildly elevated total plasma concentration of homocysteine (tHcy) as a risk factor for cardiovascular disease. The aim of this study was to investigate the relationship between the Mg dietary intake and homocysteinemia in rats, as well as the influence of ageing on these two parameters. Experiments were conducted on 4-week-old Sprague Dawley rats fed with low (0.015%), normal (0.08%) or high (0.32%) Mg diet for 4 months (n = 5 per group, young rats) or 21 months (n = 7 per group, aged rats). Total plasma concentration of Hcy and total Mg were measured at the end of the experiment by HPLC/fluorimetry and atomic absorption spectroscopy, respectively. In rats fed with normal Mg diet, ageing was associated with a 2-fold increase in homocysteinemia ($p < 0.01$) whereas it did not affect plasma total Mg level. In young rats, Mg deprivation or supplementation did not modify homocysteinemia. Conversely, in aged rats, high Mg diet decreased homocysteinemia by 64% as compared to normal diet (1.48 ± 0.48 versus 4.14 ± 0.40 µM, $p < 0.05$) whereas low Mg diet had no effect (3.95 ± 0.28 versus 4.14 ± 0.40 µM). Interestingly, plasma Hcy negatively correlated with plasma total Mg level ($r = -0.445$, $p = 0.04$) in aged rats but not in young rats. In conclusion, these results demonstrate the existence of a relationship between plasma Mg level and homocysteinemia in aged rats, and suggest that a long-term Mg supplementation is able to modify Hcy metabolism. These data are in favor of a beneficial effect of Mg supplementation for reducing the risk for cardiovascular disease in the elderly.

Introduction

On one hand, a considerable number of experimental, epidemiological and clinical studies have pointed out the role of abnormal dietary deficiency of magnesium (Mg) in the etiology of cardiovascular pathology. In addition, many studies demonstrated the beneficial effect of magnesium supplementation in chronic cardiovascular diseases [1]. Up to now, the underlying mechanisms involved in these effects are not fully understood.

On the other hand, increased plasma total homocysteine is an established independent risk factor for the development of cardiovascular disease. Indeed, epidemiological studies have shown that even mild hyperhomocysteinemia is associated with coronary artery disease, myocardial infarction, peripheral arterial disease, cerebrovascular disease or cardiac allograft vasculopathy [2]. Determinants of increased plasma homocysteine concentration remain to be elucidated.

Age is the dominant risk factor for cardiovascular diseases, and it is well-established hyperhomocysteinemia is a common phenomenon among elderly people [2]. A large part of aged subjects may have an inadequate Mg intake and a chronic latent Mg deficiency, but the link between nutritional Mg status and homocysteinemia has never been investigated. In our study, we tested the hypothesis that the magnesium intake could influence plasma homocysteine concentration in young and aged rats.

Materials and Methods

Animals

Experiments were conducted on thirty six Sprague Dawley rats aged 4-week-old. Animals were kept under a 12h-12h light/dark cycle and allowed free access to food and water. Animal care was in accordance with the French Department of agriculture guidelines.
Rats were divided in two groups: young rats (n = 15) fed with Mg diet for 4 months and aged rats (n = 21) fed with Mg diet for 21 months. In each group, animals were fed with normal Mg diet (0.080%), low Mg diet (0.015%), or high Mg diet (0.32%). The synthetic diet which was prepared in our laboratory contained the following (%): starch (42.5), casein (20), cellulose (6), sucrose (21), groundnut oil (2.5) and corn oil (2.5), mineral mixture except Mg (7) and vitamin mixtures (1). Mg was added in the form of Magnesium oxide.

Biochemical analysis

At the end of the experiment, total plasma homocysteine (tHcy) was measured by HPLC with fluorometric determination after derivation of thiols with 7-fluoro-2,1,3-benzodiazole-4 sulphonamide according to the method described by Durand *et al.* [3]. Plasma magnesium concentrations were analyzed by atomic absorption spectrophotometry.

Statistical analyses

Values are expressed as mean ± SEM. The significance of differences between groups was calculated by one-way analysis of variance following by a Tukey-Kramer unpairwise post-hoc test. Trends in the thiol values with increasing magnesium concentration were explored using the linear regression represented graphically by scatter plots. Pearson's correlation coefficient was applied to test the relation between plasma homocysteine and Mg levels. Differences were considered statically significant at $p < 0.05$.

Results

Effects of magnesium diet on body weight and plasma Mg level

Results are represented in *Table I.* Diet with different Mg content did not modify the weight of young and aged rats. As compared to rats fed with the normal Mg diet, Mg-supplemented diet did not increase the plasma Mg levels whereas the low Mg diet reduced by 42% ($p < 0.05$) and 33% ($p < 0.05$) plasma Mg levels in young and aged rats, respectively.

Table I. Effect of magnesium diet on body weight and plasma Mg level in young and in aged rats.

	Young rats			Aged rats		
	Normal Mg	High Mg	Low Mg	Normal Mg	High Mg	Low Mg
Body weight (g)	454 ± 10	500 ± 17	477 ± 26	640 ± 26	653 ± 43	580 ± 37
Plasma Mg (mmol/L)	0.67 ± 0.02	0.73 ± 0.03	0.39 ± 0.04*	0.73 ± 0.02	0.84 ± 0.01	0.49 ± 0.05*

Values are means ± SEM. Rats were fed with normal Mg (0.08%), low Mg (0.015%) or high Mg diet (0.32%) for 4 months (young rats, n = 5 per group) or 21 months (aged rats, n = 7 per group). * ($p < 0.05$) significantly different from age-matched rat fed with normal Mg diet.

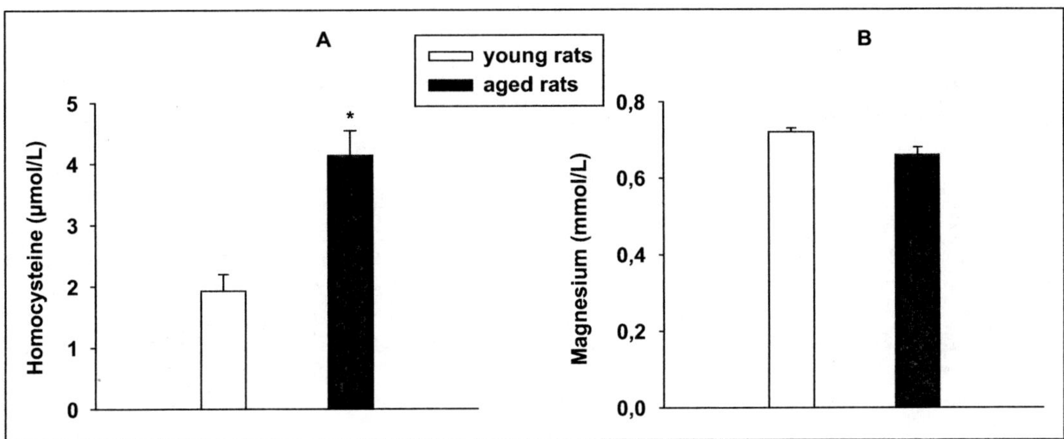

Figure 1. Effect of age on total plasma homocysteine and magnesium levels. Rats were fed with normal Mg diet for 4 months (A) (n = 15) and 21 months (B) (n = 21). * ($p < 0.05$) significantly different from young rats.

Effect of age on total plasma homocysteine and plasma Mg level

In rats fed with normal Mg diet, tHcy concentration was 2-fold higher ($p < 0.05$) in aged rats than in young rats (*Figure 1*), whereas total plasma Mg level was not affected by age.

Effects of Mg diet on homocysteinemia

As shown in *Figure 2*, low Mg or high Mg diet did not modify tHcy in young rats. Conversely, in aged rats high Mg diet decreased homocysteinemia by 64% as compared to normal Mg diet ($p < 0.05$), whereas the low Mg diet had no effect.

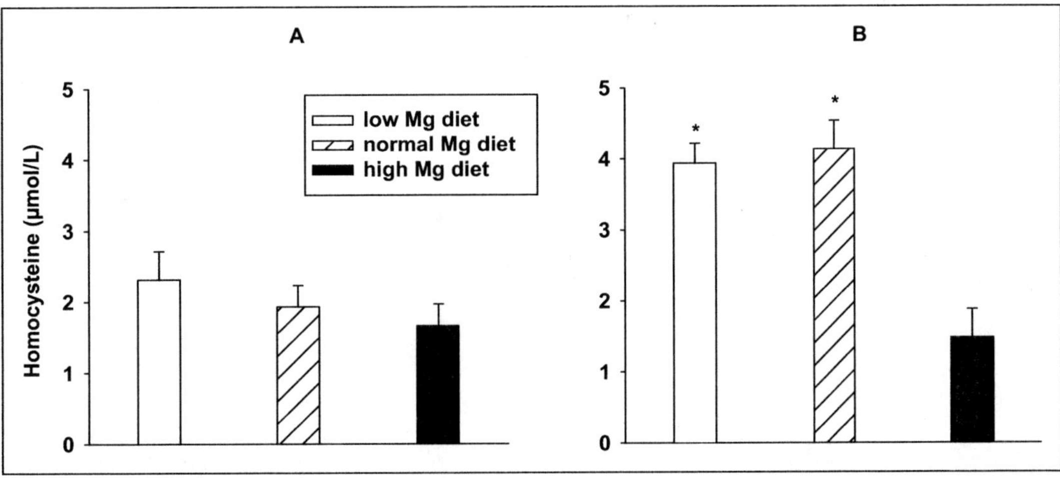

Figure 2. Effects of Mg diet on homocysteinemia. Rats were fed with normal Mg (0.08%), low Mg (0.015%) or high Mg diet (0.32%) for 4 months (A) (n = 5 per group) or 21 months (B) (n = 7 per group). Values are means ± SEM. * ($p < 0.001$) significantly different from rat fed with high Mg diet.

Correlation of homocysteinemia with plasma Mg level

Figure 3 plots the tHcy level against plasma Mg levels in young rats (*Figure 3A*) and in aged rats (*Figure 3B*). In aged rats, tHcy negatively correlated with plasma total Mg level (r = -0.445, p = 0.043). Such correlation was not observed in young rats (r = -0.440, p = 0.098).

Discussion

The present study reported that *i)* ageing is associated with increased plasma homocysteine levels in rats, *ii)* Mg-supplemented diet prevented the rise in homocysteinemia in aged rats, *iii)* plasma tHcy negatively correlated with plasma Mg levels aged rats but not in young rats.

The relationship between magnesium and cardiovascular disease is well documented, but the mechanisms involved in the deleterious effects of Mg deficiency or in the beneficial effects of Mg supplementation remain to be elucidated. In this study, we tested the hypothesis that nutritional status of Mg may influence total plasma homocysteine (tHcy) levels in rats. Comparison of homocysteinemia in young and aged rats confirms data in which ageing was associated with hyperhomocysteinemia [2]. In our study, the rise of plasma tHcy levels in aged rats occurred without any change of plasma Mg concentration. In addition, the low Mg diet did not modify tHcy levels in young and in aged rats. In their study on cultured vascular smooth muscle cells, Li *et al* [4] demonstrated that an overload of cells with tHcy induced a decrease in intracellular Mg concentration, but the effects of Mg deficiency on tHcy metabolism was not investigated. In our study, we can not exclude an effect of hyperhomocysteinemia on intracellular Mg levels since we only measured plasma Mg level which is a weak marker of cellular Mg concentrations. Nevertheless, our results are not in favor of an effect of magnesium deficiency on hyperhomocysteinemia.

Different mechanisms have been proposed to explain the beneficial effects of Mg supplementation such as modifications of cellular calcium concentration, decrease in catecholamine secretion, increase in prostacycline synthesis, inhibition of platelet aggregation [1]. In this study, we reported for the first time that Mg supplementation dramatically prevented the rise of plasma Hcy in aged rats whereas it did not affect significantly tHcy levels in young rats. This result suggests that beneficial effects of increased Mg intake are apparent only in conditions associated with abnormal tHcy status, as confirmed by the negative correlation between Mg and Hcy in aged rats. The mechanisms involved in this effect remain to be elucidated. One hypothesis could be a direct effect of magnesium on homocystein metabolism

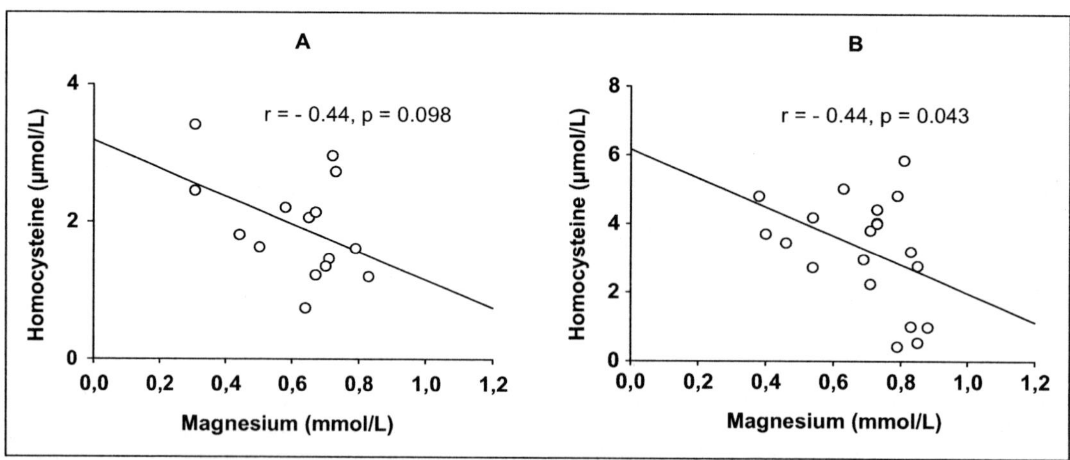

Figure 3. Correlation of plasma tHcy and plasma Mg level from young rats (A, n = 15) and aged rats (B, n = 21). The tHcy and plasma Mg levels were plotted and the correlation coefficient (r) was calculated.

since the major enzymes involved in Hcy metabolism are Mg^{2+}-dependent [4]. Interestingly, this result propose a new mechanism by which Mg supplementation exerts beneficial cardiovascular effect, since recent data suggested a direct relation between decrease in tHcy and reduction of vascular disease risk [5].

In conclusion, this study is the first to report that a long-term Mg supplementation is able to modify the rise in total plasma Hcy level associated with age. These results are in favor of a beneficial effect of Mg supplementation for preventing the risk for cardiovascular diseases in the elderly.

References

1. Chakraborti S, Chakraborti T, Mandal M, Mandal A, Das S, Ghosh S. Protective role of magnesium in cardiovascular diseases: a review. *Mol Cell Biochem* 2002; 238: 163-79.
2. Parnetti L, Bottiglieri T, Lowenthal D. Role of homocysteine in age-related vascular and non-vascular diseases. *Aging Clin Exp Res* 1997; 4: 241-57.
3. Durand P, Fortin LJ, Lussier-Cacan S, Davignon J, Blache D. Hyperhomocysteinemia induced by folic acid deficiency and methionine load-applications of modified HPLC method. *Clin Chim Acta* 1996; 252: 83-93.
4. Li W, Zheng T, Wang J, Altura BT, Altura BM. Extracellular magnesium regulates effects of vitamin B6, B12 and folate on homocysteinemia-induced depletion of intracellular free magnesium ions in canine cerebral vascular smooth muscle cells: possible relationship to (Ca2+)i, atherogenesis and stroke. *Neurosci Lett* 1999; 274: 83-6.
5. Appel LJ, Miller III ER, Jee SH, *et al.* Effect of dietary patterns on serum homocystein. Results of a randomized, controlled feeding study. *Circulation* 2000; 102: 852-7.